MODERN VIBRATIONS PRIMER

Peter M. Moretti, Ph.D., P.E.
Department of Mechanical and Aerospace Engineering
Oklahoma State University
Stillwater, Oklahoma

CRC Press
Boca Raton London New York Washington, D.C.

Cataloging-in-Publication Data is available from the Library of Congress.

International Standard Book Number 0-8493-2038-0
Printed in the United States of America 1 2 3 4 5 6 7 8 9 0
Printed on acid-free paper

Preface

This book was written to provide up-to-date material not only on fundamental methods of analysis but also on practical applications in industry. For the sake of fundamental understanding, concepts of engineering analysis are developed from the bottom up; the first few chapters develop in gradual steps what is covered in the first few pages of other books. For the sake of practical application, mathematics and physical reasoning are presented side-by-side, and later chapters focus on methods that are widely used in industry.

The material is organized so that it can be used either in an introductory course for undergraduates or in a comprehensive course for graduate students. Each chapter starts with a few "must know" elements for all students, and ends with more subtle aspects of the topic for advanced students. The instructor can use as much of each chapter as is appropriate to his course. The student knows that the first few pages of each chapter are the most important.

In order to keep up with the times, advanced topics are included with simple examples, to provide a basis for later study in evolving subject areas.

The chapters are short and focused; chapters on special topics can be included or left out, so that this introductory course will mesh appropriately with the rest of any curriculum. Material from basic engineering courses is reviewed as needed on a just-in-time basis.

This book restores some of the practical subjects included in the earliest vibrations books, which have been left out in more theoretical texts. There is basic coverage of the important topics of autonomous oscillations, flow-induced vibrations, and parametric excitation. On the other hand, coverage of subjects that are no longer central to the field has been shortened and simplified; computational procedures like Gaussian reduction have been replaced with the use of widely available computer programs.

About the Author: Professor Moretti, a native of Switzerland, was educated at the California Institute of Technology, Technische Hochschule Darmstadt in Germany, and Stanford University. He gained design experience at Interatom in Bensberg, Germany, and at Westinghouse Advanced Reactor Division in Madison, Pennsylvania. His research interests include flow-induced vibration in heat exchangers, and web flutter in paper machines, printing presses, and plastic-film drying ovens. He teaches in the School of Mechanical and Aerospace Engineering at Oklahoma State University in Stillwater and Tulsa.

Dedication:

To my teachers at Caltech, Darmstadt, and Stanford.

Contents

Part I

Simple Systems

Chapter 1

INTRODUCTION and resources

Why do we study vibrations? Because most serious dynamic problems in mechanical engineering involve vibrations. There are two main paths by which vibrations lead to trouble: either a resonance condition permits small disturbances to generate destructively large amplitudes and stresses, or else the frequent repetition of moderate stress reversals leads to premature failure from fatigue or wear. Therefore we must study both resonant and long-enduring oscillations.

1.1 Background

Some case studies in vibrations date back to the nineteenth century, but the number of "trouble jobs" increased markedly with the development of automotive engines in the early twentieth century. As speeds went up, fractured crankshafts became common; machine designers' naïve attempts to "beef up" the shafts, making them heavier, led to even more rapid failures. Solutions were not found until dynamic analysis was applied. Similar problems in big steam turbines were investigated in industry laboratories like the Westinghouse Research Center, and a vibrations discipline was defined by S. Timoshenko and J.P. Den Hartog, who assembled course textbooks in 1928 and 1934, respectively. Since then, the field has gradually expanded as instruments, computer methods, and the mathematical theory of random signals have developed; this book will expand it a little more. We now include the prevention of vibration through suppression, isolation, or balancing; the constructive use of vibrations in technical processes ranging from conveying materials to compacting them; the sources of noise; earthquake and shock loading of structures; airplane wing flutter and other self-sustaining vibrations; cyclic failure; vibration detection and control; non-linear systems; and chaos.

1.2 Objectives

The foundation of vibration analysis is mathematical modeling. At its base is the theory of linear ordinary differential equations. At the next level are the tools of matrix algebra, partial differential equations, non-linear differential equations, and probability and statistics. On top of these rests the study of actual structures, of real earthquakes, of flow-induced vibrations, and so on. The main goal of this book is to establish a solid foundation of physical understanding and mathematical representation, and to apply it to practical trouble-shooting and design.

1.3 Method

Our teaching tools towards this goal include:

- emphasizing fundamentals and de-emphasizing analytical tricks;
- developing physical reasoning hand-in-hand with the mathematics;
- starting with the most elementary theory and methods;
- incorporating realistic and practical problems;
- progressing to advanced topics like damped multi-degree-of-freedom systems;
- introducing modern applications such as experimental modal analysis;
- providing a solid basis for specialized successor courses, such as:
  - acoustics,
  - structural vibrations,
  - numerical methods,
  - instrumentation,
  - signal processing,
  - random vibrations and
  - fatigue,
  - shock loading,
  - earthquake damage,
  - active control,
  - non-linear systems, and
  - chaos;
- using both metric and English units.

1.4 References

This textbook includes the reference information needed to solve the given problems. However, every engineer should have access to a mathematical handbook with which he is familiar. Some possible choices include:

Ronald J. Tallarida, *Pocket Book of Integrals and Mathematical Formulas,* 2nd Edition, CRC Press, Boca Raton, Florida, 1992, ISBN 0-8493-0142-4;

Daniel Zwillinger, editor, *CRC Standard Mathematical Tables and Formulae,* 30th Edition, CRC Press, Boca Raton, Florida, 1996, ISBN 0-8493-2479-3; or

Lennart Råde & B. Westergren, *BETA β Mathematics Handbook: Concepts, Theorems, Methods, Algorithms, Formulas, Graphs, Tables,* 2nd Edition, CRC Press, Boca Raton, Florida, 1992, ISBN 0-8493-7758-7.

1.5 Computers

In order to obtain solutions, calculator and computer procedures are needed. In this book, our emphasis is on putting equations into canonical form so that standard mathematical procedures such as matrix inversion and eigenvalue determination can be used. This will enable the student to utilize computational software like *MATLAB.*[1]

Numerical operations are also available in the symbolic-algebra programs *Maple*[2] and *Mathematica.*[3] Parts of the former are incorporated into the word-processor *Scientific WorkPlace*[4] and the engineering tool *Mathcad.*[5]

Many matrix procedures are possible on scientific calculators.

1.6 Report Writing

Reports on engineering analyses and solutions require a combination of verbal explanations, schematic illustrations, and mathematical expressions. This book was developed using *Scientific WorkPlace,*[4] which integrates creation of text, organization of documents using the mark-up-language features of LaTeX, typesetting of mathematical equations using TeX, and mathematical operations and curve-plotting using *Maple.* Additional editing of the resulting LaTeX file was carried out using PC TeX.[6]

---

[1] The MathWorks, Inc., Natick, MA 01760-1500, `http://www.mathworks.com/`

[2] Waterloo Maple, Inc., Waterloo, Ontario N2L-5J2, Canada, `http://www.maplesoft.com/`

[3] Wolfram Research, Inc., Champaign, IL 61820-7237, `http://www.wolfram.com/`

[4] MacKichan Software, Bainbridge Island, WA 98110, `http://www.mackichan.com/`

[5] MathSoft, Inc., Cambridge, MA 02142-1521, `http://www.mathsoft.com/`

[6] Personal TeX, Inc., Mill Valley, CA 94941, `http://www.pctex.com/`

1.7 Problems

Engineering is learned by solving problems. This introductory set is intended for review of prerequisite material in the area of dynamics. Solve them as a check on your readiness for a vibrations course. If you have trouble with the units, review them in the next chapter.

Problem 1.1 *A meteor has a mass of 6000 slugs (lbf-sec^2/ft). It is moving toward the earth in the equatorial plane, where the earth's gravitational field has an acceleration of 32.09 ft/sec^2 just above the ocean surface. What is the force of gravitational attraction on the meteor just before impact?*

Problem 1.2 *Soon after, a ship of the same mass as the meteor reaches the impact point and stops for observations. What is the force of buoyancy on the ship at this time?*

Problem 1.3 *Eventually the ship steams eastward at 40 ft/sec. What is the force of buoyancy during this time?*

Problem 1.4 *A 60-pound boy on skis is pulling on the free end of a rope that is tied to a post. The maximum pull that he can exert without having the rope slip through his grasp is 12 pounds. If the ground is horizontal, calculate the maximum acceleration that he can produce.*

Problem 1.5 *Another skier, who is twice as heavy as the preceding one, unties the rope at the post and proceeds to pull the first boy toward himself. The second boy can exert a maximum pull of 24 pounds. What is the maximum acceleration each skier can experience in this pulling contest?*

Problem 1.6 *A 1750-kg automobile is approaching a crest in an otherwise level road at 50 km/hr. The crest represents a rise of 0.5 m, a length of 8 m, and constant curvature within those eight meters. How much of the car's weight is supported by the road as it passes the crest?*

Problem 1.7 *The sketch represents a circular platform (of radius $r = 4$ ft.) rotating about an axis P which itself is mounted on a rotating arm (at a radius of $R = 8$ ft.) that rotates about a fixed axis O. The smaller platform rotates with a constant clockwise angular velocity of 7.2 radians/second with respect to the line OP. The longer arm rotates with a constant counterclockwise angular velocity of 3.5 radians/second.*

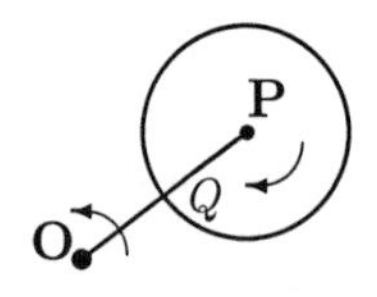

Calculate the acceleration of the point Q on the edge of the smaller platform, at the moment when Q happens to pass the line OP.

Problem 1.8 *By separation of variables and direct integration, find the solution to the equation*

$$\dot{x} + \alpha x = 0$$

where $\dot{x} \triangleq dx/dt$ *and the Initial Condition is the displacement* $x_{(0)}$. *Answer: We expect a solution in exponential form.*

Problem 1.9 *Solve the same equation by fitting a power series*

$$x = a_o + a_1 t + a_2 t^2 + a_3 t^3 + ...$$

to the equation and the Initial Condition. Answer: The result should agree with the Maclaurin-series expansion of an exponential function.

Problem 1.10 *Solve the same equation for two positive and two negative values of* α, *using a computer program like* Maple, MATLAB, *or* Mathematica. *Superposition: Since this is a linear equation, you can do this for any initial condition by making the dimensionless ratio* $x/x_{(0)}$ *the dependent variable. Similitude: If we make* (αt) *the dimensionless independent variable, all positive-*α *solutions should be identical, and all negative-*α *solutions should be identical.*

Problem 1.11 *Prepare a comparison of competing software for engineering analysis and simulation. Can you find and discuss alternatives to those already listed in this chapter? Hint: Use the Internet as a resource for commercial information.*

Chapter 2

FORMULATION OF TRANSLATIONAL SYSTEMS and review of units

The first step in solving any vibrations problem is to identify the physical system and formulate it as a mathematical equation. The simplest systems are those which can be represented as assemblies of components such as idealized "weightless" springs and concentrated "point" masses (Figure 2.1). We will carefully review the modeling of a very simple system, in order to establish a method for analyzing more complex systems.

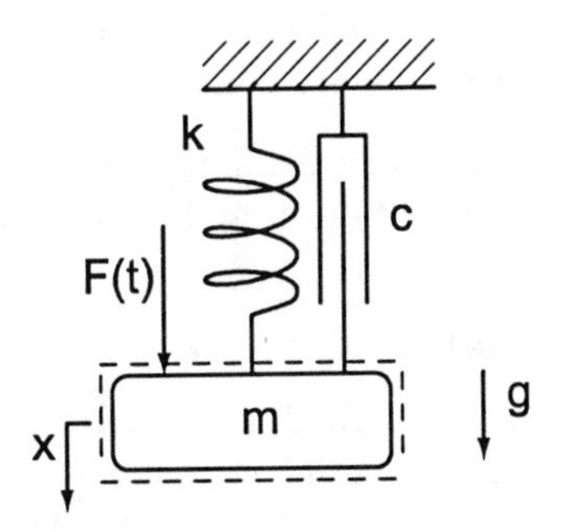

Figure 2.1: Translational System

2.1 Newton's Law

In classical mechanics, we write equations which balance forces against changes of momentum

$$F \propto \frac{d(mv)}{dt}$$

where $F \triangleq$ force, $m \triangleq$ mass, and $v \triangleq$ velocity.

In the "box method," we enclose the mass in an imaginary box and identify the forces acting through and within the box. If this box forms a "closed control volume," one which always contains identically the same mass, this equation can be expanded and then reduced to the "free-body" equation. First we expand the differentiation

$$F \propto \left(m\frac{dv}{dt} + v\frac{dm}{dt} \right)$$

Then we strike out the $v \cdot dm/dt$ term; it is zero because there is no mass flux through the control surface. Substituting $a \triangleq dv/dt$ in the other right-hand term, we arrive at the familiar

$$F \propto ma$$

It is convenient to introduce the proportionality constant g_c

$$F = \frac{ma}{g_c} \tag{2.1}$$

where g_c equals 32.174 lbm-ft/lbf-sec^2 in English units, 1 kg-m/N-s^2 in S.I. units, and 9.80665 kg-m/kg$_{\text{force}}$-s^2 in the old mks units. In this book we keep track of g_c throughout the algebraic equations as a reminder to you. Many engineers prefer to omit it, and to insert its value later, as required to make the units come out, when they finally substitute numerical values into an algebraic solution.

It is easy to make mistakes in applying force balances. To make sure we get the correct sign on each term, we use a set of three conventions:

1. that the direction of F is that of the force acting on the control volume (not the opposite force exerted by the control volume onto the environment);

2. that the coordinate directions defining positive displacement x, positive velocity v, positive acceleration a, and positive force F are all the same; and

3. that F and ma are written on opposite sides of the equation.

If any one of these conventions is violated, there will be a wrong sign on some term, a very serious "200%" error. The inquiring mind notes that violating two conventions simultaneously would be all right, since the errors would cancel. Therefore there must be three other sets of conventions which would also work. However, the set given above is the one used in most of the literature.

2.2 Lumped-Parameter System

For a mass supported from a fixed base by a spring and a damper, the equation of motion can be applied after we do two things:

- define a "control volume" enclosing the "system" to which the governing equation will be applied, as indicated by the dashed line boxing-in the mass in Figure 2.1; and
- define a coordinate system indicating the direction of positive values of displacement x, as indicated by the arrow in Figure 2.1, so that we will use a consistent convention for all the terms.

The force F in Equation 2.1 is the sum of the several forces F_{cs} acting through the control surface, plus any body force F_{cv} acting within the control volume

$$\sum F_{cs} + F_{cv} = \frac{ma}{g_c} \tag{2.2}$$

Surface forces F_{cs} include

$$F_{\text{spring}} \cong -kx \tag{2.3}$$

$$F_{\text{damper}} \cong -cv \tag{2.4}$$

and any externally applied force $F(t)$.

The body force F_{cv} typically is

$$F_{\text{gravity}} = \frac{mg}{g_c} \tag{2.5}$$

where g is a positive number if the coordinate x is taken in the direction of gravity, or negative if x is taken upwards. For brevity we write $\dot{x}$ for $v = dx/dt$ and $\ddot{x}$ for $a = dv/dx = d^2x/dt^2$. Inserting all these terms into Equation 2.2, we obtain:

$$-kx - c\,\dot{x} + F(t) + \frac{mg}{g_c} = \frac{m\,\ddot{x}}{g_c}$$

Re-arranging this into the customary sequence, we get:

$$\boxed{\frac{m}{g_c}\,\ddot{x} + c\,\dot{x} + kx = \frac{mg}{g_c} + F(t)} \tag{2.6}$$

Note that the $\ddot{x}$-term is made positive, and the $\dot{x}$- and x-terms have the same positive sign. If any of them had a different sign, we would need to be concerned: probably we made a mistake in formulating the equation. A more remote possibility would be that the system has peculiar dampers which do not retard motion but enhance it, or peculiar springs which do not restore the equilibrium position but push away from it: in short, that this is not an ordinary, stable vibrational system. The gravitational and forcing terms on the right side of the equation, on the other hand, can be positive or negative, depending which way the arrow defining the coordinate system was drawn in Figure 2.1.

Equation 2.6 is a second-order ordinary differential equation with constant coefficients. It is linear because we have chosen to represent the force in the spring as being proportional to displacement (Equation 2.3), and the force in the damper as being proportional to velocity (Equation 2.4). The former is approximately correct for many springs; the latter is usually a very crude assumption which is tolerable only if the damping term is small.

Exercise 2.1 *Show that any number of springs arranged in parallel, so that their displacements are equal but their forces add, can be replaced by a stronger equivalent spring*

$$k_{equiv} = \sum k_i$$

Exercise 2.2 *Show that any number of springs arranged in series, so that their forces are equal but their displacements add, can be replaced by a weaker equivalent spring such that*

$$\frac{1}{k_{equiv}} = \sum \frac{1}{k_i}$$

Exercise 2.3 *What additional terms do we need in Equation 2.2 if we have an open control volume, with mass passing in or out (e.g., jet propulsion).*

2.3 Caution

We have shown a single-degree-of-freedom lumped-mass problem represented by a second-order ordinary differential equation.

Even as we first drew the schematic of the system as an assembly of pure and separate components, we implicitly made assumptions: we assumed that the mass is rigid and does not flex; that the spring does not have any mass or damping; that the ground provides a rigidly fixed base, etc. These assumptions are acceptable only if the deflection within the mass is much smaller than the deflection of the spring; the mass of the spring much smaller than the that of the lumped mass, etc.

An engineer needs to be aware of the simplifications made at the start of an analysis, and remember them when interpreting the final results. This is illustrated in the following example.

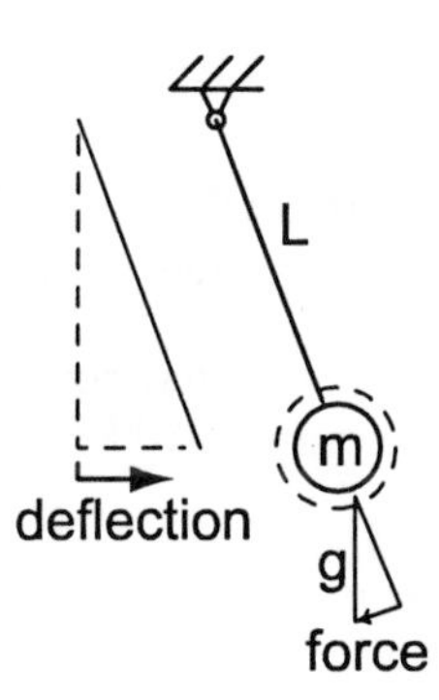

Figure 2.2: Crude Analysis of Simple Pendulum

2.4 The Simple Pendulum

A concentrated mass m hanging from a string of length L is called a simple pendulum (Figure 2.2). We can enclose the mass in a control volume and write Equation 2.1 for lateral deflections x. The restoring force F is a component of the gravitational attraction mg/g_c; from similar triangles, we can show that $Fg_c/mg = x/L$. Therefore Equation 2.1 becomes

$$-\frac{mg}{g_c} \cdot \frac{x}{L} = \frac{m\,\ddot{x}}{g_c}$$

which can be arranged as

$$\ddot{x} + \left(\frac{g}{L}\right) x = 0 \tag{2.7}$$

Although this is a useful first-order answer, the approach taken is here is *not* a good one:

1. The path measured by x is not straight, but a curve. We linearized the problem as we went along, and can no longer tell at this point how large the errors due to the linearization are. We should find a way of describing the non-linearities mathematically, so that we can estimate the approximate magnitude of the neglected terms.

2. The mass m is described as concentrated at a single point, and the inertial terms due to its rotation are neglected. We don't know what errors this simplification causes.

In the next chapter, we will develop a better approach which will let us overcome these limitations.

2.5 Generalization

We have developed the application of Newton's law to a single direction in a stationary coordinate system. We can generalize our method as follows:

1. For a concentrated mass in a stationary or inertial (i.e., uniformly translating) coordinate system, we can apply Newton's law in all spatial directions and write

$$\sum \overrightarrow{F}_{cs} + \overrightarrow{F}_{cv} = \frac{m\overrightarrow{a}}{g_c} \tag{2.8}$$

   which can be decomposed into three equations like Equation 2.2, one for each Cartesian coordinate. This will later prove to be convenient for masses that are mounted with two or three degrees of freedom.

2. If the mass is mounted within a steadily rotating system, we can use acceleration relative to that coordinate system if we add correction terms for the acceleration due to the motion of the reference system itself

$$\sum \overrightarrow{F}_{cs} + \overrightarrow{F}_{cv} = \frac{m}{g_c}\overrightarrow{\Omega} \times \left(\overrightarrow{\Omega} \times \overrightarrow{r}\right) + 2\frac{m}{g_c}\overrightarrow{\Omega} \times \overrightarrow{v}_{rel} + \frac{m\overrightarrow{a}_{rel}}{g_c} \tag{2.9}$$

   where Ω is the steady angular velocity of the reference system, the position vector $\overrightarrow{r}$ is measured from a point on the rotational axis, and v is the velocity of the concentrated mass. The inward-acting momentum-term $\left|m\Omega^2 r/g_c\right|$ can be replaced by a radially outward-acting "centrifugal force" on the left side of the equation; similarly, $|2m\Omega v_{radial}/g_c|$ acting at right angles to the radial velocity component can be replaced by an opposite "Coriolis force" on the left side of the equation.

3. In an arbitrarily moving coordinate system, additional terms for the rate of acceleration of the reference point and for the change in rate of rotation, must also be added

$$\begin{aligned}&\sum \overrightarrow{F}_{cs} + \overrightarrow{F}_{cv} \\ = \; &\frac{m}{g_c}\left(\overrightarrow{a}_{center} + \dot{\overrightarrow{\Omega}} \times \overrightarrow{r} + \overrightarrow{\Omega} \times \left(\overrightarrow{\Omega} \times \overrightarrow{r}\right) + 2\overrightarrow{\Omega} \times \overrightarrow{v}_{rel} + \overrightarrow{a}_{rel}\right)\end{aligned} \tag{2.10}$$

2.6 Review of Units

The units in these equations can be handled in three completely different ways (see Table 2.1):

(A) In older technical systems of units, the numbers obtained by weighing an object on a spring scale are interpreted as units of force (for example pounds-force "lbf"), and the units of mass (i.e., "slugs") are defined in

	UNITS	Force	Mass	Accel.	g_c
A	Tech. Engl.	*lbf*	slug	ft/sec^2	1 slug-ft/lbf-sec^2
A	"	*lbf*	slinch	in./sec^2	1 slinch-in./lbf-sec^2
B	Abs. Engl.	poundal	*lbm*	ft/sec^2	1 lbm-ft/poundal-sec^2
C	Engrg. Engl.	*lbf*	*lbm*	ft/sec^2	32.174 lbm-ft/lbf-sec^2
C	"	"	"	in./sec^2	386.088 lbm-in./lbf-sec^2
B	Metric cgs	dyne	*gm*	cm/s^2	1 gm-cm/dyne-s^2
B	S.I. mks	Newton	*kg*	m/s^2	1 kg-m/N-s^2
C	Engrg. mks	kg_{force}	*kg*	m/s^2	9.80665 kg-m/kg$_f$-s^2

Table 2.1: Force and Mass Units, with weight-scale units in *italics.*

such a way that the constant g_c turns out to be exactly one. This is convenient when the majority of problems to be solved are static-force problems on earth. It is less convenient for dynamics problems, where we have to remember that an object which weighs 1 lbf in standard gravity has a mass of about 1/32 slugs.

The traditional units of pressure are psig (lbf/in$^2_{\text{gage}}$) and psia (lbf/in$^2_{\text{abs}}$).

(B) In physicist-invented units, including Système International (S.I.), the numbers obtained by comparing weights on a balance are interpreted as units of mass (for example pounds-mass avoirdupois "lbm", or kilogram "kg"), and the unit of force (i.e. "poundal," or Newton "N") is defined such that g_c is exactly one. This is convenient in astrophysics and in particle dynamics. It is less convenient on earth, where we keep having to calculate static forces due to weight in standard gravity, and a mass of 1 kg exerts a force of about 9.8 N on its foundation.

The S.I. units of pressure are Pa$_{\text{gage}}$ and Pa$_{\text{abs}}$; 100 kPa are sometimes called a "bar." **S.I. units are required in more and more situations.**

(C) In engineers' compromise units, scale readings (as obtained here on earth in standard gravity) can be used for both mass and force: both lbm and lbf, or both kilogram$_{\text{mass}}$ and kilogram$_{\text{force}}$ (kilopond). Therefore g_c does not turn out to be a simple number (see Table 2.1). It is numerically equal (but different in units) to 32.174 ft/sec^2 or 9.80665 m/s^2, the *average* value of sea-level gravity, at a latitude of 45.5 degrees; *actual* values of the acceleration of gravity vary all over the world, from about 32 to almost 32.3 ft/sec^2.

English lbm and lbf units are still widely used in industry. Units of pressure are psig and psia in English units. In metric units, we still find kg$_f$/cm^2, sometimes called "atmospheres" abbreviated atü (for "Überdruck" or gage), atu (for "Unterdruck" or vacuum), and ata (for "absolut"). Note that 1 kg$_f$/cm^2 equals 98.0665 kPa. "Standard atmospheric

Dimension	S.I.	English
Force	1 N	≅ 0.2248 lbf
"	4.448 N	≅ 1 lbf
Mass	1 kg	≅ 2.2046 lbm (avdp.)
"	0.4536 kg	≅ 1 lbm (avoirdupois)
Length	0.3048 m	= 1 ft
"	25.4 mm	= 1 inch

Table 2.2: U.S. Industrial Unit Conversions.

pressure" is 101.325 kPa or 14.696 psia. Therefore kg_f/cm^2, bar, and standard atmospheric pressure are *almost* identical with each other.

In general, problems should be solved in whatever units the input data are given; converting units back and forth unnecessarily is not acceptable practice. However, if the input is in mixed units, a few conversions are needed (see Table 2.2). For further information, consult an authoritative reference such as the *Handbook of Chemistry and Physics.*[1]

A philosopher might note that the engineers' desire for familiar force and mass units and the physicists' desire for $g_c = 1$, could both have been fulfilled by adjusting either the length unit or the time unit. If the unit of length were equal to about 9.81 m, instead of 1/40,000,000 of some circumference of the earth, weighing-scale units could serve for both mass and force. Alternatively, if the unit of time were equal to about 0.32 seconds, instead of 1/84,600 of an average earth day, kg-force and kg-mass would be comparable on earth. Unfortunately, the early standard-setting bodies were not sophisticated enough to anticipate these issues.

Exercise 2.4 *Determine the value of g_c for a metric system using the kilogram for both force and mass, the Dekameter for length, and the second for time.*

Exercise 2.5 *Determine the value of g_c for an English system using the pound for both force and mass, the half-chain (equal to two rods or 11 yards) for length, and the second for time.*

Exercise 2.6 *Determine the value of g_c for a system using the kilogram for both force and mass, the meter for length, and a time unit which splits the day in 25 new-hours, each divided into 100 new-minutes, each divided into 100 new-seconds.*

2.7 Model Example

Homework, like all engineering work, should be laid out in such a way that it can be checked. As displayed in Figure 2.3, this means recording the following:

[1] David R. Lide, Jr., editor, *CRC Handbook of Chemistry and Physics,* 80th edition, CRC Press, Boca Raton, Florida, 1999, ISBN 0-8493-0480-6

1. PROBLEM STATEMENT: An 11-pound mass is supported by a spring of strength $k_1 = 15$ lbf/in from below, and a second spring $k_2 = 275$ lbf/ft from above. It is subject to gravity and an upward magnetic force $F_m = 33$ lbf. What is the static deflection, if the springs exerted zero force before gravity and the magnetic field were applied?

2. DIAGRAM:

3. NEWTON'S LAW:

$$\sum F = \frac{ma}{g_c}$$

$$-k_1 x - k_2 x + F_m - \frac{mg}{g_c} = \frac{m\ddot{x}}{g_c}$$

$$\frac{m}{g_c}\ddot{x} + (k_1 + k_2)\,x = \left(F_m - \frac{mg}{g_c}\right)$$

$\ddot{x} = 0$ for static equilibrium

4. ALGEBRAIC SOLUTION:

$$x = \frac{\left(F_m - \frac{mg}{g_c}\right)}{(k_1 + k_2)}$$

5. NUMBERS AND UNITS:

$$\frac{mg}{g_c} = \frac{11\ \text{lbm}}{} \left| \frac{32.1\ \text{ft}}{\text{sec}^2} \right| \frac{\text{lbf sec}^2}{32.174\ \text{lbm ft}} = 11.0\ \text{lbf weight}$$

$$k_2 = \frac{275\ \text{lbf}}{\text{ft}} \left| \frac{1\ \text{ft}}{12\ \text{in}} \right. = 22.917\ \text{lbf/in}$$

substituting into algebraic solution, on long fraction bar:

$$x = \frac{(33 - 11.0)\ \text{lbf}}{} \left| \frac{\text{in}}{(15 + 22.917)\ \text{lbf}} \right.$$

$$= +\,0.58022\ \text{in.}$$

6. FINAL ANSWER:

static x = 0.58 in. upwards

Magnetic force exceeds weight.

Springs act in parallel, even though they are on opposite sides of the mass.

Figure 2.3: Model Homework Example

1. The problem to be solved, clearly identified by means of:
   (a) a description, in words, pictures, or symbols, of what is known or given; and
   (b) the objective or purpose, i.e., what is to be found.
2. A diagram showing:
   (a) the schematic representation of the system (for example, by means of lumped masses and springs);
   (b) the nomenclature used in the analysis (labeling the mass as "m" etc.);
   (c) the control volume to which the analysis is to be applied; and
   (d) arrows indicating the coordinates and directional conventions.
3. A statement of the approach or method of solution:
   (a) the general laws or fundamental equations applied to the control volume (for example, a force/momentum balance); and
   (b) a list of the assumptions and approximations used to simplify the problem.
4. The algebraic solution, before any numerical values are entered.
5. Insertion of numbers and units into the algebraic solution in such a way that it is evident to a checker what values were used. Always write dimensional numbers together with their units. Resolve unit conversions using the "long fraction-bar" approach shown in the example.
6. Clearly identify your final solution:
   (a) round-off the final answer (but not any intermediate results) to the number of significant figures which indicate the accuracy, remembering that the accuracy of the results is limited by the accuracy of the given input data and of the assumptions;
   (b) re-examine the assumptions you made in the light of the calculated results;
   (c) report the answer in numbers and units; and
   (d) write out any conclusions, using clear wording.

Organization, neatness, logic, and clarity of solution are all important parts of a correct answer.

Problem 2.7 *On the moon, where gravitational acceleration is 5.47 ft/sec^2, what is the mass in lbm, and the weight in lbf, of a person who weighed* 165 *lbf on earth?*

Problem 2.8 *On the moon, where gravitational acceleration is* 1.67 m/s^2, *what is the mass in kg, and the weight in N, of a person who weighed* 735 *N on earth?*

Problem 2.9 *A* 5-*kg mass is supported by a spring of strength* $k_1 = 2.6$ *kN/m from below, and a second spring of* $k_2 = 4.0$ *kN/m from above. It is subject to gravity and an upward magnetic force* $F_m = 150$ *N. What is the static deflection, if the springs exerted zero force before both gravity and the magnetic field were applied? Hint: Follow the procedure of the Model Homework Example, Figure 2.3. The parameters have similar values, so the answer should have the same order of magnitude.*

Problem 2.10 *In the same problem, change the configuration so that the two springs are in series. How much do the equivalent spring constant and the static deflection change?*

Problem 2.11 *Develop a differential equation for a mass constrained by a spring and a damper connected in series. Hint: Rather than summing forces as a function of* x *and its derivatives, add up displacements as a function of the force* f *and its integrals.*

Problem 2.12 *What is the equivalent spring constant* k *at the tip of a uniform cantilever beam, as a function of length* L *and stiffness* EI? *Reminder: If location* x *is measured from the base of the cantilever, a force* F *applied at the tip* $x = L$ *leads to*

$$\begin{aligned}
Shear &= F \\
Moment &= F(L-x) \\
\frac{d^2y}{dx^2} &= \frac{F(L-x)}{EI} \\
\frac{dy}{dx} &= \frac{F}{EI}\left(Lx - \frac{x^2}{2}\right) \\
y &= \frac{F}{EI}\left(\frac{Lx^2}{2} - \frac{x^3}{6}\right) \\
y_{tip} &= \frac{FL^3}{3EI} \\
\left[\frac{dy}{dx}\right]_{tip} &= \frac{FL^2}{2EI}
\end{aligned}$$

Problem 2.13 *What is the equivalent spring constant* k *at the center of a uniform cantilever beam, as a function of length* L *and stiffness* EI? *Hint:*

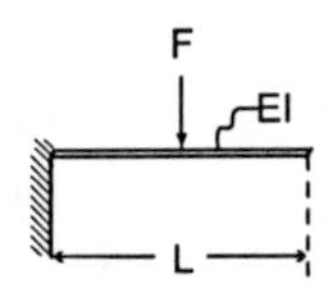

The outer half of the cantilever is unstressed and extends without curvature. The inner half is stressed like a cantilever of length $L/2$.

Problem 2.14 *What is the equivalent spring constant* k *at the tip of a six-foot long cantilever beam, made of solid square one-inch steel bar? Reminder: Young's modulus* E *for steel is approximately* 28×10^6 *psi; the bending "moment of inertia"* I *for a square cross-section of width and height* b *is* $b^4/12$. *Approximate answer:* 20 *lbf/in.*

Problem 2.15 *What is the equivalent spring constant for an object suspended from a rubber band with a spring constant of* 15 *lbf/in, if the other end of the rubber band is attached to the tip of a cantilever with a spring constant of* 20 *lbf/in.?*

Problem 2.16 *In the previous problems, what is the equivalent spring constant if a second rubber band of spring constant* 15 *lbf/in is mounted as a helper-spring between the ceiling and the tip of the cantilever?*

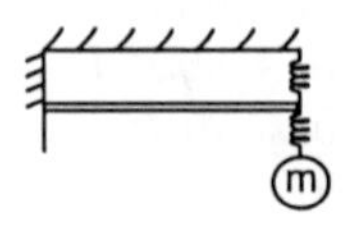

Problem 2.17 *In the previous problem, what is the equivalent spring constant if the second "helper" rubber band is connected directly from the ceiling to the suspended object?*

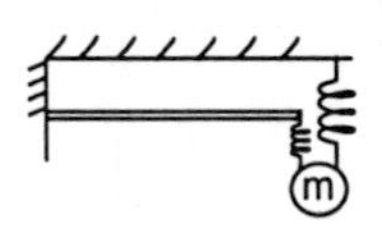

Problem 2.18 *What is the equivalent spring constant* k *for lateral deflection for a platform supported by four uniform pillars built-in both at the bottom (into the ground) and at the top (into the platform), as a function of length* L *and stiffness* EI*?*

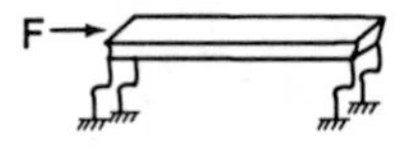

Hint: If location x *is measured from the ground, the shear in each pillar is*

$$
\begin{aligned}
\text{Shear} &= F/4 \\
\text{Moment} &= \frac{F}{4}\left(\frac{L}{2} - x\right)
\end{aligned}
$$

Problem 2.19 *What is the equivalent spring constant k for lateral deflection for a table-like platform supported by six uniform legs built-in at top and bottom, each 0.8 m tall and made of solid round 50-mm steel rods? Reminder: Young's modulus E for steel is approximately 200×10^3 MPa; the bending "moment of inertia" I for a circular cross-section of diameter D is $\pi D^4/64$. Approximate answer: 9000 kN/m.*

Problem 2.20 *What is the equivalent spring constant k at the center of a simply supported uniform beam, as a function of length L and stiffness EI? Hint: From symmetry, each half of the beam behaves like a cantilever extending $L/2$ from the center, and loaded at its tip by the support reaction $F/2$.*

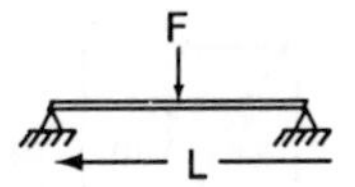

Problem 2.21 *What is the equivalent spring constant k at the center of a uniform beam built-in at both ends, as a function of length L and stiffness EI? Hint: This beam can be divided into segments which have the same loading as cantilevers and simply-supported beams.*

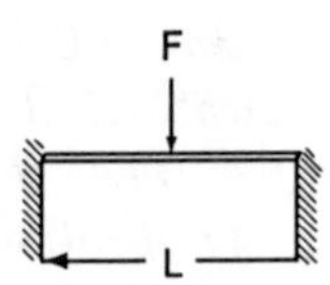

Problem 2.22 *What is the equivalent spring constant k at the center of a uniform beam built-in at one end and hinged at the other, as a function of length L and stiffness EI? Hint: This beam's deflection can be obtained by superposing the response of a cantilever to a downward load F in the middle, and the response to an appropriate upward force at the tip to bring the pinned end back to its proper location.*

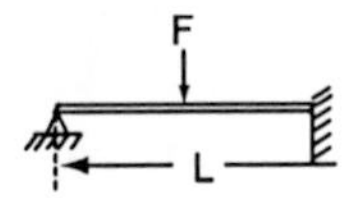

Problem 2.23 *What is the equation of motion of the pulley problem pictured here:*

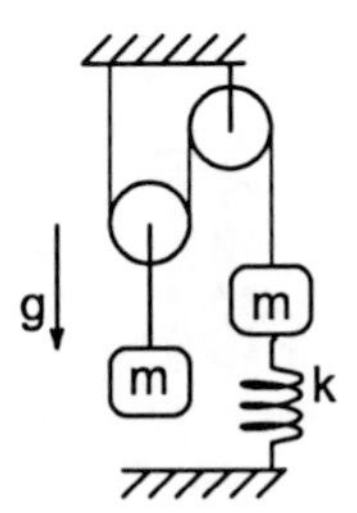

Hint: To solve this problem systematically, assign a coordinate to each mass, enclose each mass in a control volume, and write Newton's law for each control volume in terms of its mass m_n*, its vertical coordinate* x_n*, the spring constant* k*, and the string tension* T*. This will give you two differential equations in three unknowns,* x_1*,* x_2*, and* T*. The third equation is the conservation-of-string equation, relating* x_1 *and* x_2 *so that the length the string remains constant. All three equations can be combined into one differential equation for* x_1 *or* x_2*.*

Problem 2.24 *A mass is held away from a wall by two rigid links, pinned at each end, and supported by a diagonal spring* k *which is angled 30 degrees from the vertical.*

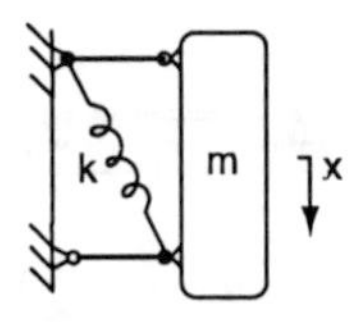

When the links are horizontal, what is the effective spring constant for vertical motion of the mass?

Problem 2.25 *For your bicycle, estimate the spring constant of one tire when it is supporting about half your weight. Hint: For a flexible tire, the support force is the gage pressure in the tire times the flattened area of contact with the ground, so you need to find a way of estimating the change of contact area resulting from vertical displacement. Does the air pressure change significantly with displacement? Is the tire a linear spring?*

Problem 2.26 *Devise an experiment to determine the spring constant of a bicycle tire. Can you also determine if the deflection of the rim, spokes, fork, frame, and saddle are comparable to the tire deflection?*

Chapter 3

FORMULATION OF ROTATIONAL SYSTEMS and review of second moments

Many engineering problems involve rotation, and it is not realistic to speak of a single "point" mass. But if we idealize the mass system as rigid, we can write similar differential equations for rotation θ as we did for translation x, by formulating a rotational equivalent mass called "moment of inertia."

3.1 Newton's Law Revisited

Let us begin by examining a very simple rotational system. If a concentrated "point" mass m is placed at the end of a rigid, weightless lever of length r (Figure 3.1) and a torque T is applied to the lever, we observe that a small displacement x of the mass can also be expressed as $r\theta$, a velocity $\dot{x}$ as $r\,\dot{\theta}$, an acceleration $\ddot{x}$ as $r\,\ddot{\theta}$, and the force F at the mass as T/r. Substituting these into Equation 2.1 $F = ma/g_c$, we obtain $T/r = (mr/g_c)\,\ddot{\theta}$, which we can solve

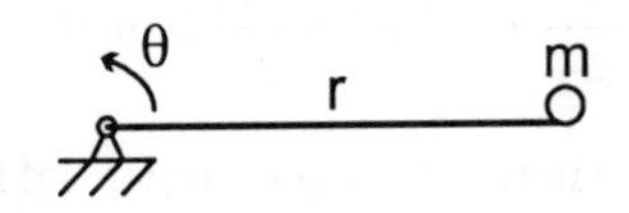

Figure 3.1: Rotational System

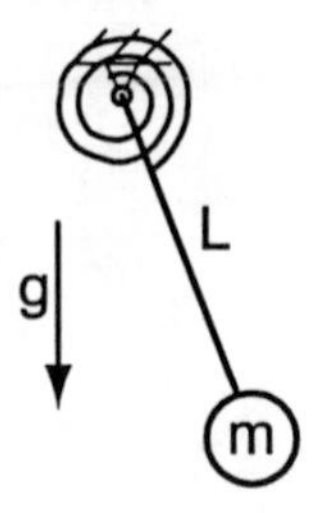

Figure 3.2: Simple Pendulum with Torsion Spring

for T to obtain Newton's law in units of torque:

$$T = \frac{(mr^2)\ddot{\theta}}{g_c} \tag{3.1}$$

We identify (mr^2) as the rotational equivalent mass or "moment of inertia"; it is a function of both the magnitude of the mass and the distance r from pivot to mass. Since this radius r is squared, this moment is also called the "second moment of the mass" about the pivot; it is designated by the symbol J_{pivot}.

3.2 Torsional Systems

If we add a torsional spring (a clock spring or a torsion bar), a torsional damper, and torsional excitation to Figure 3.1, we can write a differential equation that sums up torques similar to the way Equation 2.6 sums up forces:

$$\boxed{\frac{J_{\text{pivot}}}{g_c}\ddot{\theta} + c_T\dot{\theta} + k_T\theta = T(t)} \tag{3.2}$$

where $c_T \triangleq$ torsional damping constant, $k_T \triangleq$ torsional spring constant, and $T(t) \triangleq$ any externally applied torque.

Example: A pendulum with a concentrated mass is called a "simple pendulum." If we also add a torsion spring k_T (Figure 3.2) attached such that its force is zero when the pendulum is vertical, and evaluate the restoring moment—force times lever arm—due to gravity, the equations above reduce to

$$\frac{mL^2}{g_c}\ddot{\theta} + k_T\theta + \frac{mg}{g_c}L\sin\theta = 0 \tag{3.3}$$

We can interpret the last term as either the tangential force component $(mg/g_c)\sin\theta$ times the radial length, or else as the vertical force mg/g_c times the horizontal lever arm $L\sin\theta$. Unlike our crude analysis in the previous chapter (page 13), this equation is not yet linearized. Recalling the Maclaurin series expansion

$$\begin{aligned} \sin\theta &= \theta - \frac{\theta^3}{3!} + \frac{\theta^5}{5!} - \dots \\ &\cong \theta, \text{ if } \frac{\theta^3}{6} \ll \theta \end{aligned} \tag{3.4}$$

we recognize that we can linearize the equation by replacing $\sin\theta$ by θ. This leads to the same gravitational term as the crude analysis of Section 2.4, but now we know the magnitude of the terms we have neglected when we linearized. The fractional error in the last term of the linearized equation is $\frac{\theta - \sin\theta}{\sin\theta} \cong \frac{\theta^2}{6}$: if θ remains within ± 6 degrees (0.1 radians) of the vertical, the error in the gravitational term remains within 0.2%; if θ increases to 57 degrees (1 radian) from the vertical, the error in the last term increases to 16%. Because it is important to know the limitations of the analysis, it is always best to delay linearization as long as possible.

3.3 Rigid Bodies

In Chapter 2, where we had pure straight-line translation x, a body was treated as a concentrated "point" mass assuming that all parts of it moved as one; i.e., that it was rigid. Thus the translation x of any point in the body was also the translation of any other point. In mathematical terms: if a rigid body is made up of a large number of small masses dm, it can be represented by the single concentrated mass $M = \iiint dm$, where the integration stands for the summation of all the infinitesimal parts of the body.

Similarly, in problems where we have pure rotation θ about a point, we can sum the second moments of all the small masses dm in a rigid body to obtain the total moment of inertia

$$\boxed{J_{\text{pivot}} = \iiint r^2 dm} \tag{3.5}$$

where r is the distance of each mass dm from the axis of rotation.

Example: For a flywheel of uniform thickness b and outside radius R, the mass is the sum of a series of infinitesimal rings of width b, circumference $2\pi r$, and thickness dr:

$$M = \rho b \int_0^R (2\pi r)\, dr = \rho b \pi R^2$$

The second moment about the central axis of the disk is:

$$J_{zz} = \rho b \int_0^R (2\pi r)\, r^2 dr = \rho b \frac{\pi}{2} R^4 = 0.5 R^2 M$$

In translation, a rigid flywheel behaves like a concentrated mass M located at the center of gravity. In rotation, it behaves as if that mass were an average distance of $\sqrt{0.5 \cdot R^2}$ from the axis. We can define that average distance as the "radius of gyration"

$$r_{gyr}^2 \triangleq \frac{J}{M} = \frac{\iiint r^2 dm}{\iiint dm} \tag{3.6}$$

which turns out to be $0.707R$ for any uniform-disk flywheel of radius R.

Exercise 3.1 *Find the second moment of a disk with a hole in the center.*

Exercise 3.2 *Show that a thin hoop has $J_{zz} \approx 1.0MR^2$. What is the order of magnitude of the error due to neglecting the thickness of the hoop?*

Since sectional second moments are of importance also in the torsion of rods and bending of beams, values for different shapes can be found in many engineering texts[1] and handbooks.[2]

- Sectional second moments for torsion $J_z \triangleq \int r^2 dA = \int \left(x^2 + y^2\right) dA$ can be adapted to find the polar moments J_{zz} of uniform-thickness bodies, whether circular or not.
- Sectional second moments in bending $I_x \triangleq \int y^2 dA$ and $I_y \triangleq \int x^2 dA$, can be adapted to find the moments of inertia $I_{xx} \triangleq \iiint \left(y^2 + z^2\right) dm$ or $I_{yy} \triangleq \iiint \left(x^2 + z^2\right) dm$ of flat, thin shapes ($z \ll x$ or y) about axes within their own plane.

Exercise 3.3 *Compare the $J_{zz} = 0.5MR^2$ obtained above for a uniform circular flywheel with the sectional moment of a circle in a strength-of-materials book.*

Exercise 3.4 *Show that a butterfly valve (a thin circular disk rotating symmetrically about an axis within its own plane) such as is used as an automotive-engine throttle, has a moment of inertia $I_{xx} \cong 0.25MR^2$. What is the order of magnitude of the error due to the thickness of the disk?*

In sectional moments, we can show from Pythagoras' theorem $r^2 = \left(y^2 + x^2\right)$ that $J_{zz} = I_{xx} + I_{yy}$; this is approximately true also for flat, thin shapes ($z \ll x$ or y). Since the integrals for one or the other of these may be much easier to obtain, depending on shape boundaries, we can use this relationship for developing hard-to-get second moments.

Exercise 3.5 *Show that a thin circular hoop has $I_{xx} = I_{yy} \cong 0.5MR^2$.*

Exercise 3.6 *Find I_{xx} and I_{yy} for a thin rectangular plate by integration, and obtain J_{zz} from them.*

---

[1] *e.g.*, Warren C. Young, *Roark's Formulas for Stress and Strain*, 6th Edition, McGraw-Hill, New York, 1989, ISBN 0-07-072541-1.

[2] *e.g.*, Ray E. Bolz and G.L. Tuve, editors, *CRC Handbook of Tables for Applied Engineering Science*, 2nd Edition, CRC Press, Boca Raton, Florida, 1973, ISBN 0-8493-0252-8.

3.4 Center of Gravity

Any rigid-body rotation about some particular pivot can also be represented at any instant as the combination of rotation about some other point, plus a translation of that point. (Conversely, any combination of rotation and translation can be represented as pure rotation about the "instantaneous center of rotation.") In rigid-body dynamics, it is often convenient to represent a motion as the sum of rotation about the center of gravity (c.g.) plus translation of the c.g.

The coordinates of the c.g. in Cartesian coordinates can be obtained by taking the first moments of the mass

$$\overline{x} = \frac{\iiint x dm}{\iiint dm}; \overline{y} = \frac{\iiint y dm}{\iiint dm}; \overline{z} = \frac{\iiint z dm}{\iiint dm} \quad (3.7)$$

Exercise 3.7 *Find the mass, the location of the c.g., and some moments of inertia of a thin plate in the shape of a right triangle.*

Exercise 3.8 *Find the mass, the location of the c.g., and some moments of inertia of a thin plate in the shape of a half-circle.*

Exercise 3.9 *Find the mass, the location of the c.g., and some moments of inertia of a thin plate in the shape of a ninety-degree sector of a circle.*

Exercise 3.10 *Find the mass, the location of the c.g., and some moments of inertia of a thin plate in the shape of a ninety-degree segment of a circle (i.e., a sector minus a right triangle).*

3.5 Parallel-Axis Theorem

The mass of a body is the same, no matter what coordinates are chosen; The location of its c.g. is in the same place, no matter in what coordinates it is expressed. But the second moment of a body depends on the location of the axis about which it turns. Steiner's parallel-axis theorem transfers the second moment about the c.g., to the second moment about any other pivot:

$$J_{\text{pivot}} = J_{\text{c.g.}} + Md^2 \quad (3.8)$$

where d is the distance from the axis of rotation to the center of gravity. If we know the second moment about any axis through the c.g., we can find the second moment about any other parallel axis. Conversely, if we know the second moment about any axis, we can find the second moment about the parallel axis through the c.g., and from that the second moment about any other parallel axis.

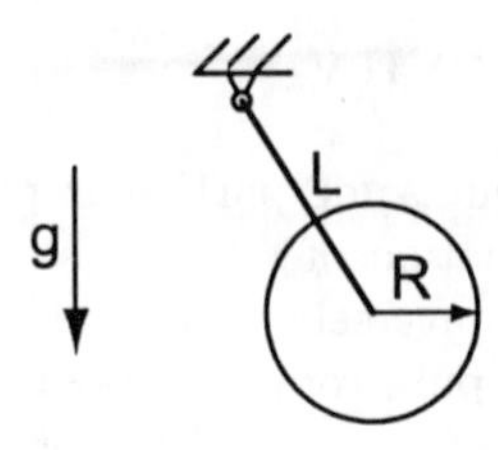

Figure 3.3: Compound Pendulum

Example: The mass of a yardstick of cross-sectional area A is

$$M = \rho A \int_0^L dx = \rho AL$$

The first moment of mass is

$$\rho A \int_0^L x dx = 0.5LM$$

so that the c.g. must be at $\overline{x} = 0.5L$. The second moment for a thin stick about a pivot at one end is

$$J_{\text{pivot}} = \rho A \int_0^L x^2 dx = ML^2/3$$

and the second moment for a thin stick about its center is

$$J_{c.g.} = 2\rho A \int_0^{L/2} x^2 dx = ML^2/12$$

We can verify that for a pivot at the end

$$J_{\text{pivot}} = J_{\text{c.g.}} + Md^2 = ML^2/12 + M\left(L/2\right)^2 = ML^2/3$$

We can also obtain the second moment about another pivot, for example at the one-foot mark ($L/3$ from the end or $L/6$ from the center) by writing $J_{\text{pivot}} = J_{\text{c.g.}} + Md^2 = ML^2/12 + M\left(L/6\right)^2 = ML^2/9$.

Exercise 3.11 *In this example, what is the order of magnitude of the error due to the width of the yardstick?*

3.6 The Compound Pendulum

Let us study a pendulum which consists of a mass M in the form of a flat, solid disk of radius R, supported by a "weightless" rod measuring length L from pivot to the mass' center of gravity (Figure 3.3). The torque in the governing Equation 3.1 is the force due to gravity multiplied by the lever-arm length upon which it acts:

$$\frac{-MgL\sin(\theta)}{g_c} = \frac{J_{\text{pivot}}}{g_c}\ddot{\theta} \tag{3.9}$$

where $J_{\text{pivot}} = J_{\text{c.g.}} + ML^2$; for a flat disk, $J_{\text{c.g.}} = 0.5MR^2$ so that we get

$$\left(\frac{0.5MR^2}{g_c} + \frac{ML^2}{g_c}\right)\ddot{\theta} + \frac{MgL}{g_c}\sin\theta = 0 \tag{3.10}$$

Compare this with the solutions for the simple pendulum in the previous chapter (page 13) and in Section 3.2.

1. This is a nonlinear equation. We can linearize it by substituting θ for $\sin\theta$. We know what kind of error this introduces from the Maclaurin series expansion (Equation 3.4); the error is small if $\theta^2/6 \ll 1$.

2. The effect of the size of the mass is now quantified. It is negligible if $R^2 \ll L^2$.

For a compact mass, this reduces to our Equation 2.7 for the simple pendulum. However, using the rotational-system approach, we now know what the effects of the linearization and of the point-mass assumption are. On the one hand, the effect of the size of the mass is less than 0.5% if $R < 0.1L$. In the other extreme case—a flywheel mounted slightly eccentrically—the length L is negligible for the moment-of-inertia term (but important in the gravitational term) if $L \ll R$.

3.7 Summary

Rotational motion of a body is conveniently described in either of two coordinate systems:

- either as a combination of translation of the c.g. plus rotation—the variables are $x_{c.g.}$ and θ;

- or as pure rotation about a pivot or other instantaneous center of rotation—the variables are θ and the location of the center of rotation.

For the second description, a single differential equation can be written by summing up all the torques about the instantaneous center of rotation. The coefficient of $\ddot{\theta}$ is the moment of inertia, which can be computed by means of a second-moment integral about the given pivot axis. Also, if the location of the c.g. is known, the transfer theorem lets us compute moment of inertia about any pivot axis from the moment of inertia about any other parallel axis, by way of the parallel axis through the c.g.

Problem 3.12 *Find the moment of inertia of a two-pound yardstick when it pivots at the x-inch marking, where x is chosen according to the middle initial of your name:*

init.	x	*init.*	x	*init.*	x	*init.*	x	*init.*	x
A.	0	*F.*	5	*L.*	10	*Q.*	15	*V.*	20
B.	1	*G.*	6	*M.*	11	*R.*	16	*W.*	21
C.	2	*H.*	7	*N.*	12	*S.*	17	*X.*	22
D.	3	*I./J.*	8	*O.*	13	*T.*	18	*Y.*	23
E.	4	*K.*	9	*P.*	14	*U.*	19	*Z.*	24

Problem 3.13 *Find the governing equation for the rocking of a one-half-circular solid cylinder on a flat surface.*

Problem 3.14 *Find the governing equation for the rocking of a thin pipe, longitudinally sawed in half, on a flat surface.*

Problem 3.15 *Find the governing equation for the rocking of a solid circular cylinder of radius R_1 inside a tube with internal radius R_2.*

Problem 3.16 *Find the governing equation for the motion of a mass $M = 100$ lbm at the end of a rigid crank $R = 1$ ft., attached to a steel cylindrical torsion bar $D = 1$ inch, of length $L = 1$ yard; check it for dimensional consistency. Hint: solve it in algebraic terms first, before plugging in numbers and units. Reminder: The shear modulus G of steel is about 11×10^6 psi; and the sectional "polar moment of inertia" J of a solid circle is $\pi D^4/32$.*

Problem 3.17 *Estimate the moment of inertia of a bicycle wheel, using the approximate dimensions of your own bicycle. Hint: Estimate (or weigh or look up in a mail-order catalog) the mass of each part of a wheel—tire, tube, rim, nipples, and spokes—and apply the appropriate radius of gyration.*

Chapter 4

UNDAMPED FREE VIBRATION and static deflection

In the simplest vibration analyses, damping is neglected, and the motion is the result of initial conditions. These assumptions lead to the first of the Standard Form equations we will study.

4.1 Governing Equations

Neglecting damping, and in the absence of external excitation, the force balance for a translational system (Equation 2.6) reduces to

$$\frac{m}{g_c}\ddot{x} + kx = 0 \tag{4.1}$$

This simple equation has a form which shows up frequently in other problems of engineering and physics. For example, from the torque balance of Equation 3.2, free vibrations in undamped rotational problems are governed by

$$\frac{J}{g_c}\ddot{\theta} + k_T\theta = 0 \tag{4.2}$$

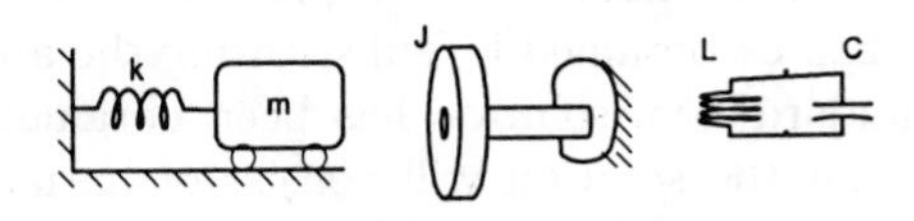

Figure 4.1: Undamped Translation, Rotation, and Electric Circuits

Similarly, electrical tank circuits (see Figure 4.1) lead to a voltage balance

$$L\frac{d\Im}{dt} + \frac{1}{C}\int \Im dt + \text{constant} = 0 \tag{4.3}$$

or an electric-current balance

$$C\frac{dE}{dt} + \frac{1}{L}\int E dt + \text{constant} = 0 \tag{4.4}$$

both of which can be transformed into second-order differential equations, without integral terms, by differentiating each term with respect to time.

Exercise 4.1 *Find a reference on analog computers, and determine how similar equations can be generated using active components like integrators.*

4.2 Standard Form

Each of the second-order differential equations above has two constant coefficients. In order to reduce the number of constants, it is customary to divide by the coefficient of the highest-order term, leading to

$$\ddot{x} + \left(\frac{kg_c}{m}\right)x = 0 \tag{4.5}$$

$$\ddot{\theta} + \left(\frac{k_T g_c}{J}\right)\theta = 0 \tag{4.6}$$

$$\ddot{I} + \left(\frac{1}{LC}\right)I = 0 \tag{4.7}$$

$$\ddot{E} + \left(\frac{1}{LC}\right)E = 0 \tag{4.8}$$

Each of these equations now has only one normalized parameter, which is given the name ω_n^2; it is taken as a square for convenience, because we will be taking the square root later. Thus the Standard Form of all these equations is

$$\boxed{\ddot{x} + \left(\omega_n^2\right)x = 0} \tag{4.9}$$

By comparing this equation with the previous four equations, we see that the definition of the normalized parameter is $\left(\omega_n^2\right) \triangleq (kg_c/m)$, or $(k_T g_c/J)$, or $(1/LC)$, respectively.

Next, we will solve the Standard Form Equation 4.9; any of the other equations' solutions can then be obtained by substituting the appropriate definition for ω_n^2, after the standard-form solution has been obtained. The alert reader already anticipates that the solution will consist of harmonic motion; nevertheless, let us carry out a fractional analysis, as well as examine exponential solutions, as a rehearsal for meeting the challenges of later chapters.

4.3 Fractional Analysis

As a first step in studying a differential equation, we can obtain partial information about its solution by inspecting the non-dimensional coefficients in the equation.[1] But ω_n^2 is not dimensionless: it must have units of $\sec^{-2}$ in order for the units of the two terms in the equation to agree. We could fix this by re-scaling the time coordinate to a dimensionless time

$$t^* \triangleq \omega_n t \tag{4.10}$$

then

$$\ddot{x} = \frac{d^2x}{dt^2} = \omega_n^2 \frac{d^2x}{d(\omega_n t)^2} = \omega_n^2 \frac{d^2x}{d(t^*)^2} = \omega_n^2 x''$$

Substituting this in the Standard Form of the differential equation and dividing both terms by ω_n^2 results in

$$x'' + x = 0 \tag{4.11}$$

This form of the equation does not have any coefficients! We must conclude that all undamped second-order systems are similar, and ω_n is the time-scaling factor for obtaining similitude. From a mathematician's viewpoint, the coefficientless equation is the defining equation for trigonometric functions, just like Bessel's equation is the defining equation for Bessel functions. Since engineers are not comfortable with using dimensionless time coordinates, we will stay with our Standard Form Equation 4.9. However, it is useful to remember that we must expect the dimensionless product $\omega_n t$ to keep occurring in the solutions.

4.4 Exponential Solution

Equation 4.9 is an Ordinary Differential Equation (O.D.E.), because the dependent variable $x(t)$ is a function of only one independent variable, time t. This O.D.E. has the following properties:

- Because x always appears in the first power, the equation is **linear**. Therefore its solutions are superposable: the sum of any two solutions is also a solution. This is an important feature, because it means that we can write generic solutions, rather than solve each problem individually.
- Because ω_n^2 is not a function of t, the equation has only **constant coefficients.** Therefore we will have no solutions caused by fluctuating parameters.

[1] S.J. Kline, *Similitude and Approximation Theory,* reissued by Springer-Verlag, Berlin/Heidelberg, 1986, ISBN 0-38716518-5.

- Because there is no non-x term—such as a forcing term $F(t)$ on the right-hand side of the equation—the equation is **homogeneous**. It has no solutions caused by external influences. That leaves only solutions caused by the Initial Conditions.
- Because the equation is homogeneous and linear **with constant coefficients**, we expect solutions that have an exponential-function form.
- Because the equation is **second-order**, we expect to find two parameters, determined by two Initial Conditions, in the solution.

Let us try a solution in the form of an exponential function

$$x = Ae^{st} \tag{4.12}$$

where A and s are unknown constants. Then

$$\ddot{x} = As^2 e^{st} = s^2 x$$

We immediately see the usefulness of the exponential form: differentiation results in similar functions of time, so that we can arrange to cancel terms incorporating this solution and its derivatives. Substituting into Equation 4.9, we get

$$As^2 e^{st} + \left(\omega_n^2\right) Ae^{st} = 0$$

or

$$A(s^2 + \omega_n^2)e^{st} = 0$$

Because e^{st} is zero only when $st = -\infty$, and $A = 0$ is the trivial solution where x is always zero, our solution must satisfy the characteristic equation

$$\left(s^2 + \omega_n^2\right) = 0 \tag{4.13}$$

As expected, there are two non-trivial solutions for the value of s

$$s_{1,2} = \pm\sqrt{-\omega_n^2} = \pm\sqrt{-1}\omega_n = \pm i\omega_n$$

and Equation 4.12 becomes

$$x = A_1 e^{i\omega_n t} + A_2 e^{-i\omega_n t} \tag{4.14}$$

We have assumed here that the coefficient $\left(\omega_n^2\right)$ in Equation 4.9 was a positive number, as is usually the case. Therefore ω_n is a real number, and $\pm i\omega_n t$

are imaginary exponents, which we find difficult to interpret. Exponential functions with imaginary arguments suggest oscillating solutions, through Euler's formulas

$$\begin{aligned} e^{i\omega t} &= \cos\omega t + i\sin\omega t \\ e^{-i\omega t} &= \cos\omega t - i\sin\omega t \end{aligned} \tag{4.15}$$

or conversely

$$\begin{aligned} \sin\omega t &= \tfrac{1}{2i}\left(e^{i\omega t} - e^{-i\omega t}\right) \\ \cos\omega t &= \tfrac{1}{2}\left(e^{i\omega t} + e^{-i\omega t}\right) \end{aligned} \tag{4.16}$$

For simplicity, we will not try to convert our exponential solution into sinusoidal form, but will start over again, looking directly for sinusoidal solutions.

Exercise 4.2 *Compare the Maclaurin series expansions of sine, cosine, and exponential functions.*

4.5 Sinusoidal Solution

Let us now try the solution

$$x = A\sin(\omega t) + B\cos(\omega t) \tag{4.17}$$

where A, B, and ω (without a subscript) are unknown constants. Then

$$\begin{aligned} \dot{x} &= A\omega\cos(\omega t) - B\omega\sin(\omega t) \\ \ddot{x} &= -A\omega^2\sin(\omega t) - B\omega^2\cos(\omega t) \\ &= -\omega^2 x \end{aligned}$$

We see that sinusoidal solutions, like exponential functions, have the desirable property of yielding similar functions when differentiated. Substituting into Equation 4.9, we find that the terms can be made to cancel, even for A and $B \neq 0$, if:

$$\omega^2 = \omega_n^2 \tag{4.18}$$

Therefore the constant ω, the angular frequency (in radians-per-second) of the solution of an unforced, undamped system is identically equal to the system parameter ω_n (actually, the sign is $\pm$ after you take the square root, but only the positive value makes sense, since cause-and-effect considerations tell us that ωt must be increasing with the passage of time). Therefore the non-trivial solution is

$$\boxed{x = A\sin(\omega_n t) + B\cos(\omega_n t)} \tag{4.19}$$

Exercise 4.3 *Noting that $s = \pm i\omega_n$, and using Euler's formulas above, convert the solution from the sinusoidal form of Equation 4.19 to the exponential form of Equation 4.14. In other words: what are the values of A_1 and A_2 in Equation 4.14, in terms of the A and B of Equation 4.19?*

4.6 Initial Conditions

A and B in Equation 4.19 are still unknown; they must be determined from Initial Conditions. Taking Equation 4.19 and its derivative

$$\dot{x} = A\omega_n \cos(\omega_n t) - B\omega_n \sin(\omega_n t)$$

and evaluating them at $t = 0$, at which time $\sin(\omega_n t) = 0$ and $\cos(\omega_n t) = 1$, we get:

$$x(0) = B \tag{4.20}$$

and

$$\dot{x}\,(0) = A\omega_n \tag{4.21}$$

Therefore the constant B is the initial displacement $x(0)$, and A is obtained by dividing the initial velocity $\dot{x}\,(0)$ by the system parameter ω_n; the solution is now fully determined. It is a sinusoidal oscillation which can also be expressed in the alternative form

$$\boxed{x = C\sin(\omega_n t + \phi)} \tag{4.22}$$

where $C = \sqrt{A^2 + B^2}$ and $\phi = \arctan(B/A)$. Examining this solution, we see that the maximum amplitude of the motion is C; the period of a complete oscillation (when $\omega_n t$ goes through 2π radians, or 360 degrees) is $\tau_n = 2\pi/\omega_n$, and its inverse is the frequency $f_n = \omega_n/2\pi$.

By differentiating this solution, we learn that

$$\left|\dot{x}_{\max}\right| = \omega_n C = \omega_n \left|x_{\max}\right| \tag{4.23}$$

$$\left|\ddot{x}_{\max}\right| = \omega_n^2 C = \omega_n^2 \left|x_{\max}\right| \tag{4.24}$$

Of course, $x_{\max}$ and $\dot{x}_{\max}$ occur at different times, since sines and cosines peak in different parts of the cycle.

The name "natural frequency" is used for both f_n and ω_n; the difference is that f_n is expressed in units of Hertz (cycles-per-second), while ω_n is in angular units of radians-per-second. Since this difference does not show up in a dimensional check (cycles and radians are both dimensionless), we can *avoid numerical errors by distinguishing between these symbols in our algebraic expressions.*

The word "amplitude" is also ambiguous. Mathematicians call C the amplitude and $2C$ the "double amplitude" or "peak-to-peak amplitude." Experimentalists traditionally call $2C$ the "amplitude" and C the "half-amplitude."

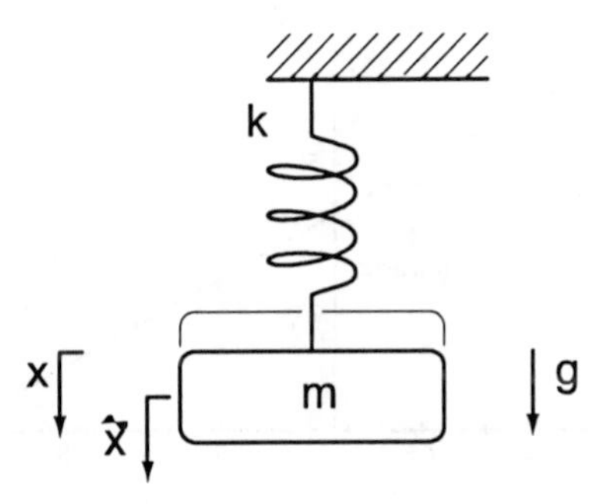

Figure 4.2: Static Deflection

4.7 Static Deflection

In obtaining Equation 4.1, we neglected not only damping and the forcing function $F(t)$, but also the effect of gravity. Adding back in the gravitational term from the original Equation 2.6, the governing equation becomes:

$$\frac{m}{g_c}\ddot{x} + kx = \frac{mg}{g_c} \tag{4.25}$$

This differential equation is non-homogenous, but only in a trivial way, because the right-hand term is a constant. We can represent x as the sum of a constant or "DC" value $\overline{x}$ and a fluctuating or "AC" value $\widetilde{x}$. Substituting these components for x in the equation (but leaving out $\ddot{\overline{x}}$, because the derivative of a constant must equal zero):

$$\frac{m}{g_c}\ddot{\widetilde{x}} + k(\widetilde{x} + \overline{x}) = \frac{mg}{g_c} \tag{4.26}$$

Since the constant terms must cancel against each other, and the fluctuating terms against other fluctuating terms, we really have two equations now:

$$k\overline{x} = \frac{mg}{g_c} \tag{4.27}$$

$$\frac{m}{g_c}\ddot{\widetilde{x}} + k\widetilde{x} = 0 \tag{4.28}$$

The equilibrium Equation 4.27 tells us that the static deflection $\overline{x} = mg/kg_c = g/\omega_n^2$. This relationship turns out to be handy in the field: if we can observe the static deflection $\overline{x}$ of a mass on its supporting spring, we can obtain the natural frequency without measuring m or k. Solving for ω_n and f_n, and inserting numerical values for 2π and g_c, we can develop the rule of thumb:

$$\boxed{f_n = \frac{1}{2\pi}\sqrt{\frac{g}{\overline{x}}} \cong \sqrt{\frac{250\,\text{mm}}{\overline{x}_{\text{mm}}}}\,\text{Hertz} \cong \sqrt{\frac{10\,\text{inches}}{\overline{x}_{\text{inches}}}}\,\text{Hertz}} \tag{4.29}$$

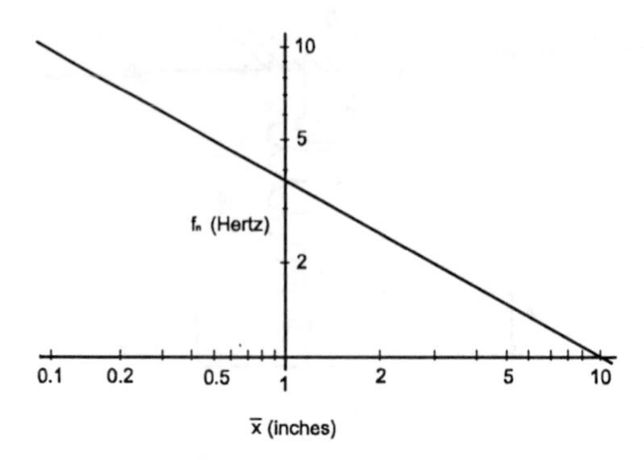

Example: What is the natural frequency of a car that settles about ten inches into its suspension when it is lowered from a lift onto the ground? From Equation 4.29, the bouncing frequency must be about 1 Hz (60 c.p.m.).

The significance of Equation 4.28 is that it is a homogeneous equation again, identical to Equation 4.1 except that the variable x has been replaced by $\widetilde{x} \triangleq (x - \overline{x})$; the coefficients are the same and the natural frequency therefore is also unchanged. This means that the solutions we obtained for Equation 4.9 are still valid for Equation 4.25 if we transform to the coordinate $\widetilde{x}$, shifting the origin by $\overline{x}$. This demonstrates that linear equations are unaffected by biasing constants (forces which push steadily in one direction) except for a shift in the equilibrium point.

4.8 Stability

We wonder what would happen if (ω_n^2) were not a positive number, but happened to be a negative number? The exponents s_1 and s_2 of the solution Equation 4.12 would then turn out to be a pair of real numbers, one positive and one negative, so that we would have one exponentially growing term and one exponentially decaying term. Unless the Initial Conditions were such that the constant A_1 turned out to be *exactly* zero, x would diverge and grow towards infinity. We conclude that a negative (ω_n^2) indicates unstable equilibrium.

If we tried to fit a harmonic-motion solution, Equation 4.17, to this case, we would end up with imaginary numbers for ω in the arguments of the sine and cosine, which we would find difficult to interpret. But we now see that whenever we get imaginary numbers in the arguments of exponential solutions, we should have tried sines and cosines; and whenever we get imaginary numbers in the arguments of sines and cosines, we should have tried an exponential function instead.

4.9 Summary

We have obtained a general solution for the undamped homogenous second-order Ordinary Differential Equation; it is given in Equations 4.19 and 4.22. So far, we have learned two ways of obtaining the system constant ω_n

1. From Newton's law, obtain the differential equation and compare it with the Standard Form Equation 4.9 to find the value of (ω_n^2); or

2. From the static deflection, find f_n from Equation 4.29, remembering that $\omega_n = 2\pi f_n$.

Finally, the constants A and B in Equation 4.19 are determined from the Initial Conditions, as demonstrated in Equations 4.20 and 4.21.

We are now able to find the response of an undamped system to Initial Conditions by finding the natural frequency; fitting the solution to initial velocity and displacement; and predicting the ensuing maximum amplitude C in Equation 4.22.

Problem 4.4 *A mass oscillates harmonically with an amplitude of* 0.1 *inches and a frequency of* 20 *Hertz. What are the values of the maximum velocity in in/sec and the maximum acceleration in g? Hint: "Harmonically" means that the motion has the form of a sine wave (Equation 4.22); differentiate it twice and recall that* $\omega = 2\pi f$. *The unit of acceleration "g"* $\triangleq$ 32.174 *ft/sec*^{2}.

Problem 4.5 *An instrument mounted on an automotive engine detects a sinusoidal vibration at* 1200 *cycles-per-minute with a maximum acceleration of* 4 *g. What is the amplitude in millimeters? The unit of acceleration "g"* $\triangleq$ 9.80665 *m/s*^{2}.

Problem 4.6 *A heavy piece of machinery is set on four rubber mounts, one at each corner. Each mount is observed to compress about* 0.375 *inches. Estimate the natural frequency of the machine for vertical motion. Report your result in Hertz.*

Problem 4.7 *An airplane settles* 150 *mm into its landing-gear springs when the airplane is at rest. What is the natural frequency* f_n *for vertical motion of the airplane?*

Problem 4.8 *A* 2.0*-pound mass is attached to a spring of stiffness* 15 *pounds/inch. What is the natural frequency in Hertz?*

Problem 4.9 *A* 1.0*-pound mass is attached to a spring of stiffness* 15 *pounds/inch. What is the natural frequency in Hertz? Approximate answer:* 12 *Hz.*

Problem 4.10 *A* 2000*-kilogram mass is attached to a spring such that the natural frequency is one Hertz. What is the spring constant?*

Problem 4.11 *A* 75 *kg man stands at the tip of a spring-board, where the spring constant is* 5.0 *KN/m. What is the bouncing frequency in Hz?*

Problem 4.12 *A 50 kg woman stands at the tip of a spring-board, where the spring constant is 5.0 N/mm. What is the bouncing frequency in Hz?*

Problem 4.13 *A thin rod, 36 inches long, is suspended by a pivot which is x inches from one end, where x is chosen according to the first letter of your middle name:*

init.	x	*init.*	x	*init.*	x	*init.*	x	*init.*	x
A	0	F	5	L	10	Q	15	V	20
B	1	G	6	M	11	R	16	W	21
C	2	H	7	N	12	S	17	X	22
D	3	I,J	8	O	13	T	18	Y	23
E	4	K	9	P	14	U	19	Z	24

Find the natural frequency for small oscillation as a pendulum. Express the results in Hertz. Hint: Solve the problem generically for a rod of length L and a c.g.-to-pivot distance d, before plugging in numbers and units. Useful reminder: the second moment of a thin rod about its center of gravity is $mL^2/12$. Answers range from 0 to 0.7 Hertz.

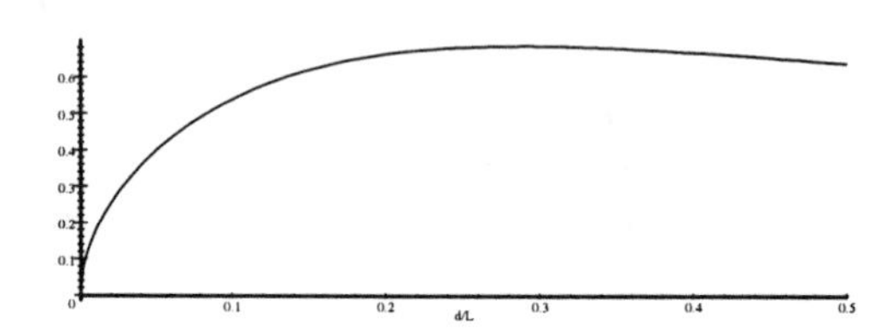

Problem 4.14 *A thin ring (for example, a barrel band or a hula hoop) is suspended from a point at its edge and swings back and forth in its own plane. Treating it as a compound pendulum, find the natural frequency in terms of the diameter D and gravity g.*

Problem 4.15 *A thin ring (for example, a barrel band or a hula hoop) is suspended from a point at its edge and swings back and forth at right angles to its own plane (not in its own plane). Treating it as a compound pendulum, find the natural frequency in terms of the diameter D and gravity g. Hint: we expect a lower second moment and a higher natural frequency than in the preceding problem.*

Problem 4.16 *The post to which the hinges of a swinging gate are attached, has tilted towards the opposite post by about five degrees. If the gate is a uniform panel one meter square, estimate the time it takes to swing shut from a partially-open position, under the influence of gravity.*

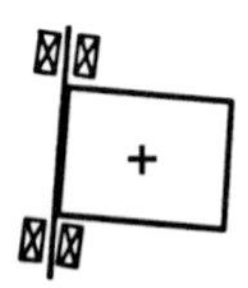

Hint: If the initial opening angle is not too great, we can linearize the governing equation. The time from a maximum displacement to the closed position will turn out to be one-fourth of the characteristic period $\tau = 2\pi/\omega_n$.

Problem 4.17 *Fasten a small weight to different locations on the rim of your bicycle's front wheel, and time the period of small oscillations. From your measurement, calculate the moment of inertia of the rim with and without the added weight.*

Chapter 5

ENERGY METHODS FOR NATURAL FREQUENCY with an introduction to Hamiltonian methods

One of the most important tasks of the engineer is to find the appropriate mathematical representation of a physical problem. In previous chapters, the governing equations were developed from Newton's law. In complex situations, it is frequently easier to obtain the mathematical model and determine the natural frequency from Conservation of Energy.

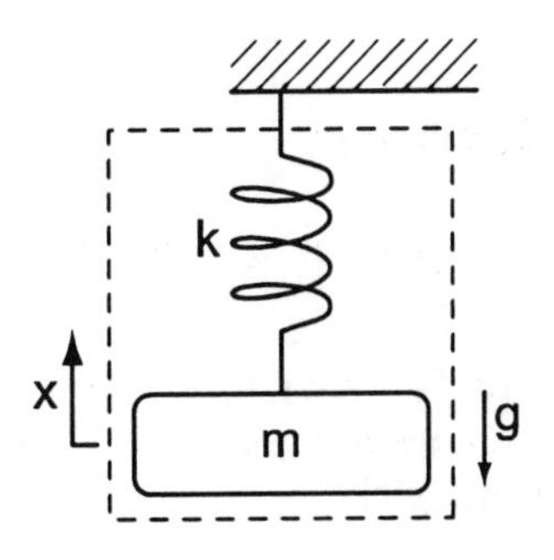

Figure 5.1: Conservative System

5.1 Conservative Systems

If a system is conservative—that is, if there is no damping term which dissipates mechanical energy into heat—the total mechanical energy stored within an autonomous system remains constant. We recognize two forms of mechanical energy: kinetic energy "T", and potential energy "U". Therefore

$$T + U = \text{constant} \tag{5.1}$$

or

$$\boxed{\frac{d}{dt}(T + U) = 0} \tag{5.2}$$

where

$$T = \frac{1}{2}\frac{m}{g_c}\left(\dot{x}\right)^2 \text{ and/or } \frac{1}{2}\frac{J}{g_c}\left(\dot{\theta}\right)^2, \text{ respectively}$$

and

$$U = \frac{mg}{g_c}h \text{ and/or } \frac{1}{2}kx^2 \text{ or } \frac{1}{2}k_T\theta^2, \text{ respectively}$$

When we apply this to a simple mass and spring as in Figure 5.1, Equation 5.2 becomes

$$\frac{d}{dt}\left(\frac{1}{2}\frac{m}{g_c}\left(\dot{x}\right)^2 + \frac{mg}{g_c}x + \frac{1}{2}kx^2\right) = 0$$

which simplifies to

$$\frac{m}{g_c}\dot{x}\ddot{x} + kx\dot{x} = -\frac{mg}{g_c}\dot{x}$$

If we divide all terms by $\dot{x}$, we get Equation 4.25 again (except for the minus sign in the gravitational term, because we happened to take x positive upward this time). In this simple problem, there was no real advantage in using the energy method. It is very useful, however, in complex systems that have many masses or many springs which are constrained to move simultaneously.

5.2 Complex Systems

The solution of one-degree-of-freedom systems with many components can be broken down into five simple steps:

1. Write an expression for T, adding up the all the kinetic energy terms, and using whatever coordinates are geometrically convenient. In the example of Figure 5.2, this would yield

$$T = \frac{1}{2}\frac{m_1}{g_c}\left(\dot{x}_1\right)^2 + \frac{1}{2}\frac{m_2}{g_c}\left(\dot{x}_2\right)^2 + \frac{1}{2}\frac{J_1}{g_c}\left(\dot{\theta}\right)^2$$

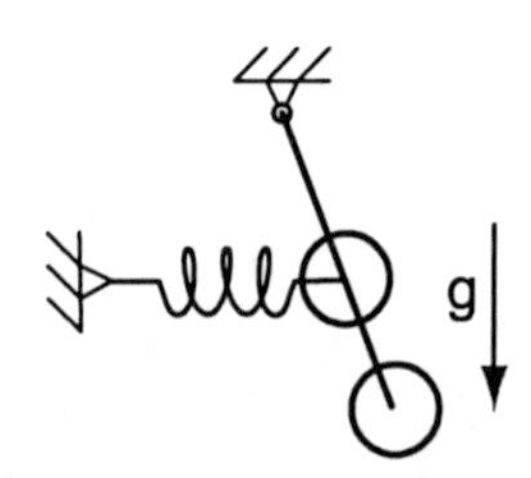

Figure 5.2: Complex System

2. Write an expression for U, adding up all the potential energy terms, using whatever coordinates are easiest and least error-prone. Continuing our example

$$U = \frac{m_1 g}{g_c} h_1 + \frac{m_2 g}{g_c} h_2 + \frac{1}{2} k x_1^2$$

3. Write kinematic equations relating the several coordinates. In our example

$$\begin{aligned}
x_1 &\cong l_1 \sin(\theta) \\
&\cong l_1 \theta \text{ forsmall } \theta, \text{ since } \sin(\theta) = \left(\theta - \frac{1}{6}\theta^3 + \frac{1}{120}\theta^5 - \ldots \right) \\
\dot{x}_1 &\cong l_1 \dot{\theta} \\
\dot{x}_2 &\cong l_2 \dot{\theta} \\
h_1 &= l_1 (1 - \cos(\theta)) \\
&\cong l_1 \left(\frac{1}{2}\theta^2 \right) \text{ forsmall } \theta, \text{since } \cos(\theta) = \left(1 - \frac{1}{2}\theta^2 + \frac{1}{24}\theta^4 - \ldots \right) \\
h_2 &= l_2 (1 - \cos(\theta)) \cong l_2 \left(\frac{1}{2}\theta^2 \right)
\end{aligned}$$

4. Convert T and U to any one of the coordinates. In our example

$$\begin{aligned}
T &\cong \frac{1}{2} \left[\frac{m_1 l_1^2 + m_2 l_2^2 + J_1}{g_c} \right] \dot{\theta}^2 \\
U &\cong \frac{1}{2} \left[\frac{m_1 g l_1 + m_2 g l_2}{g_c} + k l_1^2 \right] \theta^2
\end{aligned}$$

5. Find the derivatives and plug into Equation 5.2 to obtain the system equation. In our example

$$\begin{aligned}
\frac{dT}{dt} &= \left[\frac{m_1 l_1^2 + m_2 l_2^2 + J_1}{g_c} \right] \dot{\theta}\ddot{\theta} \\
\frac{dU}{dt} &= \left[\frac{m_1 g l_1 + m_2 g l_2}{g_c} + k l_1^2 \right] \dot{\theta}\, \theta
\end{aligned}$$

and therefore

$$\left\{ \left[\frac{m_1 l_1^2 + m_2 l_2^2 + J_1}{g_c} \right] \ddot{\theta} + \left[\frac{m_1 g l_1 + m_2 g l_2}{g_c} + k l_1^2 \right] \theta \right\} \dot{\theta} = 0$$

Note that the expressions in the square brackets come from T and U respectively; we could call them the total "equivalent mass" and the total "equivalent spring" relative to the coordinate θ

$$\begin{aligned} \frac{M_{\text{equiv}}}{g_c} &= \left[\frac{m_1 l_1^2 + m_2 l_2^2 + J_1}{g_c} \right] \\ K_{\text{equiv}} &= \left[\frac{m_1 g l_1 + m_2 g l_2}{g_c} + k l_1^2 \right] \end{aligned}$$

If we had chosen a different coordinate in Step 4—for example x_1—these expressions would have been different: *equivalent masses and springs must be defined with respect to a particular coordinate.* However, their ratio, which we will call ω_n^2 as before, will be the same no matter which coordinate we had chosen to use, because the natural frequency is a property of the system itself.

Exercise 5.1 *Work the example using coordinate* x_2.

5.3 Short-cut Method Using Equivalent Elements

For linear systems, we can save ourselves the trouble of going through the differential calculus, if we recognize that the same patterns occur again and again in different problems. T normally takes the form

$$T = \frac{1}{2} \left[\frac{M_{\text{equiv}}}{g_c} \right] \dot{x}^2 \tag{5.3}$$

as we saw in the example above. In this quadratic form, we call whatever shows up in the brackets: "M_{equiv} relative to the coordinate x" (or θ or whatever coordinate that shows up in the time-derivative next to the brackets—if it is an angular coordinate, the equivalent mass will have the form and dimensions of moment-of-inertia, of course). U often takes the form

$$U = \frac{1}{2} [K_{\text{equiv}}] x^2 \tag{5.4}$$

In this quadratic form, we call whatever shows up in these brackets: "K_{equiv} relative to the coordinate x" (or θ or other coordinate next to the brackets). As long as the equivalent masses and springs are all referred to the same coordinate x, we immediately can write the differential equation

$$\frac{M_{\text{equiv}}}{g_c} \ddot{x} + K_{\text{equiv}} x = 0 \tag{5.5}$$

Normalizing this equation and comparing it with Equation 4.9, we learn right away that

$$\left(\omega_n^2\right) = \left(\frac{K_{\text{equiv}} g_c}{M_{\text{equiv}}}\right) \tag{5.6}$$

What if T or U do not fall into the expected form? The quadratic form for U implies that there is a linear spring element law like Equation 2.3. It comes from the definition of work as a force acting through a distance: $U = \int_0^x F(\xi)d\xi$; if $F(\xi) = k\xi$, then $U = \frac{1}{2}kx^2$. If U takes some other form than the quadratic, then the spring must follow a non-linear function of the deflection coordinate. For example, the compound pendulum of Figure 3.3 gives us the expected quadratic form $T = \frac{1}{2}\left[\frac{J_{\text{pivot}}}{g_c}\right]\dot{\theta}^2$, but an irregular $U = \frac{mg}{g_c}R(1-\cos\theta)$. To proceed with the short-cut method, we would have to force U into the quadratic form $\frac{1}{2}\left[\frac{mgR}{g_c}\right]\theta^2$ by substituting the Maclaurin series expansion $\left(1 - \frac{1}{2}\theta^2 + \frac{1}{24}\theta^4 - \ldots\right)$ for $\cos\theta$ and restricting ourselves to small θ so that $\frac{1}{24}\theta^4 \ll \frac{1}{2}\theta^2$; this has the practical result of linearization. If we did not wish to restrict ourselves to small angles θ, we could go back from the short-cut methods to the full-length method of the previous section and differentiate $(1-\cos\theta)$; we would then get the non-linear differential equation $J_{\text{pivot}}\ddot{\theta} + mgR\sin\theta = 0$.

One special case of a non-quadratic form is work against a steady gravity term: $U = \frac{mg}{g_c}x$, as in Figure 4.2. Differentiation here gives us $\frac{mg}{g_c}\dot{x}$; after we put this into Equation 5.2 and divide by $\dot{x}$, this yields a constant on the right-hand side as in Equation 4.25.

5.4 Caution

The Conservation of Energy method has two limitations:

- If there is a dissipative element (damper) in the system, mechanical energy is not conserved—some of it is converted to thermal energy. We may still use this method to obtain mass and spring coefficients, and then figure out the damping term later, as shown in Chapter 10.

- If there is a boundary which is not fixed in an inertial coordinate system, the vibrating system might not be isolated, and might be exchanging energy with the environment. Then the energy in the system might periodically increase and decrease during the course of an oscillatory cycle. This situation is difficult to recognize; in case of doubt it is best to avoid the Conservation of Energy method and use Lagrange's equation instead.

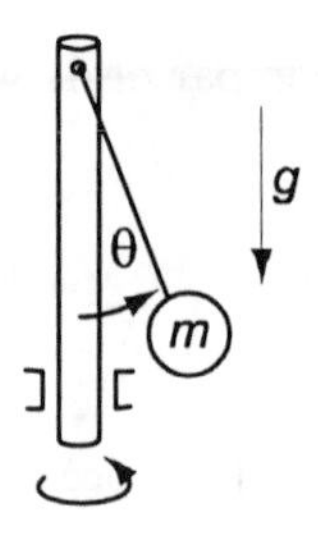

Figure 5.3: Flyball Governor

5.5 A Pathological Case

In order to learn the limits of the Conservation of Energy method, let us look at a case where it gives the wrong answer. Our problem is a simplified version of a flyball governor: a vertical shaft is driven at constant angular velocity $\dot{\Phi}$ and supports the hinge of a lever which holds a concentrated mass at its tip (Figure 5.3). As the shaft spins, the whirling lever is pulled out and upward by the centrifugal force, and down and inward by gravity. The energy method appears to be easy to apply, because the velocity of the concentrated mass can be described by two components, $l\,\dot{\theta}$ and $l\sin(\theta)\,\dot{\Phi}$, which are at right angles to each other, so that the total velocity squared is the sum of the squares of these components. Therefore kinetic energy $T = \frac{1}{2}\frac{m}{g_c}l^2\left(\dot{\theta}^2 + \dot{\Phi}^2\sin^2(\theta)\right)$. Potential energy is $U = \frac{mg}{g_c}l\,(1-\cos(\theta))$. Differentiating these (remembering that $\dot{\Phi}$ is a constant and θ is the variable) and plugging into Equation 5.2 yields

$$\frac{m}{g_c}l^2\,\ddot{\theta}\dot{\theta} + \frac{m}{g_c}l^2\,\dot{\Phi}^2\sin(\theta)\cos(\theta)\,\dot{\theta} + \frac{mg}{g_c}l\left(\sin(\theta)\,\dot{\theta}\right) = 0$$

which simplifies to

$$\ddot{\theta} + \dot{\Phi}^2\sin(\theta)\cos(\theta) + \frac{g}{l}\sin(\theta) = 0$$

This equation is *wrong!* It passes superficial inspection: for $\dot{\Phi}=0$ it reduces to the simple pendulum equation. But when we try to solve for the equilibrium $\overline{\theta}$, we get

$$\dot{\Phi}^2\sin(\overline{\theta})\cos(\overline{\theta}) + \frac{g}{l}\sin(\overline{\theta}) = 0$$

which has two solutions: either $\sin(\overline{\theta}) = 0$ or else $\cos(\overline{\theta}) = g/l\,\dot{\Phi}^2$. The first of these, $\overline{\theta} = 0$, obviously does not cover all the cases we would like to consider;

and the second calls for $\overline{\theta}$ outside of the acceptable range of $0 \leq \overline{\theta} < \frac{\pi}{2}$. Where did we go wrong?

Is this a conservative system? Yes, in the sense that it has no dampers to dissipate mechanical energy. But it is *not* an isolated system: mechanical energy flows in and out by way of the driven shaft during each cycle of oscillation. Every time the mass moves outward, the angular momentum increases and the shaft must exert torque and put energy into the system in order to maintain constant angular velocity; every time the mass moves inward the shaft absorbs energy. On the average, the energy in the system is maintained, but at any instant it may enter or leave the control volume.

Warning: Do not assume Conservation of Energy unless the system is not only free of internal losses, but also isolated. If the forces or torques exerted through the control surface act through displacements or rotations of the environment, energy may pass in or out.

We can still solve this problem by going back to Newton's law, remembering that rotating coordinate systems require consideration of centrifugal and Coriolis forces. However, there is another energy method which does not depend on system isolation: Lagrange's equation.

Exercise 5.2 *Solve for $\overline{\theta}$ in the problem above, by balancing centrifugal force and gravity.*

5.6 Lagrange's Equation

In Chapter 2, we set up dynamics equations using Newton's law, Equation 2.1. In this chapter, we showed that we could set up the same equations—often much more easily—using Conservation of Energy, Equation 5.2. There is a third way of setting up the equations of classical mechanics, from Hamilton's principle that $(T - U)$ is minimized throughout a motion. Euler showed that such a variational principle leads to a differential equation.[1] For non-dissipative systems, the equation resulting from Hamilton's principle is

$$\frac{d}{dt}\left(\frac{\partial (T-U)}{\partial v}\right) - \frac{\partial (T-U)}{\partial x} = 0 \tag{5.7}$$

For the moment, we have returned to writing v for $\dot{x}$, to emphasize that v and x are treated as separate state variables here. In other words, we treat T as $T(x, v)$ rather than $T(x, t)$. T is often a function only of v, but can be a function of both x and v in some coordinate systems. U is characteristically a function of x but not of v, so that we can leave out $\partial U / \partial v$

$$\boxed{\frac{d}{dt}\left(\frac{\partial T}{\partial v}\right) - \frac{\partial T}{\partial x} + \frac{dU}{dx} = 0} \tag{5.8}$$

[1] Richard Courant and D. Hilbert, *Methods of Mathematical Physics,* Volume I, Wiley, New York, reprinted 1989, ISBN 0-471-50447-5.

This is called Lagrange's equation. It can handle complicated boundary constraints, so we can try it on the problem of the preceding section

$$
\begin{aligned}
T &= \frac{1}{2}\frac{m}{g_c}l^2\left(\dot{\theta}^2 + \dot{\Phi}^2 \sin^2(\theta)\right) \\
\left[\frac{\partial T}{\partial\left(\dot{\theta}\right)}\right]_{\theta=const.} &= \frac{m}{g_c}l^2\left(\dot{\theta}\right) \\
\frac{d}{dt}\left[\frac{\partial T}{\partial\left(\dot{\theta}\right)}\right] &= \frac{m}{g_c}l^2\left(\ddot{\theta}\right) \\
\left[\frac{-\partial T}{\partial \theta}\right]_{\dot{\theta}=const.} &= \frac{-m}{g_c}l^2\left(\dot{\Phi}^2 \sin(\theta)\cos(\theta)\right) \\
U &= \frac{mg}{g_c}l\left(1-\cos(\theta)\right) \\
\frac{dU}{d\theta} &= \frac{mg}{g_c}l\left(\sin(\theta)\right)
\end{aligned}
$$

Inserting these into Lagrange's equation

$$\frac{m}{g_c}l^2\,\ddot{\theta} - \frac{m}{g_c}l^2\,\dot{\Phi}^2 \sin(\theta)\cos(\theta) + \frac{mg}{g_c}l\sin(\theta) = 0$$

which simplifies to

$$\ddot{\theta} - \dot{\Phi}^2 \sin(\theta)\cos(\theta) + \frac{g}{l}\sin(\theta) = 0$$

Despite the superficial resemblance to our previous solution, there is a very big difference: the opposite sign in the middle term! We will have to wait for a later chapter to attempt to solve a non-linear differential equation like this one, but we can search for an equilibrium $\overline{\theta}$ by means of the equation

$$-\dot{\Phi}^2 \sin(\overline{\theta})\cos(\overline{\theta}) + \frac{g}{l}\sin(\overline{\theta}) = 0$$

which has a solution $\cos(\overline{\theta}) = g/l\,\dot{\Phi}^2$, if $g \leq l\,\dot{\Phi}^2$. If $g > l\,\dot{\Phi}^2$, we must resort to the other possible solution, $\sin(\overline{\theta}) = 0$. Apparently $\overline{\theta} = 0$ represents a stable equilibrium when $g > l\,\dot{\Phi}^2$. but an unstable equilibrium when $g < l\,\dot{\Phi}^2$.

5.7 Summary

In the previous chapter, we developed two methods for obtaining the characteristic frequency of a system: Newton's law and observation of static deflection. Now we have two more: Conservation of Energy and Lagrange's equation.

Conservation of Energy is useful for isolated systems with many masses and springs; it also allows us to introduce the concept of "equivalent" masses and springs. We can extend it to the fundamental modes of continuous systems—like beams and fluid-filled cavities—by assuming approximate mode shapes.

Lagrange's equation must be used in place of Conservation of Energy when T is not purely a function of velocity alone, but also depends in some way on displacement, as we showed in an example of a non-dissipative, but not isolated system.

Problem 5.3 *Given a pendulum as in Figure 3.2, spring-assisted but biased such that θ_o is the angle from the vertical at which the spring exerts zero force. From the Energy Method, find the governing equation and the equilibrium position.*

Problem 5.4 *Develop the governing equation for the natural frequency of a mass suspended by a rope of overall stiffness k_{rope} running through a (weightless) pulley attached to a ceiling hook of stiffness k_{hook} and then running back down to the floor where it is fastened.*

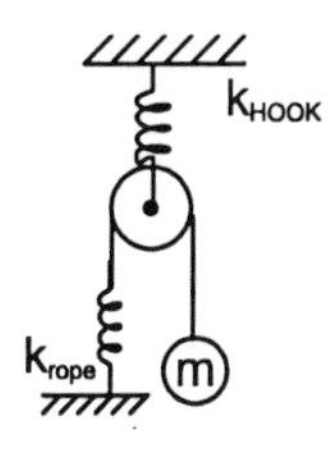

Problem 5.5 *What is the natural frequency of a flywheel J mounted on a shaft of diameter D, when the shaft is set on knife-edges so it can roll back and forth, and an unbalance mass m is attached to the flywheel at radius R? Always start by drawing a diagram!*

Problem 5.6 *A vertical shaft, driven at constant speed of Ω radians/second by a large motor, supports a turntable on which a compact mass m moves in a radial slot. A spring k restrains the mass; the force of the spring is zero when the mass is at radius $r = r_o$. Find the natural frequency of the mass, oscillating in and out along the radial slot, as a function of m, k, Ω, and r_o.*

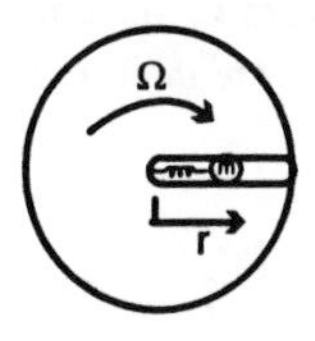

Hint: Use either Newton's law or Lagrange's equation. The resulting equation should be linear.

Problem 5.7 *A vertical shaft, driven at constant speed of Ω radians/second by a large motor, supports a turntable on which a compact mass m moves in a radial slot. A spring k restrains the mass; the force of the spring is zero when the mass is at radius $r = 0$. Find the natural frequency of the mass, oscillating along the radial slot, as a function of m, k, and Ω. Answer: The equilibrium is stable at $r = 0$ for $\Omega^2 < \frac{kg_c}{m}$, is indeterminate when $\Omega^2 = \frac{kg_c}{m}$, and is unstable for $\Omega^2 > \frac{kg_c}{m}$.*

Problem 5.8 *A vertical shaft, driven at constant speed of Ω radians/second by a large motor, supports a turntable to which a pendulum is attached, with the pivot at radius R. The pendulum is a simple pendulum in the plane of the turntable, with mass m and length L. Find the natural frequency of the pendulum, oscillating forward and back, as a function of R, L, and Ω. (A device like this is used in some piston engines.)*

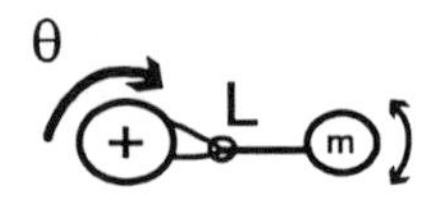

Hint: The absolute velocity of the mass has the components $\Omega \cdot R + L \cdot \left(\Omega + \dot{\theta}\right)\cos\theta$ *and* $L \cdot \left(\Omega + \dot{\theta}\right)\sin\theta$.

Problem 5.9 *A vertical shaft, driven at constant speed of Ω radians/second by a large motor, supports a turntable to which a pendulum is attached, with the pivot at radius R. The pendulum is a simple pendulum out-of-plane to the turntable, with mass m and length L. Find the natural frequency of the pendulum as a function of R, L, and Ω. Neglect gravity. (This is a highly idealized form of an helicopter blade-flapping problem.) Answer: The natural frequency is proportional to the rotational frequency.*

Problem 5.10 *Design an experiment for determining the moment of inertia of a bicycle wheel by hanging it horizontally from three identical vertical strings fastened to the rim at the bottom and attached to three hooks in the ceiling, and timing small rotational oscillations around a vertical axis. Sketch a suitable arrangement. What is the moment-of-inertia as a function of measured period, wheel mass, rim diameter, and string length? Prepare an error analysis estimating the effect of uncertainties in active string length, effective attachment radius, etc.*

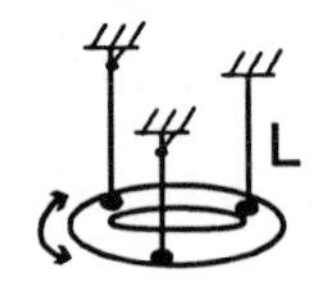

Problem 5.11 *In "blueprinting" engines for racing, pistons are matched for cylinder clearance and weight, and connecting rods for weight and c.g.-location. Devise a procedure for locating the center of gravity and also determining the radius of gyration (around the axis through the c.g.) of connecting rods, from a measurement of the natural frequencies of the rod suspended as a pendulum from a knife edge, first from within the little end and secondly from within the big end.*

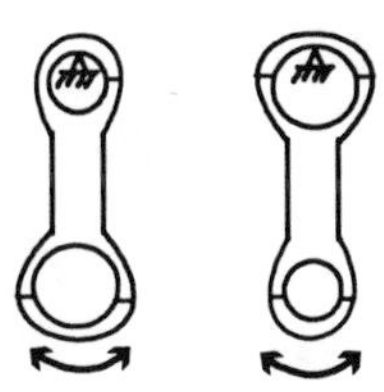

Problem 5.12 *To determine whether a one-design-rule sailboat is within specifications, not only the dimensions, but also the weight-distribution must be tested, because a boat with less mass in the ends will be faster through the waves. Devise a procedure for determining the moment of inertia of a sailboat with respect to rocking-horse motion, by suspending it in slings of different length, and measuring the period of the resulting pendulum.*

Chapter 6

APPROXIMATIONS FOR DISTRIBUTED SYSTEMS and hydrodynamic inertia

The Conservation of Energy method can be extended to situations that are not strictly single-degree-of-freedom systems, if we have a pretty good idea what is the "mode shape" (the relative motion of all parts of the system).

6.1 Rayleigh's Method

We can often take a good guess at the relative amplitudes of the different parts of a multi-degree-of-freedom system, or the shape of the oscillatory deflection of a continuously flexible system. Having made that guess, we can then go through the steps of summing up T and U in terms of a reference coordinate as we did for rigidly one-degree-of-freedom systems.

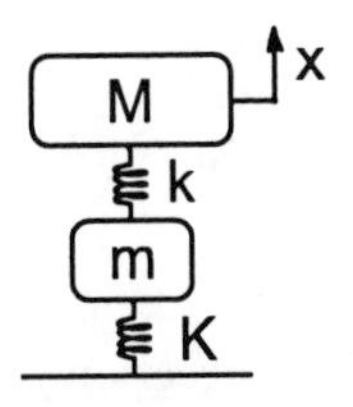

Figure 6.1: Quarter-Car System

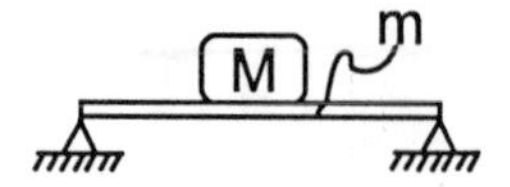

Figure 6.2: Beam Spring

6.1.1 Two-Lump System

For example, if we wish to know the effect of the unsprung mass on the fundamental frequency of a car, we can assume that (at low frequencies) the unsprung mass will move proportionately to the sprung mass, as it would for static-loading deflection. Let us say that a quarter of a car (Figure 6.1) has a mass of $M = 500$ kg and is suspended on a spring with an effective spring constant of $k = 20$ KN/m or N/mm. From $\omega_n^2 = kg_c/M$, our rough first estimate of the natural frequency of the bouncing car is 1.0066 Hz or, truncating to significant digits, 60.4 c.p.m.

If we consider that the flexibility of the tire adds another spring of about $K = 10k = 200$ KN/m in series, we can correct the effective spring constant to 18.18 KN/m, and make an improved second estimate of natural frequency 0.9597 Hz.

But if we also consider that the axle, brake, and wheel constitute an unsprung weight of $m = M/10 = 50kg$, we find ourselves with a two-degree-of-freedom problem. Since the unsprung mass is much smaller than the main mass, we can guess that, at low frequencies, the unsprung weight moves proportionately to the deflection of the main mass, in the same proportion that it would for slowly applied loading. From consideration of the spring constants, that proportion is

$$x_m/x_M \cong 18.18 \div 200 \cong 0.09$$

that is, the unsprung weight moves about 9% of the motion of the main mass. We can now write

$$T \cong \frac{1}{2}\left(\frac{M}{g_c}\right)\dot{x}_{M1}^2 + \frac{1}{2}\left(\frac{m}{g_c}\right)\dot{x}_m^2 = \frac{1}{2}\left[\frac{M}{g_c} + \frac{m}{g_c}\left(\frac{x_m}{x_M}\right)^2\right]\dot{x}_M^2$$

$$U \cong \frac{1}{2}k\left(x_M - x_m\right)^2 + \frac{1}{2}Kx_m^2 = \frac{1}{2}\left[k\left(1 - \frac{x_m}{x_M}\right)^2 + K\left(\frac{x_m}{x_M}\right)^2\right]\dot{x}_M^2$$

For this guessed mode, $M_{\text{equiv}} = 500.4$ kg; $k_{\text{equiv}} = 18.18$ KN/m; and $f_n = 0.9593$ Hz, or 57.6 c.p.m. Since the correction to our first estimate is not large, it seems that further refinements to our guessed mode probably won't make much difference, and we feel justified to quit at this point.

6.1.2 Bending Beams

Similarly, if we wish to know the effect of the mass m of a simply supported beam (Figure 6.2), holding up a concentrated mass M at its center, we must assume a mode shape for the beam. We know that the support conditions require zero displacements at the ends The simplest shape we can guess is the parabola

$$\frac{y(x)}{y_{\text{center}}} \cong 4\left(\frac{x}{L}\right) - 4\left(\frac{x}{L}\right)^2 \tag{6.1}$$

which has zero values for y at $x = 0$ or L. We can insert this function $y(x)$ into the integral for kinetic energy

$$\begin{aligned} T &= \frac{1}{2}\frac{M}{g_c}\dot{y}^2_{\text{mass}} + \frac{1}{2}\int_0^L \left(\dot{y}(x)\right)^2 \left(\frac{m}{Lg_c}\right) dx \\ &= \frac{1}{2}\left[\frac{M}{g_c} + \frac{m}{g_c}\int_0^1 \left(\frac{y(x)}{y_{\text{center}}}\right)^2 d\left(\frac{x}{L}\right)\right]\dot{y}^2_{\text{center}} \\ &\cong \frac{1}{2}\left[\frac{M}{g_c} + \frac{m}{g_c}\int_0^1 \left(4\xi - 4\xi^2\right)^2 d\xi\right]\dot{y}^2_{\text{center}} \end{aligned} \tag{6.2}$$

and into the integral for potential energy in beams

$$\begin{aligned} U &= \frac{1}{2}\int_0^L EI\left(\frac{d^2y}{dx^2}\right)^2 dx \\ &= \frac{1}{2}\left[\frac{EI}{L^3}\int_0^1 \left(\frac{d^2\left(y/y_{\text{center}}\right)}{d\left(x/L\right)^2}\right)^2 d\left(\frac{x}{L}\right)\right] y^2_{\text{center}} \\ &\cong \frac{1}{2}\left[\frac{EI}{L^3}\int_0^1 (8)^2\, d\xi\right] y^2_{\text{center}} \end{aligned} \tag{6.3}$$

In each equation, the integral depends only on the shape of the guessed mode. Evaluating the integrals for our guessed parabolic mode-shape function, we get

$$\begin{aligned} T &\cong \frac{1}{2}\left[\frac{M}{g_c} + \frac{8}{15}\frac{m}{g_c}\right]\dot{y}^2_{\text{center}} \\ U &\cong \frac{1}{2}\left[\frac{64EI}{L^3}\right] y^2_{\text{center}} \end{aligned}$$

From these, we can identify equivalent mass and spring constant, and obtain the natural frequency

$$2\pi f_n = \omega_n \cong \sqrt{\frac{\frac{64EI}{L^3}}{\frac{M}{g_c} + \frac{8}{15}\frac{m}{g_c}}} \tag{6.4}$$

According to the estimate assuming this deflection shape, 8/15 of the mass of a uniform beam-spring should be added to the concentrated mass it is supporting.

6.2 Short-Cut Method Using Maximum Energy

In our example above, deflection y along the beam varies as a function of both space and time; since the mode-shapes at any moment are similar to those at any other moment, we were able to eliminate time-variation from our integrals. A shorthand way of eliminating time is to write $T_{\max}$ for the kinetic energy at the moment that velocity is maximum and potential energy is zero, and $U_{\max}$ for the moment that deflection is maximum and velocity is zero. We can then write

$$T + U = \text{constant} = \boxed{T_{\max} = U_{\max}} \tag{6.5}$$

Thus we will write maximum kinetic energy as a function of $\dot{x}^2_{\max}$ and set it equal to maximum potential energy (which occurs at a different time) as a function of $x^2_{\max}$. If we know that the system is governed by linear laws, we can assume harmonic motion, with the consequence that

$$\boxed{\dot{x}^2_{\max} = \omega^2 x^2_{\max}} \tag{6.6}$$

which we use to eliminate the velocity from $T_{\max}$. When we set $T_{\max} = U_{\max}$ and solve for ω^2, $x^2_{\max}$ will cancel out in linear problems.

We can demonstrate this by repeating the beam problem of the previous section; for variety, we will try a different guessed mode-shape which also meets our end conditions of $y = 0$ for $x = 0$ and L

$$y_{\max} = C \sin\left(\pi \frac{x}{L}\right) \tag{6.7}$$

where the constant $C = y_{\text{center}}$. Therefore, Equation 6.2 becomes

$$\begin{aligned} T_{\max} &\cong \frac{1}{2}\frac{M}{g_c}\omega^2 C^2 + \frac{1}{2}\int_0^L \omega^2 C^2 \sin^2\left(\frac{\pi x}{L}\right)\frac{m}{Lg_c}dx \\ &\cong \omega^2 \frac{1}{2}\left[\frac{M}{g_c} + \int_0^\pi \sin^2\xi \frac{m}{\pi g_c} d\xi\right] C^2 \\ &\cong \omega^2 \frac{1}{2}\left[\frac{M}{g_c} + \frac{1}{2}\frac{m}{g_c}\right] C^2 \end{aligned}$$

and Equation 6.3 becomes

$$\begin{aligned} U_{\max} &\cong \frac{1}{2}\int_0^L EI \left(\frac{-\pi^2}{L^2} C \sin\left(\frac{\pi x}{L}\right)\right)^2 d\left(\frac{x}{L}\right) \\ &\cong \frac{1}{2}\left[EI \frac{\pi^4}{L^3}\int_0^1 \sin^2\xi d\xi\right] C^2 \\ &\cong \frac{1}{2}\left[\frac{\pi^4}{2}\frac{EI}{L^3}\right] C^2 \end{aligned}$$

Setting these equal to each other yields an expression for ω^2, while C cancels out

$$2\pi f_n = \omega_n \cong \sqrt{\frac{\frac{\pi^4}{2}\frac{EI}{L^3}}{\frac{M}{g_c} + \frac{1}{2}\frac{m}{g_c}}} \tag{6.8}$$

Caution: The invocation of linear systems and harmonic response reminds us that these approximate methods are limited to problems with substantially linear governing equations.

6.3 Comparison of Different Guesses

The burning question is, of course, how sensitive this approach is to good or bad guesses of the mode shape. Comparing our two guesses, parabola or sinewave, for the mode shape of the loaded beam in the preceding two sections, we see that there is some difference. For $M \gg m$, Equation 6.4 gives $2\pi f_n \cong 8.00\sqrt{EIg_c/ML^3}$, but Equation 6.8, $6.98\sqrt{EIg_c/ML^3}$. And for $M \ll m$, the parabolic guess gives $2\pi f_n \cong 10.95\sqrt{EIg_c/mL^3}$, but the sinewave, $9.87\sqrt{EIg_c/mL^3}$. Which values are better?

The answer is that the *lower* value for ω_n is always better, *if both guesses for the mode shape meet the physical restraints* the environment places on the system (i.e., that $y = 0$ at both $x = 0$ and $x = L$). The reason is that a bad guess places an artificial additional constraint on the motion on the system, requiring its parts to move differently in our mathematical model than they naturally would want to in the physical prototype; this artificial additional constraint stiffens up the mathematical model, and raises the fundamental frequency obtained. Therefore, Rayleigh estimates of fundamental frequency are always high, and the lowest value obtained is closest to the truth. One approach to getting better answers is to solve the problem for many different guesses at the mode shape, and picking the one that gives the lowest answer. Another is trying to start with a good guess in the first place.

Exercise 6.1 *Repeat the analysis for a beam with a concentrated mass at its center, with the assumed mode-shape*

$$\frac{y(x)}{y_{center}} \cong \frac{8}{3}\left(\frac{x}{L}\right) - \frac{8}{3}\left(\frac{x}{L}\right)^3$$

6.4 Static Deflection

For many systems such as beams with a single span, the shape of the static deflection distribution can be used as a very good guess at the fundamental mode shape. In our frequency analysis, we ignored static deflection of the beam, because the deflection due to steadily biasing forces does not affect natural frequency, as we showed in Section 4.7; to get a homogeneous differential equation we use the equilibrium deflection as our starting point for dynamic deflections. Now we will borrow the static deflection shape as a guess for the shape of the dynamic deflections. From strength-of-materials books[1] we can see that the static deflection of a centrally loaded beam follows the shape

$$\begin{aligned} \text{for } \left(\frac{x}{L}\right) &\le \frac{1}{2}, \quad \frac{y(x)}{y_{\text{center}}} = 3\left(\frac{x}{L}\right) - 4\left(\frac{x}{L}\right)^3 \\ \text{for } \left(\frac{x}{L}\right) &\ge \frac{1}{2}, \quad \frac{y(x)}{y_{\text{center}}} = -1 + 9\left(\frac{x}{L}\right) - 12\left(\frac{x}{L}\right)^2 + 4\left(\frac{x}{L}\right)^3 \end{aligned} \tag{6.9}$$

which ought to be a good guess if $M \gg m$. On the other hand, a uniformly loaded beam follows the shape

$$\frac{y(x)}{y_{\text{center}}} = \frac{16}{5}\left(\frac{x}{L}\right) - \frac{32}{5}\left(\frac{x}{L}\right)^3 + \frac{16}{5}\left(\frac{x}{L}\right)^4 \tag{6.10}$$

which ought to be a good guess for $M \ll m$. If M and m are in the same order of magnitude, one could use the combined static deflection:

$$\frac{y(x)}{y_{\text{center}}} = \frac{\left(3\left(\frac{x}{L}\right) - 4\left(\frac{x}{L}\right)^3\right)M + \left(2\left(\frac{x}{L}\right) - 4\left(\frac{x}{L}\right)^3 + 2\left(\frac{x}{L}\right)^4\right)m}{M + \frac{5}{8}m} \quad \text{for } \left(\frac{x}{L}\right) \le \frac{1}{2}$$

Exercise 6.2 *Compute and compare solutions obtained with these different guesses, for $M = m$.*

6.5 Equivalent Mass of Springs

As we did with the simple beam, we can use Rayleigh's method to estimate the effect of the mass of other kinds of springs.

Example: For a flywheel mounted on a long shaft of length L acting as a torsion spring, if we assume that a torsion spring twists uniformly

$$\begin{aligned} \frac{\theta(x)}{\theta_{\text{end}}} &= \frac{x}{L} \\ \frac{d\theta}{dx} &= \frac{\theta_{\text{end}}}{L} \end{aligned}$$

with an overall spring constant k_T and an overall mass moment of inertia J_{shaft},

[1] *e.g.*, Warren C. Young, *Roark's Formulas for Stress and Strain,* 6th Edition, McGraw-Hill, New York, 1989, ISBN 0-07-072541-1.

we find that

$$\begin{aligned} T &= \frac{1}{2}\left[\frac{J_{\text{wheel}}}{g_c}\right]\dot{\theta}_{\text{end}}^{2}+\frac{1}{2}\int_0^L\left[\frac{J_{\text{shaft}}}{Lg_c}\right]\left(\frac{x}{L}\,\dot{\theta}_{\text{end}}\right)^2 dx \\ &= \frac{1}{2}\left[\frac{J_{\text{wheel}}+\frac{1}{3}J_{\text{shaft}}}{g_c}\right]\dot{\theta}_{\text{end}}^{2} \\ U &= \frac{1}{2}\int_0^L [k_T L]\left(\frac{\theta_{\text{end}}}{L}\right)^2 dx = \frac{1}{2}\left[k_T\right]\theta_{\text{end}}^2 \\ \omega_n^2 &= \left[\frac{k_T g_c}{J_{\text{wheel}}+\frac{1}{3}J_{\text{shaft}}}\right] \end{aligned}$$

so that 1/3 of the moment of inertia of the shaft should be added to the moment of inertia of the flywheel.

Exercise 6.3 *Show that, for a coil spring, about 1/3 of its mass m should be added to the concentrated moving mass M at its end. Hint: If we have a coil spring with many turns and an overall length of L, a mass of m, and a spring constant k, we can treat it like an elastic rod with a mass-per-unit-length of $\mu = m/L$ and an extensional spring-constant per unit length $\kappa = kL$. The energy in the elastic rod is*

$$\begin{aligned} T &= \frac{1}{2}\int_0^L \mu\cdot u^2 dx \\ U &= \frac{1}{2}\int_0^L \kappa\left(\frac{du}{dx}\right)^2 dx \end{aligned}$$

Note that U for a stretched rod involves the first derivative of the mode, not the second derivative like a bending beam. Assume an uniform x-direction deflection $u_{(x)}/u_{end} = x/L$.

Exercise 6.4 *Repeat the problem, using a static-deflection-like mode-shape*

$$\frac{u_{(x)}}{u_{end}} = \frac{M\frac{x}{L}+m\left(\frac{x}{L}-\frac{x^2}{2L^2}\right)}{\left(M+\frac{m}{2}\right)}$$

The answer will be complicated—the assumed deflection affects both the equivalent mass and the effective spring constant—but should approach the previous answer if $m \ll M$.

6.6 Lamb's Hydrodynamic Inertia

When a body in contact with a fluid moves, some of the mass of the fluid moves along with it, and we should include this mass in our dynamic equation for the body. Let us consider a heat-exchanger tube that vibrates like a beam. Obviously, the mass of the liquid inside the tube should be added to the mass of the beam; but how much equivalent mass should be attributed to the mass of liquid outside the tube which must be pushed out of the way when the tube accelerates?

If we know the flow pattern of the fluid outside the tube, we can integrate the kinetic energy in the flow field which is due to the lateral velocity of the tube, express it in terms of the reference velocity (of the tube), and compare it with the form

$$T_{\text{fluid}} = \frac{1}{2}\left[\frac{m_{\text{equiv}}}{g_c}\right]\dot{x}_{\text{ref}}^2$$

so that the induced equivalent mass due to the surrounding fluid is equal to twice the kinetic energy induced in the surrounding fluid by a unit velocity of the body. Therefore the *total* equivalent mass of the tube is the sum of the mass of the tube itself, plus the mass of the enclosed fluid, plus the induced hydrodynamic inertia due to the surrounding fluid.

The simplest guess at the flow pattern around a moving body is potential flow. The analytical solution for two-dimensional potential-flow velocity is the gradient $\vec{\nabla}$ of a harmonic function Φ fitting the boundaries conditions,[2] which is available for many simple flow geometries. For example, the two-dimensional flow around a circular cylinder of radius R moving at unit velocity through a large pool of stagnant fluid can be described by the potential function Φ (and the stream function Ψ which is orthogonal to it)

$$\begin{aligned}
\Phi &= \frac{-R^2\cos\theta}{r} + \text{constant} \\
\Psi &= \frac{R^2\sin\theta}{r} + \text{constant} \\
v_r &= \frac{\partial\Phi}{\partial r} = \frac{1}{r}\frac{\partial\Psi}{\partial\theta} = \left(\frac{R}{r}\right)^2\cos\theta \\
v_\theta &= \frac{1}{r}\frac{\partial\Phi}{\partial\theta} = -\frac{\partial\Psi}{\partial r} = \left(\frac{R}{r}\right)^2\sin\theta
\end{aligned}$$

We can check that, along the surface $r = R$, each point moves to the right at unit velocity. Integrating the kinetic energy throughout the space $r > R$ outside the surface, we obtain an induced hydrodynamic mass per unit length of

$$\frac{m_{\text{ext.fluid}}}{L} = \rho_{\text{fluid}}\pi R^2$$

which, purely by coincidence, is the mass of fluid displaced by the cylinder. We can express the hydrodynamic inertia as the product of a dimensionless

[2] The existence of a potential function Φ indicates that the flow is irrotational, which is true for initially irrotational and substantially inviscid flows; if it is a harmonic function $\nabla^2\Phi = 0$, the flow is also incompressible.

coefficient times the fluid mass displaced by the submerged body; for a circular cylinder, this coefficient is 1.0 exactly. However, displacement is not necessarily a good way to think about hydrodynamic inertia, because the hydrodynamic inertia of a thin ribbon (with negligible displacement) is the same as a circular or elliptical cylinder of the same width.

If there are boundaries nearby, harmonic-function solutions can be found by conformal transformation: the induced hydrodynamic mass will generally be greater, because a smaller amount of fluid moves at a higher velocity, and the velocity term is squared to obtain kinetic energy. Therefore, the dimensionless hydrodynamic inertia coefficient for a circular cylinder in a confined space has been shown to be greater than 1.0 in the literature.[3]

Potential flow can also be used to guess at the flow around a three-dimensional body. For example, the flow around a submerged sphere yields a dimensionless hydrodynamic inertia coefficient of 0.5; that means that a balloon or a spherical submersible must overcome an added 50% of mass, in addition to its own, when it accelerates. Coefficients for other ellipsoidal bodies are given in the literature.[4]

It is important to note that the mass effect of the fluid is not a result of viscosity; hydrodynamic inertia appears even in our inviscid flow analysis. Viscosity in the fluid will mainly add dissipative terms that enter into the damping coefficient. As a secondary consequence, the boundary layers and flow separation might cause us to change our guess at the flow field. It is also important to note that our hydrodynamic inertia term is independent on any superposed mean flow, if our assumption of potential flow is good. If viscosity effects are significant, this is no longer true, because the viscous-flow equations are not linear and flows cannot be superposed.

6.7 Southwell's Method

For realistically complex geometries, obtaining the potential function by analytical means becomes very tedious. Fortunately, it can be obtained quite easily by a finite-difference method, which was developed for conduction problems governed by the equation $\nabla^2 T = 0$; the potential function Φ is governed by the same partial differential equation $\nabla^2 \Phi = 0$.

Let us write the potential function in terms of discrete values $\Phi_{n,m}$ on a square grid, where the first integer subscript n is the grid-location in the x-direction, and the second subscript m is the grid-location in the x-direction The governing equation $\nabla^2 \Phi = 0$ in two-dimensional Cartesian coordinates can be converted from a differential form to an approximately equivalent difference

[3] P.M. Moretti and R.L. Lowery, "Hydrodynamic Inertia Coefficients for a Tube Surrounded by Rigid Tubes," *Trans. ASME, J. of Pressure Vessel Technology,* vol. 98, series J, no. 3 (Aug. 1976), pp. 190–193.

[4] Sir Horace Lamb, *Hydrodynamics,* 6th edition (1932) reissued by Dover Publications, New York, ISBN 0-48660256-7, and by Cambridge University Press, United Kingdom, ISBN 0-52145868-4.

equation

$$
\begin{aligned}
0 &= \nabla^2\Phi = \frac{\partial^2\Phi}{\partial x^2} + \frac{\partial^2\Phi}{\partial y^2} \\
0 &\cong \frac{\frac{\Phi_{n+1,m}-\Phi_{n,m}}{\Delta x} - \frac{\Phi_{n,m}-\Phi_{(n-1,m)}}{\Delta x}}{\Delta x} + \frac{\frac{\Phi_{n,m+1}-\Phi_{n,m}}{\Delta y} - \frac{\Phi_{n,m}-\Phi_{n,m+1}}{\Delta y}}{\Delta y} \\
&\cong \frac{\Phi_{n+1,m} - 2\Phi_{n,m} + \Phi_{n-1,m}}{(\Delta x)^2} + \frac{\Phi_{n,m+1} - 2\Phi_{n,m} + \Phi_{n,m+1}}{(\Delta y)^2}
\end{aligned}
$$

If we have chosen a square grid $\Delta x = \Delta y$, we can solve for the central value of the function

$$
\Phi_{(n,m)} \cong \frac{\Phi_{(n+1,m)} + \Phi_{(n-1,m)} + \Phi_{(n,m+1)} + \Phi_{(n,m-1)}}{4}
$$

In other words, each point is the average of its adjacent points in a square grid (a similar result can be obtained for three-dimensional Cartesian coordinates). If we know the constraints on Φ at the boundaries, we can fill in all the interior points by first guessing at them, and then randomly improving on them, until the numbers "relax" into stable values.

To illustrate this with an example, let us look at the hydrodynamic induced mass on a thin web (for example, a continuous sheet of paper) billowing up and down in a square duct (for example, the drying oven of a printing press). If the span is long compared to the width of the web (the so-called slender-body assumption), the air being pushed out to the way moves mostly around the sides of the strip of paper, and not towards the end of the span, so that we can assume two-dimensional flow (this is not a good assumption if the waves in the moving paper are short compared to the width of the web).

To impose the reference condition of unit velocity (upward) at the web, we require a unit vertical gradient $\partial\Phi/dy$. To impose fixed boundaries at the wall, we specify a zero gradient at the wall, most conveniently by having phantom grid-points behind each wall and always assigning them values which "mirror" the real grid-points opposite them.

This problem, because it is two-dimensional, can also be solved by looking for a stream function Ψ which is orthogonal to the potential function Φ, and also meets the condition $\nabla^2\Psi = 0$. This is conceptually easier to follow because lines along $\Psi =$ constant are streamlines. The outer boundaries are streamlines themselves, and can be assigned constant values. There must be a unit value for $\partial\Psi/dx$ along the web in order to impose unit velocity. The symmetry line is also a stream line, so that the outer boundary must have the same value of Ψ as the center of the web. Therefore the boundary values can be inked in, once and for all. Arbitrary initial guesses are pencilled in for the other grid points, and the relaxation process is begun by erasing random grid values and replacing them with the average of adjacent points. After a number of iterations, the numbers settle into steady values for more and more digits of accuracy.

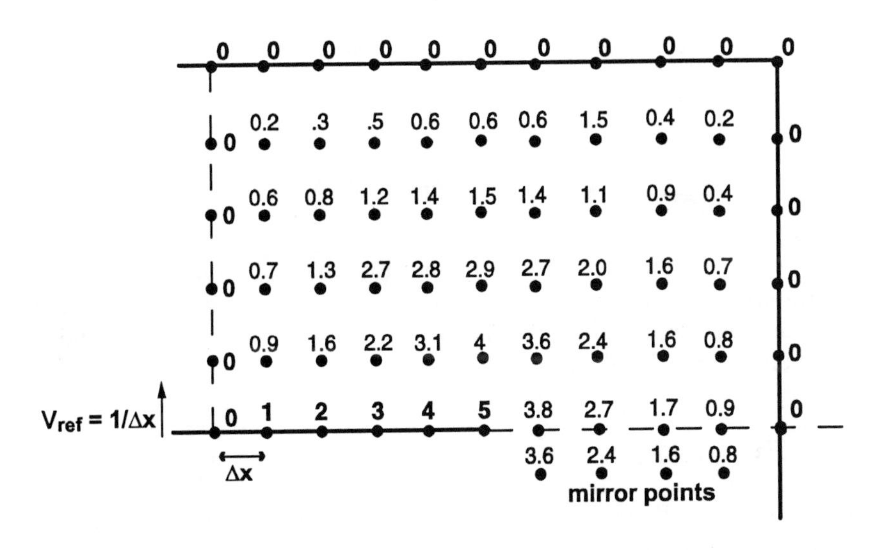

Figure 6.3: Southwell's Method

Once we have all the values of either Φ or else Ψ, we can find values for vertical velocity v_y and horizontal velocity v_x from the definition of Φ which requires that the gradient of the potential function is the velocity $\overrightarrow{v} = \overrightarrow{\nabla}\Phi$, and the definition of Ψ which requires that the stream function is orthogonal to Φ

$$v_x = \frac{\partial \Phi}{\partial x} = \frac{\partial \Psi}{\partial y}$$
$$v_y = \frac{\partial \Phi}{\partial y} = \frac{-\partial \Psi}{\partial x}$$

By the finite-difference approximation we can obtain a value for one of the velocity component

$$v_x = \frac{\Psi_{n,m+1} - \Psi_{n,m}}{\Delta y}$$

between every vertical pair of grid-points, and a value for the other of the components

$$v_y = \frac{-(\Psi_{n+1,m} - \Psi_{n,m})}{\Delta x}$$

between every horizontal pairs of grid-points. Associating each value with a square box Δx by Δy (different boxes for v_y and v_x, and some near the boundaries only half as big as others, so they only count half!), we can write for the kinetic energy per unit length (for a unit reference velocity of the web)

$$\begin{aligned}\frac{T}{L} &\cong \frac{1}{2}\frac{\rho_{fluid}\Delta x\Delta y}{g_c}\left(\sum v_x^2+\sum v_y^2\right)\\ &\cong \frac{1}{2}\frac{\rho_{fluid}}{g_c}\left(\sum(\Psi_{n,m+1}-\Psi_{n,m})^2+\sum(\Psi_{n+1,m}-\Psi_{n,m})^2\right)\end{aligned}$$

which we multiply by two to get the equivalent mass-per-unit-length.

If we are concerned about the accuracy of this method, we can halve the grid dimension, rework the problem for four times as many points, and see if there is any change in the answer.

6.8 Summary

If we assume a particular mode shape, we can treat a complex problem as a quasi-one-degree-of-freedom problem. This allows us to make allowance for the mass of a spring, the mass of liquid in a shell-and-tube heat exchanger, and the mass of air around a thin, wide web of paper or plastic.

We have also demonstrated the use of a finite-difference method for solving an elliptical partial differential equation.

6.9 Review

We have learned six ways of finding the natural frequency or time scale of a system:

1. From Newton's law,
   (a) writing a force/momentum balance for a control volume to obtain the governing differential equation, and reducing it to Standard Form; or
   (b) directly from static deflection.
2. From Conservation of Energy,
   (a) writing $dT/dt + dU/dt = 0$; or
   (b) recognizing equivalent mass and spring terms in T and U, respectively; or
   (c) writing $T_{\max} = U_{\max}$ and $\dot{x}_{\max} = \omega_n x_{\max}$.
3. From Hamilton's principle,
   (a) writing Lagrange's equation.

You should now be able to choose an appropriate methods (more than one will usually work) and determine ω_n, f_n, and τ_n for a simple system.

Problem 6.5 *Demonstrate Rayleigh's method for estimating the fundamental frequency ω_n of a uniform cantilever of length L, mass-per-unit-length μ (total mass $\mu \cdot L$), and stiffness EI. Use the assumed deflection shape*

$$\begin{aligned} y &= \frac{x^3}{L^3} \cdot y_{tip} \\ \frac{dy}{dx} &= \frac{3x^2}{L^3} \cdot y_{tip} \\ \frac{d^2y}{dx^2} &= \frac{6x}{L^3} \cdot y_{tip} \end{aligned}$$

which will give a slightly high answer.

Problem 6.6 *Find the equivalent mass of a cantilever acting as a spring at its tip, assuming the deflection shape corresponding to a static load at the tip*

$$\begin{aligned} y &= \frac{F_{tip} x^2}{6EI} (3L - x) \\ y_{tip} &= \frac{F_{tip} L^3}{3EI} \\ k &\triangleq \frac{F_{tip}}{y_{tip}} = \frac{3EI}{L^3} \end{aligned}$$

Express the equivalent mass as a fraction of the cantilever's mass $m = \mu L$.

Problem 6.7 *A long, flat plate of width $2c$ oscillates out-of-plane in a fluid-filled duct, $2a \times 2b$, with a gap $(b - c)$ on each side.*

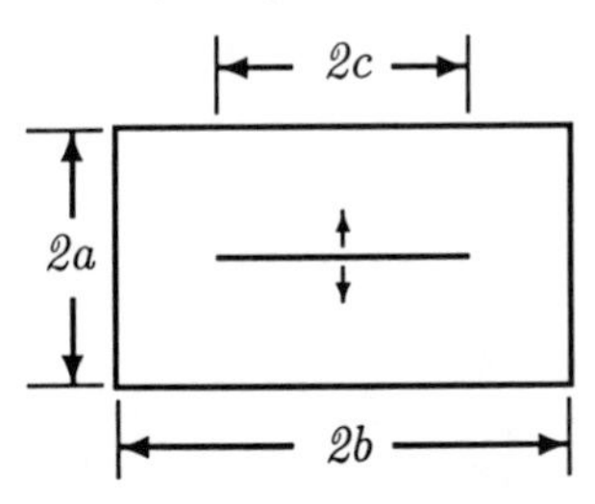

Estimate the added mass per unit length μ_{hydro} in terms of ρ_{fluid} and width $2c$. Select the parameters for your calculation according to the first letter of your first name:

	b/c	a/c		b/c	a/c		b/c	a/c		b/c	a/c		b/c	$a/$
A	1.6	0.6	F	1.7	0.6	L	1.8	0.6	Q	1.9	0.6	V	2.0	0.
B	1.6	0.8	G	1.7	0.8	M	1.8	0.8	R	1.9	0.8	W	2.0	0.
C	1.6	1.0	H	1.7	1.0	N	1.8	1.0	S	1.9	1.0	X	2.0	1.
D	1.6	1.2	IJ	1.7	1.2	O	1.8	1.2	T	1.9	1.2	Y	2.0	1.
E	1.6	1.4	K	1.7	1.4	P	1.8	1.4	U	1.9	1.4	Z	2.0	1.

Helpful hints: Use hard, vellum-like paper and a soft eraser. Draw the boundaries, gridpoints, and fixed values of the stream function in ink, but the values to be "relaxed" in faint pencil. Recognize symmetries, so you have to "relax" only one-fourth of the grid-points.

Chapter 7

PERIODIC FORCE EXCITATION OF UNDAMPED SYSTEMS and review of numerical Fourier analysis

In previous chapters, the behavior of undamped systems without excitation was investigated. We will now study the effect of a periodic forcing function $F(t)$. (Figure 7.1)

7.1 Governing Equations

We would like to solve Equation 2.6, leaving out only the damping term

$$\frac{m}{g_c}\ddot{x} + kx = F(t) \tag{7.1}$$

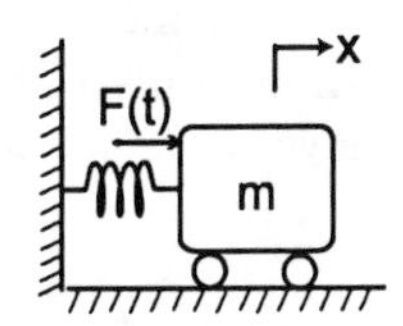

Figure 7.1: System with Excitation Force

(We have also left out the gravitational term, but that is because we learned in Section 4.7 that we can get rid of constant forces by means of a coordinate transformation.) To study a periodic $F(t)$, we will let it be a sinusoidal function of arbitrary amplitude F_o and arbitrary angular frequency ω_{ex}

$$\frac{m}{g_c}\ddot{x} + kx = F_o \sin(\omega_{ex} t) \tag{7.2}$$

As usual, we normalize to make the first coefficient unity

$$\begin{aligned} \ddot{x} + \left(\frac{kg_c}{m}\right) x &= \frac{F_o g_c}{m} \sin(\omega_{ex} t) \\ &= \left(\frac{F_o}{k}\right)\left(\frac{kg_c}{m}\right) \sin(\omega_{ex} t) \end{aligned}$$

and defining ω_n as before, write this in Standard Form

$$\boxed{\ddot{x} + \left(\omega_n^2\right) x = \left(\tfrac{F_o}{k}\right)\left(\omega_n^2\right)\sin(\omega_{ex} t)} \tag{7.3}$$

This is the form of the equation we will solve; later, when we have the solution, we can substitute the appropriate expressions for the Standard-Form parameters $\left(\omega_n^2\right)$ and $\left(\frac{F_o}{k}\right)$.

7.2 Complete Solution

The solution $x(t)$ of Equation 7.3 can have two kinds of terms: "particular integral" terms x_p that balance the particular forcing function; and "complementary function" terms x_h that balance off against each other—just like the terms of the homogeneous-equation solution—and serve to satisfy the Initial Conditions

$$x = x_p + x_h \tag{7.4}$$

For the left-hand terms to balance off against each other, x_h must contain the system parameter ω_n and takes the same form as Equation 4.19

$$x_h = A \sin\left(\omega_n t\right) + B \cos\left(\omega_n t\right) \tag{7.5}$$

but we cannot evaluate A and B from the Initial Conditions yet, because the particular-solution terms x_p may have some effect.

In order for the particular-solution terms x_p to balance off against the right-hand forcing terms, they must have a similar form containing the excitation parameter ω_{ex}; we will try

$$\begin{aligned} x_p &= X \sin(\omega_{ex} t - \theta) \qquad (7.6) \\ \ddot{x}_p &= -X\omega_{ex}^2 \sin(\omega_{ex} t - \theta) \end{aligned}$$

Plugging these into Equation 7.3, we get

$$-X\omega_{ex}^2 \sin(\omega_{ex}t - \theta) + \left(\omega_n^2\right) X \sin(\omega_{ex}t - \theta) = \left(\frac{F_o}{k}\right) \left(\omega_n^2\right) \sin(\omega_{ex}t)$$

To balance this equation, we must make $\theta = 0$; factoring out the maximum amplitude X and $\sin(\omega_{ex}t)$

$$X\left(\omega_n^2 - \omega_{ex}^2\right) \sin(\omega_{ex}t) = \left(\frac{F_o}{k}\right) \left(\omega_n^2\right) \sin(\omega_{ex}t)$$

which we can solve for X; we write it in the dimensionless form

$$\boxed{\frac{X}{(F_o/k)} = \frac{\omega_n^2}{(\omega_n^2 - \omega_{ex}^2)} = \frac{1}{1-\left(\frac{\omega_{ex}}{\omega_n}\right)^2}} \tag{7.7}$$

Inserting $\theta = 0$ and this value of X into our trial solution, Equation 7.6, we obtain x_p. Combining this and x_h from Equation 7.5 into Equation 7.4, we get the complete solution

$$x = \left(\frac{F_o}{k}\right) \left[\frac{1}{1-\left(\frac{\omega_{ex}}{\omega_n}\right)^2}\right] \sin(\omega_{ex}t) + A\sin(\omega_n t) + B\cos(\omega_n t) \tag{7.8}$$

We can now fit this to the Initial Conditions; first we take the derivative

$$\dot{x} = \left(\frac{F_o}{k}\right) \left[\frac{1}{1-\left(\frac{\omega_{ex}}{\omega_n}\right)^2}\right] \omega_{ex}\cos(\omega_{ex}t) + A\omega_n\cos(\omega_n t) - B\omega_n\sin(\omega_n t) \tag{7.9}$$

then we evaluate x and $\dot{x}$ at $t = 0$, remembering that $\sin(0) = 0$ and $\cos(0) = 1$; therefore the Initial Conditions are

$$\begin{aligned} x(0) &= B \\ \dot{x}(0) &= \left(\frac{F_o}{k}\right) \left[\frac{1}{1-\left(\frac{\omega_{ex}}{\omega_n}\right)^2}\right] \omega_{ex} + A\omega_n \end{aligned}$$

We can solve these equations for A and B and insert them into Equation 7.8 to obtain the complete solution. For example, the zero-initial-state solution, $x(0) = 0$ and $\dot{x}(0) = 0$, is

$$x = \left(\frac{F_o}{k}\right) \left[\frac{1}{1-\left(\frac{\omega_{ex}}{\omega_n}\right)^2}\right] \sin(\omega_{ex}t) - \left(\frac{F_o}{k}\right) \left[\frac{1}{1-\left(\frac{\omega_{ex}}{\omega_n}\right)^2}\right] \left(\frac{\omega_{ex}}{\omega_n}\right) \sin(\omega_n t) \tag{7.10}$$

We have demonstrated the general procedure for solving non-homogeneous linear equations; it has three steps:

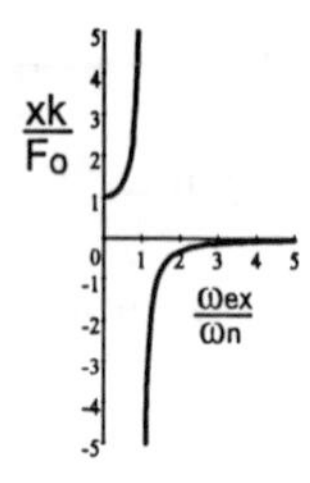

Figure 7.2: Response to periodic exculation

1. Obtain the complementary solution x_h, but do not fit the constants from Initial Conditions yet. (The reason we examine x_h first is to prepare us to look out for repeated roots, as we will see in a later example.)

2. Obtain the particular solution x_p.

3. Write the complete solution $x = x_p + x_h$, and solve for the constants A and B which came with x_h so that the complete solutions fits the Initial Conditions.

We will now study the particular solution to Equation 7.3 in more detail.

7.3 Amplification Factor

Let us examine the particular solution x_p. Equations 7.6 and 7.7 combine to

$$x_p = X_{\max} \sin(\omega_{ex} t) = \left(\frac{F_o}{k}\right) \left[\frac{1}{1 - \left(\frac{\omega_{ex}}{\omega_n}\right)^2}\right] \sin(\omega_{ex} t) \tag{7.11}$$

From Equation 7.7 this we learn that the maximum amplitude $X_{\max}$ of the response to forced excitation depends on two input parameter: excitation strength (F_o/k) and frequency ratio $f_{ex}/f_n = \omega_{ex}/\omega_n$. The former, (F_o/k), has the dimension of x, and the magnitude of the deflection due to a steady force F_o. $X_{\max}$ is directly proportional to it. This is a consequence of the linearity of the differential equation: superposing a second force $F_o \sin(\omega_{ex} t)$, thereby doubling the input, must double the output.

The dependence of $X_{\max}$ on (ω_{ex}/ω_n) is more complex; it is plotted in Figure 7.2:

- at very low excitation frequencies, the response approximates the static deflection, $X_{\max} = F_o/k$;

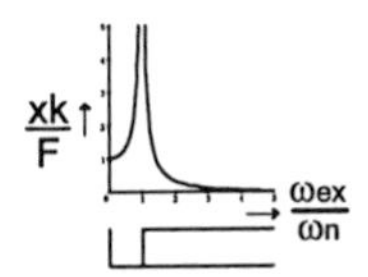

Figure 7.3: Undamped Amplitude Response: linear plot

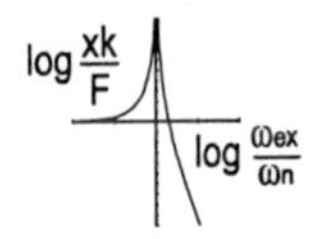

Figure 7.4: Undamped Amplitude Response: log-log plot

- at all sub-resonant excitation frequencies, the motion is in the same direction as the force;

- as the frequency ratio rises towards $\omega_{ex}/\omega_n = 1$, the response approaches ∞;

- at very high excitation frequencies, the response approaches zero;

- at all super-resonant excitations, the motion is in the *opposite* direction from the force;

- as the frequency ratio decreases towards $\omega_{ex}/\omega_n = 1$, the response approaches $-\infty$.

The situation becomes easier to understand if we plot the absolute magnitude of $(X_{\max}k/F_o)$ and fix the sign by letting θ be 180 degrees (π radians) for the larger values of (ω_{ex}/ω_n), as in Figure 7.3. The magnitude can also be plotted on log-log coordinates in Figure 7.4, which shows that, at large values of (ω_{ex}/ω_n), $(X_{\max}k/F_o)$ decreases asymptotically towards $(\omega_{ex}/\omega_n)^{-2}$.

The important lesson of these plots is that extraordinarily large response amplitudes are possible when ω_{ex} is near ω_n.

We are startled by a paradox: just below and just above resonance we get opposite responses. What happens at resonance itself? Is x_p in or out of phase with the excitation when ω_{ex} is exactly ω_n? We will investigate this in the next section.

7.4 Resonance

At resonance itself, when $\omega_{ex} = \omega_n$, the subscripts become superfluous and we can write Equation 7.3 as

$$\ddot{x} + (\omega^2)\, x = \left(\frac{F_o}{k}\right)(\omega^2)\sin(\omega) \tag{7.12}$$

Following our procedure of Section 7.2, we first we obtain the form of the complementary solution

$$x_h = A\sin(\omega t) + B\cos(\omega t)$$

Secondly, we try the particular solution of Equation 7.6

$$x_p = X_{\max}\sin(\omega t - \theta) = C\sin(\omega t) + D\cos(\omega t)$$

Hold on—this is a repeated-roots situation: x_p has the same form as x_h! If we try to use it in Equation 7.12, the x_p terms will cancel out against each other just like x_h, and we can never get a solution that balances the excitation term. In cases like this, we remember that we should try t-multiplied solutions, so we replace the trial form for x_p with

$$\begin{aligned} x_p &= Ct\sin(\omega t) + Dt\cos(\omega t) \qquad (7.13)\\ \dot{x}_p &= C\sin(\omega t) + C\omega t\cos(\omega t) + D\cos(\omega t) - D\omega t\sin(\omega t)\\ \ddot{x}_p &= 2C\omega\cos(\omega t) - C\omega^2 t\sin(\omega t) - 2D\omega\sin(\omega t) - D\omega^2 t\cos(\omega t) \end{aligned}$$

Plugging these into Equation 7.12, we get many cancellations; what remains is

$$(-2D\omega)\sin(\omega t) + (2C\omega)\cos(\omega t) = \left(\frac{F_o}{k}\omega^2\right)\sin(\omega t)$$

Since the coefficients of the sine have to cancel off against each other, and the coefficients of the cosine likewise, we really have two equations here; they yield

$$\begin{aligned} D &= -\frac{1}{2}\frac{F_o}{k}\omega \qquad (7.14)\\ C &= 0 \end{aligned}$$

so that Equation 7.13 becomes

$$x_p = -\frac{1}{2}\frac{F_o}{k}\omega t\cos(\omega t) \tag{7.15}$$

Therefore, the resonant solution is neither in-phase nor opposed to the excitation function, but is 90 degrees ($\pi/2$ radians) out of phase with it. The t-multiplier indicates linear growth of the amplitude of the oscillation. This makes sense: the fact that the solution is out-of-phase results in work being done by the exciting force against the displacement of the mass, at a rate which is proportional to the amplitude. Thus the power injected into the system increases linearly with time, and the energy accumulated within the system increases quadratically with time, approaching infinity, until the system breaks down and no longer follows the linear governing equation. In many practical cases where the periodic excitation persists for only a short time, the rate of growth is vital because it determines whether destructive amplitudes are reached or not.

This resolves the paradox with which we started this section, but it creates a new one: if the particular solution x_p for $\omega_{ex} = \omega_n$ is the steadily growing function of Equation 7.15, how can we have a completely different kind of solution like Equation 7.11 for an immediately adjoining solution, even with just an infinitesimal difference between ω_{ex} and ω_n?

7.5 Beats

To get a correct comparison between two solutions, one at and another near resonance, we need to look at complete solutions $x = x_p + x_h$. First let us examine resonance. If we add the complementary solution to Equation 7.15 and adjust A and B to yield the zero-initial-state solution, we get:

$$x = -\frac{1}{2}\frac{F_o}{k}\omega t\cos(\omega t) + \frac{1}{2}\frac{F_o}{k}\sin(\omega t) \tag{7.16}$$

This is not fundamentally different from Equation 7.15. The second, complementary term of the solution compensates for the first, particular-integral term, in order to get zero initial velocity. After a few cycles have elapsed, the fast-growing first term completely overshadows the second term, so that the long-term solution is still the steady growth of amplitude we have already seen at resonance (Figure 7.5).

What about the complete solution nearby, when $\omega_{ex} = (\omega_n \pm \varepsilon)$? Restating Equation 7.10 for this case:

$$x = \left(\frac{F_o}{k}\right)\left[\frac{1}{1-\left(\frac{\omega_n \pm \varepsilon}{\omega_n}\right)^2}\right]\left[\sin\left(\left(\omega_n \pm \varepsilon\right)t\right) - \left(\frac{\omega_n + \varepsilon}{\omega_n}\right)\sin\left(\omega_n t\right)\right]$$

If $\varepsilon \ll \omega_n$, this can be closely approximated by:

$$x \cong \left(\frac{F_o}{k}\right)\left(\frac{-\omega_n}{2\varepsilon}\right)\left[\sin\left(\omega_n t \pm \varepsilon t\right) - \sin\left(\omega_n t\right)\right]$$

Using trigonometric formulas (from the references suggested in Section 1.4) we can expand $\sin(\omega_n t \pm \varepsilon t)$ into $(\sin(\omega_n t)\cos(\varepsilon t) \pm \cos(\omega_n t)\sin(\varepsilon t))$ and find

$$x \cong \left(\frac{F_o}{k}\right)\left(\frac{-\omega_n}{2\varepsilon}\right)\left[\left(\cos\left(\varepsilon t\right) - 1\right)\sin\left(\omega_n t\right) \pm \sin\left(\varepsilon t\right)\cos\left(\omega_n t\right)\right]$$

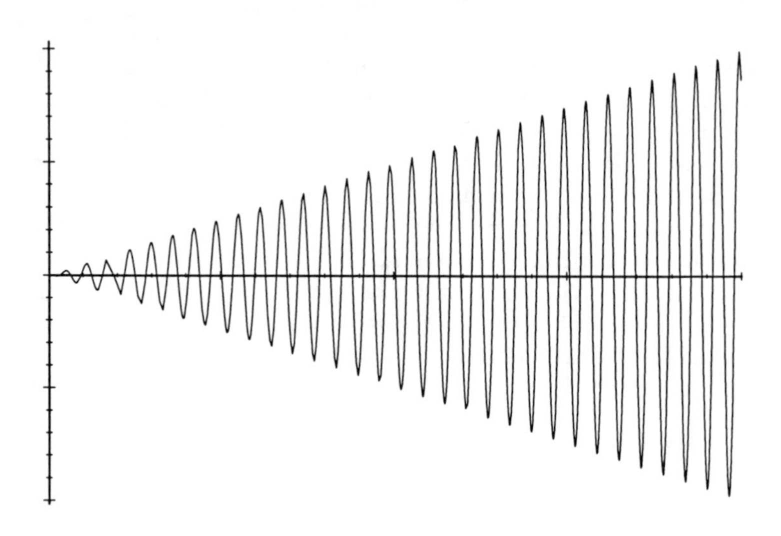

Figure 7.5: Complete Solution at Resonance

where the functions of εt are slowly varying parameters (because ε is a small number) and the functions of $\omega_n t$ are relatively rapid oscillations. Therefore, x has the form $C \sin(\omega t - \phi)$, where C and ϕ are slowly varying functions of time, and ω is in the vicinity of ω_n and ω_{ex}. Using trigonometric relations again, we can obtain the value of C:

$$\begin{aligned} C &\cong \left(\frac{F_o}{k}\right)\left(\frac{-\omega_n}{2\varepsilon}\right)\sqrt{(\cos(\varepsilon t) - 1)^2 + \sin^2(\varepsilon t)} \\ &\cong \left(\frac{F_o}{k}\right)\left(\frac{-\omega_n}{2\varepsilon}\right)\sqrt{2 - 2\cos(\varepsilon t)} \\ &\cong \left(\frac{F_o}{k}\right)\left(\frac{-\omega_n}{2\varepsilon}\right) 2\sin\left(\frac{1}{2}\varepsilon t\right) \end{aligned}$$

For $\varepsilon \ll \omega$ we can interpret the absolute value of C as a slowly varying envelope of the rapid oscillation ωt, with get the response of Figure 7.6, which periodically rises from zero to twice the amplitude of x_p and then falls back to zero again.

This phenomenon is called "beating" and occurs whenever two nearly identical frequencies of oscillation are added together. One can hear it when one of two strings for the same note on a piano are a little bit out of tune: the more they are apart, the faster the beating occurs. Beating between the response to a periodic excitation and an initial-condition solution is more unusual, but can be useful in the laboratory: it tells us just how far off from resonance a newly

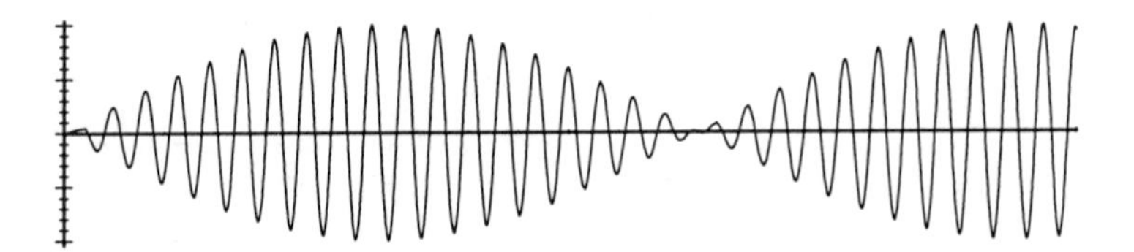

Figure 7.6: Complete Solution near Resonance

added harmonic excitation is.

The initial growth of the amplitude is $\frac{dC}{dt} = \frac{1}{2}\frac{F_o}{k}\cos\left(\frac{1}{2}\varepsilon t\right)$; this is identical with the growth of the resonant solution. Therefore, it is impossible to distinguish between resonance and near-resonance by observing the first few moments of response to a periodic excitation.

Exercise 7.1 *For simplicity, the derivation above has been a little sloppy. For greater rigor, write the excitation frequency as* $\left(\omega + \frac{1}{2}\varepsilon\right)$ *and the natural frequency as* $\left(\omega - \frac{1}{2}\varepsilon\right)$*, and repeat the derivation. This should lead to the same envelope function of the form* $\sin\left(\frac{1}{2}\varepsilon t\right)$*, but will prove that the rapid oscillation is at a frequency between* ω_{ex} *and* ω_n*.*

7.6 Base Excitation

Our solutions for periodic excitation can also be adapted to excitation by ground motion. If we let the ground motion be $y(t) = Y_o \sin(\omega_{ex} t)$, the governing equation is

$$\frac{m}{g_c}\ddot{x} + k\left(x - Y_o \sin(\omega_{ex} t)\right) = 0 \tag{7.17}$$

or

$$\frac{m}{g_c}\ddot{x} + kx = (kY_o)\sin(\omega_{ex} t) = (F_{equiv})\sin(\omega_{ex} t) \tag{7.18}$$

so that all our equations apply if we substitute Y_o for (F_o/k).

7.7 Fourier Analysis

Most periodic excitations do not have a simple sinusoidal wave-form. However, we can represent any periodic function as the sum of a series of sinusoidal functions. Since we are dealing with linear differential equations, superposition applies: we can obtain the response to each of these harmonic components, and then add them up to obtain the total response to the entire excitation.

The representation of the arbitrary periodic function of frequency ω is the Fourier series expansion

$$\begin{aligned} f(t) = \overline{f} + a_1 \cos \omega t + a_2 \cos 2\omega t + a_3 \cos 3\omega t + a_4 \cos 4\omega t + \ldots + \\ + b_1 \sin \omega t + b_2 \sin 2\omega t + b_3 \sin 3\omega t + b_4 \sin 4\omega t + \ldots \end{aligned} \tag{7.19}$$

where

$$\begin{aligned} \overline{f} &= \frac{\int_0^{2\pi} f(t) d(\omega t)}{\int_0^{2\pi} d(\omega t)} = \frac{1}{2\pi} \int_0^{2\pi} f(t) d(\omega t) = \frac{a_o}{2} \\ a_n &= \frac{\int_0^{2\pi} f(t) \cos(n\omega t)\, d(\omega t)}{\int_0^{2\pi} \cos^2(n\omega t)\, d(\omega t)} = \frac{1}{\pi} \int_0^{2\pi} f(t) \cos(n\omega t)\, d(\omega t) \\ b_n &= \frac{\int_0^{2\pi} f(t) \sin(n\omega t)\, d}{\int_0^{2\pi} \sin^2(n\omega t)\, d} = \frac{1}{\pi} \int_0^{2\pi} f(t) \sin(n\omega t)\, d(\omega t) \end{aligned} \tag{7.20}$$

Exercise 7.2 *Prove these relationships by multiplying both sides of the definition of the Fourier series expansion above, by* $\cos(n\omega t)$ *for some value of* n*, and integrating both sides over one complete cycle of the fundamental frequency, and evaluating each term*

$$\begin{aligned} &\int_0^{2\pi} f(t) \cos(n\omega t)\, d\omega t \\ &= \int_0^{2\pi} \left(\tfrac{a_o}{2} + a_1 \cos \omega t + b_1 \sin \omega t + a_2 \cos 2\omega t + b_2 \sin 2\omega t + \ldots\right) \cos(n\omega t)\, d\omega t \end{aligned}$$

Hint: remember orthogonality relationships like $\int_0^{2\pi} \cos(m\omega t) \cos(n\omega t)\, d\omega t = 0$ *for* $m \neq n$*. Recall also that* $\int_0^{2\pi} \cos^2(n\omega t)\, d\omega t = \pi$*, etc.)*

These integrals can be evaluated for any continuous periodic function; for example, a square wave of unit amplitude decomposes to

$$f(t) = \frac{4}{\pi} \sin \omega t + \frac{4}{3\pi} \sin 3\omega t + \frac{4}{5\pi} \sin 5\omega t + \ldots + \frac{4}{(2n-1)\pi} \sin(2n-1)\omega t + \ldots$$

if the rise of the wave is taken as $t = 0$. If we take the origin elsewhere, we will get cosine terms, but the total amplitude at any particular frequency will come out the same. We note that the coefficients decrease for the higher harmonics. Smooth functions are represented by coefficients which decrease more rapidly: most periodic functions found in nature can be represented reasonably accurately by the first few terms of their Fourier series expansion.

Exercise 7.3 *Find the Fourier representation of a triangular wave.*

To solve a vibration problem with arbitrary periodic excitation, we first decompose the excitation function into its harmonic components, as shown above. Then we write the particular solution for each excitation frequency, using Equations 7.6 and 7.7. The sum of these solutions is the Fourier series expansion of the complete particular solution.

Example: What is the response of an undamped system to a square wave of amplitude F_{sqr} if $\omega_{ex}/\omega_n = 0.5$? Using the expansion of the square wave above, we can obtain the solution term-by-term from the particular solution for each individual sinusoidal excitation

$$F(t) = \frac{4F_{\text{sqr}}}{\pi}\sin\omega t + \frac{4F_{\text{sqr}}}{3\pi}\sin 3\omega t + \ldots + \frac{4F_{\text{sqr}}}{(2n-1)\pi}\sin(2n-1)\omega t + \ldots$$

$$x_p = \frac{16F_{\text{sqr}}}{3\pi k}\sin\omega t - \frac{16F_{\text{sqr}}}{15\pi k}\sin 3\omega t \ldots - \frac{4F_{\text{sqr}}}{(2n-1)\pi k}\left|\frac{1}{1-(n-0.5)^2}\right|\sin n\omega t$$

Note that the system "filters out" frequencies which are well away from resonance; on the other hand, if any one of the harmonics coincided with resonance, unbounded growth at that frequency would be expected.

7.8 Numerical Fourier Analysis

In many cases, a periodic function has been measured, processed through an analog-to-digital (A/D) converter, and stored as a series of amplitude values. If these data points are spaced at equal time intervals, and if the period is an even multiple m of the time interval, numerical integration for decomposition of the tabulated function into Fourier coefficients is particularly easy:

$$\overline{f} \cong \frac{1}{m}\sum_{1}^{m} f_i = \frac{a_o}{2}$$

$$a_n \cong \frac{\sum f_i \cos n\omega t_i}{\sum \cos^2 n\omega t_i} = \frac{2}{m}\sum_{1}^{m} f_i \sin n\omega t_i$$

$$b_n \cong \frac{\sum f_i \sin n\omega t_i}{\sum \sin^2 n\omega t_i} = \frac{2}{m}\sum_{1}^{m} f_i \sin n\omega t_i$$

Using a spreadsheet program (for example, *Lotus 1-2-3* or *Microsoft Excel*) a table is set up, with the following columns:

1. The index numbers of all the data points (but not listing any data point twice: remember that the point at the very end of the interval is the same as the point at the very beginning—only one of these two points with identical values should be included so that we don't input the same number twice); there should be m data points if the period is divided into m intervals.

2. The time associated with each data point; these entries should advance by the time interval, which is the period of the observed function divided by m; at the foot of this column we can write the total elapsed time, which is the fundamental period τ of the system.

3. The recorded values f_i of the data points; at the foot of this column we can write the average value $\overline{f} = \frac{1}{m}\sum_0^m f_i$

4. The value of ωt_i at each data point; this starts with zero at $t = 0$ and increases by $2\pi/m$ radians or $360/m$ degrees with each entry.

5. The value of $\cos \omega t_i$ for each data point; at the foot of this column we can write $\frac{1}{m}\sum_0^m \cos^2 \omega t_i$ to verify that it equals 1/2

6. The product of the third and fifth column, $f_i \cos \omega t_i$; at the foot of this column we can write $a_1 = \sum_1^m f_i \cos \omega t_i \div \sum_1^m \cos^2 \omega t_i = \frac{2}{m}\sum_1^m f_i \cos \omega t_i$

7. The value of $\sin \omega t_i$ for each data point; at the foot of this column we can write $\frac{1}{m}\sum_0^m \sin^2 \omega t_i$ to verify that it equals 1/2

8. The product of the third and seventh column, $f_i \sin \omega t_i$; at the foot of this column we can write $b_1 = \sum_1^m f_i \sin \omega t_i \div \sum_1^m \sin^2 \omega t_i = \frac{2}{m}\sum_1^m f_i \sin \omega t_i$

9. The value of $2\omega t_i$ at each data point.

10. The value of $\cos 2\omega t_i$ for each data point; at the foot of this column we can write $\frac{1}{m}\sum_0^m \cos^2 2\omega t_i$ to verify that it equals 1/2

11. The product of the third and tenth column, $f_i \cos 2\omega t_i$; at the foot of this column we can write $a_2 = \sum_1^m f_i \cos 2\omega t_i \div \sum_1^m \cos^2 2\omega t_i = \frac{2}{m}\sum_1^m f_i \cos 2\omega t_i$

12. The value of $\sin 2\omega t_i$ for each data point; at the foot of this column we can write $\frac{1}{m}\sum \sin^2 2\omega t_i$ to verify that it equals 1/2

13. The product of the third and twelfth column, $f_i \sin 2\omega t_i$; at the foot of this column we can write $b_2 = \sum_1^m f_i \sin 2\omega t_i \div \sum_1^m \sin^2 2\omega t_i = \frac{2}{m}\sum_1^m f_i \sin 2\omega t_i$

14. The value of $3\omega t_i$ at each data point.

15. The value of $\cos 3\omega t_i$ for each data point; at the foot of this column we can write $\frac{1}{m}\sum \cos^2 3\omega t_i$ to verify that it equals 1/2

16. The product of the third and sixteenth column, $f_i \cos 3\omega t_i$; at the foot we can write $a_3 = \sum_1^m f_i \cos 3\omega t_i \div \sum_1^m \cos^2 3\omega t_i = \frac{2}{m}\sum_1^m f_i \cos 3\omega t_i$

17. The value of $\sin 3\omega t_i$ for each data point; at the foot of this column we can write $\frac{1}{m}\sum \sin^2 3\omega t_i$ to verify that it equals 1/2

18. The product of the third and eighteenth column, $f_i \sin 3\omega t_i$; at the foot we can write $b_3 = \sum_1^m f_i \sin 3\omega t_i \div \sum_1^m \sin^2 3\omega t_i = \frac{2}{m}\sum_1^m f_i \sin 3\omega t_i$

If the data points are not quite equally spaced, we can adjust the numerical integration to weight each point appropriately for the time-interval it represents. The usual problem is that the period is not an even multiple of the sampling interval, which affects the proper weighting of the first and/of last point.

Exercise 7.4 *Numerically evaluate the Fourier coefficients for a square wave. Compare the result with the analytical solution in the previous section. In your spreadsheet, check that the squares of the terms in each of the sine and cosine columns add up to* $m/2$.

7.9 Summary

The undamped system with periodic excitation can be put into the Standard Form Equation 7.3; the particular solution is the sinusoidal response of Equation 7.11, with amplitudes plotted in Figure 7.2. For linear systems, the simultaneous excitation by several sinusoidal functions can be solved by superposing the responses to the input functions. Smooth periodic functions can be described concisely by stating the mean value, the amplitude of the fundamental, and the relative amplitudes of higher harmonics.

Problem 7.5 *A harmonic force with a "half-amplitude" (see page 36) of* 0.3 *lbf and a frequency of* 10 *Hz acts on an initially stationary* 2.0 *lbm mass attached to a spring of stiffness* 15 *lbf/inch. What is the amplitude of the response? Is it in- or out-of-phase with the excitation force?*

Problem 7.6 *A* 1.0*-pound mass is attached to a spring of stiffness* 15 *lbf/inch. The foundation begins to shake with a "double amplitude" (see page 36) of* 5×10^{-3} *inch. What is the amplitude of the response? Is it in- or out-of-phase with the ground motion?*

Problem 7.7 *Write the response for an undamped system excited by a square wave with a fundamental frequency which is two-thirds of the natural frequency. Sketch the resulting response.*

Problem 7.8 *Numerically evaluate the Fourier coefficients for a triangular wave.*

Compare the result with an analytical solution in the previous section. In your spreadsheet, check that the squares of the terms in each of the sine and cosine columns add up to $m/2$.

Problem 7.9 *Analytically find the Fourier representation of a saw-tooth wave.*

Problem 7.10 *Numerically evaluate the first few terms of the Fourier representation of a saw-tooth wave.*

Problem 7.11 *Find the Fourier representation, including the average value and the first three harmonics, of the following datapoints of force F (in lbf) as a function of time t (in seconds):*

t	F	t	F	t	F	t	F	t	F
0.0	6.00	0.5	5.51	1.0	2.07	1.5	−3.19	2.0	−8.03
0.1	12.51	0.6	1.00	1.1	1.72	1.6	−2.37	2.1	−8.95
0.2	15.93	0.7	−1.04	1.2	0.00	1.7	−1.98	2.2	−6.73
0.3	15.19	0.8	−0.63	1.3	−2.05	1.8	−3.00	2.3	−1.25
0.4	11.03	0.9	0.95	1.4	−3.27	1.9	−5.42	2.4	6.00

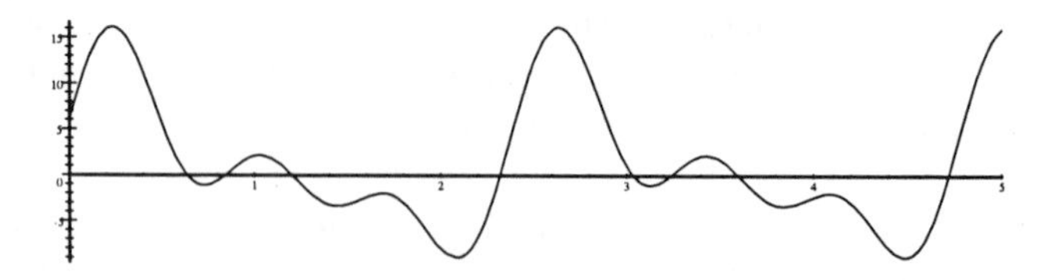

Approx. answer: $1+3\cos\left(\frac{\pi t}{1.2}\right)+2\cos\left(\frac{2\pi t}{1.2}\right)+6\sin\left(\frac{\pi t}{1.2}\right)+5\sin\left(\frac{2\pi t}{1.2}\right)+4\sin\left(\frac{3\pi t}{1.2}\right)$

Problem 7.12 *Find the first seven Fourier coefficients for the following observed force F (in Newtons) as a function of time t (in seconds):*

t	F	t	F	t	F	t	F	t	F
0.0	7.00	1.0	8.66	2.0	−5.20	3.0	−0.83	4.0	0.90
0.2	6.36	1.2	11.00	2.2	−10.88	3.2	2.90	4.2	2.00
0.4	5.20	1.4	11.21	2.4	−13.00	3.4	4.00	4.4	4.20
0.6	4.83	1.6	8.10	2.6	−11.02	3.6	3.00	4.6	6.22
0.8	6.10	1.8	2.00	2.8	−6.20	3.8	1.45	4.8	7.00

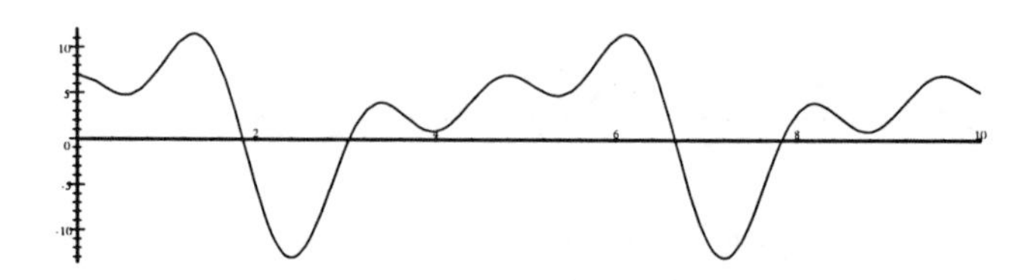

Approx. answer: $2+6\cos\left(\frac{\pi t}{2.4}\right)-5\cos\left(\frac{2\pi t}{2.4}\right)+4\cos\left(\frac{3\pi t}{2.4}\right)+3\sin\left(\frac{\pi t}{2.4}\right)-\sin\left(\frac{3\pi t}{2.4}\right)$

Chapter 8

UNBALANCE EXCITATION and rotating shafts

In many cases, the source of the excitation is not an externally imposed force, but an unbalanced mass. We can convert the effect of that mass to an equivalent force. (Figure 8.1)

8.1 Governing Equation

If a spring-supported machine of total mass M includes a smaller mass m which moves relative to it with a (half-)amplitude e and an angular frequency ω_{ex} so that its displacement is $(x + e \cdot \sin \omega_{ex} t)$, we can write Newton's law

$$-kx = \frac{(M-m)}{g_c} \ddot{x} + \frac{m}{g_c} \left(\ddot{x} - e \cdot \omega_{ex}^2 \cdot \sin \omega_{ex} t \right) \tag{8.1}$$

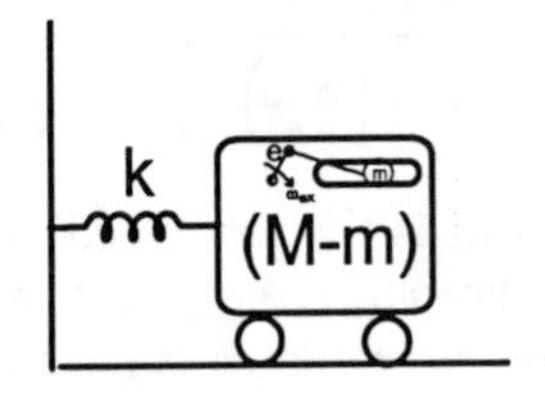

Figure 8.1: Mass Excitation

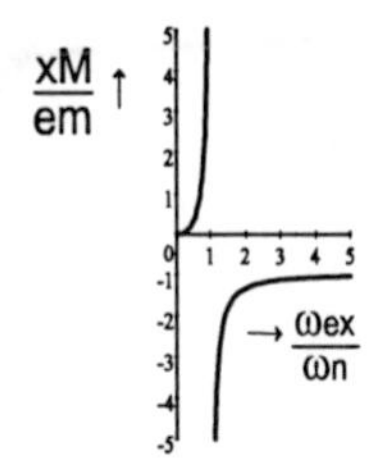

Figure 8.2: Unbalance Response

which we can reorganize to

$$\frac{M}{g_c}\ddot{x} + kx = \left(\frac{em\omega_{ex}^2}{g_c}\right)\sin\omega_{ex}t \tag{8.2}$$

which has the same form as Equation 7.2 if we replace F_o in that equation with

$$F_{\text{equiv}} \triangleq \frac{em\omega_{ex}^2}{g_c} \tag{8.3}$$

so that

$$\frac{F_{\text{equiv}}}{k} = \frac{em}{M}\frac{\omega_{ex}^2}{\omega_n^2} \tag{8.4}$$

and the amplitude of our solution in the form of Equation 7.6

$$x_p = X\sin(\omega_{ex}t - \phi) \tag{8.5}$$

is found by replacing the force in Equation 7.7 with the equivalent force

$$\frac{Xk}{F_{\text{equiv}}} = \frac{XM}{em}\frac{\omega_n^2}{\omega_{ex}^2} = \frac{1}{1-\left(\frac{\omega_{ex}}{\omega_n}\right)^2} \tag{8.6}$$

$$\frac{XM}{em} = \frac{\left(\frac{\omega_{ex}}{\omega_n}\right)^2}{1-\left(\frac{\omega_{ex}}{\omega_n}\right)^2} = \frac{1}{\left(\frac{\omega_n}{\omega_{ex}}\right)^2 - 1} \tag{8.7}$$

As shown in Figure 8.2, this is the a "flipped" version of the response curve described by Equation 7.7. For $\omega_{ex}/\omega_n \ll 1$, X approaches zero, because the equivalent exciting force goes towards zero for small ω_{ex}^2. For $\omega_{ex}/\omega_n \gg 1$, $XM = -em$, so that the center of gravity of the combined masses stays stationary.

8.2 Rate of Growth

At $\omega_{ex}/\omega_n = 1$, we have resonance. From Equation 7.15 combined with Equation 8.4 we can show that the growth of amplitude at resonance is determined by

$$x_p = -\frac{1}{2}\frac{em}{M}\omega t\cos(\omega t) \tag{8.8}$$

This equation warns us that we must not stay at or near resonance for too long a time if we want to avoid large amplitudes and stresses.

The domestic illustration is spin-drying in washing machines: if the drum does not accelerate fast enough (because the load is large and waterlogged) or the unbalance is too great (because the load is not distributed uniformly), the deflections become too large and the limit switch shuts off the machine. But once the drum makes it past resonance, the amplitude settles down to the more moderate equilibrium value (and decreases as the load becomes dryer and lighter). Many other fast-spinning devices, from ultracentrifuges in biochemistry laboratories to steam turbines in power plants, also operate beyond resonance, and have to be sped through the critical speeds.

8.3 Unbalance

The source of the term *em* need not be a translating relative motion; it could be a rotating unbalance of which one component acts on our one-degree-of-freedom system. A typical way to construct a motor-driven "shaker" for the intentional excitation of a system, is to gear two unbalanced flywheels together in such a way that the unbalances cancel in one direction, but add in the other.

In the mass-excitation equations, the terms e and m always occur together. We cannot distinguish between a slight eccentricity e of the entire mass M on the one hand, and a small added mass m mounted at a large radius e on the other. That means that we can compensate for a small error in centering the hub of a flywheel, by means of a small mass added to the rim.

8.4 Crank-Slider Harmonics

Reciprocating mechanisms may produce more than one excitation frequency. If a slider or piston is moved by means of a connecting rod and a crank, the motion of the slider will not be sinusoidal, even if the crank were rotating perfectly uniformly. The motion of a piston reverses more suddenly near top dead center than it does near bottom dead center. The relative displacement d of a slider is

$$d \cong e\left(\cos\left(\omega_{ex}t\right) + \frac{1}{4}\frac{e}{L}\cos\left(2\omega_{ex}t\right)\right) \tag{8.9}$$

where e is the eccentricity of the crank (i.e., the stroke is $2e$) and L is the length of the connecting rod. For example, if $L/2e = 1.5$, then there is a second harmonic which has an amplitude 1/12 as large as that of the fundamental. This is not insignificant, because the second harmonic will have a velocity which is 1/6 as great as that of the fundamental motion, and an acceleration (and associated force) which is 1/3 as great as the fundamental one.

Exercise 8.1 *Show that the relationship between crank angle and piston displacement is*

$$d = e\left(\cos\theta + \frac{L}{e}\sqrt{1-(e/L)^2\sin^2\theta}\right) \tag{8.10}$$

where $\theta = \sin\omega t$. *Hint: From trigonometry, show that* $d = e\cdot\cos\theta + L\cdot\cos\phi$, *where* ϕ *is the angle made by the connecting rod. It is determined by* $L\cdot\sin\phi = e\cdot\sin\theta$. *Using Pythagoras' theorem, eliminate* ϕ.

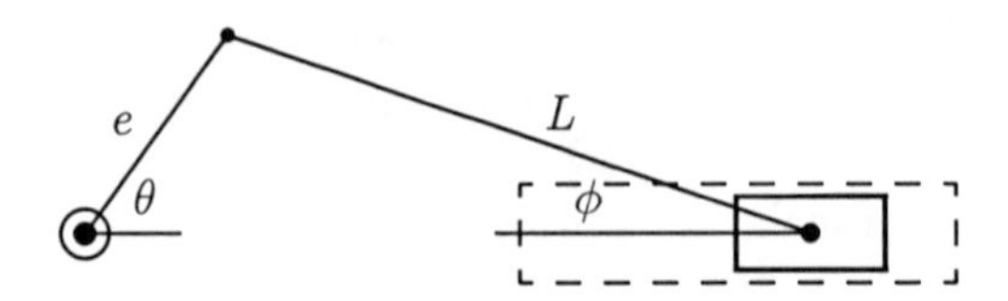

Exercise 8.2 *Expand Equation 8.10 into a Fourier series, obtaining at least one term more than Equation 8.9, for example* $\cos(4\omega_{ex}t)$. *Methods: You may either use a mathematical handbook (Section 1.4) to expand the square root by the binomial theorem and to make trigonometric substitutions for the powers of sines, obtaining a power series in* (e/L) *and/or a Fourier series with even cosine terms; or else you may use a symbolic-algebra computer program like* Maple *or* Mathematica *(Section 1.5), to expand Equation 8.10 into a Fourier Series and/or evaluate integrals like*

$$a_2 = \frac{L}{e}\cdot\frac{1}{\pi}\int_0^{2\pi}\sqrt{1-\left(\frac{e}{L}\right)^2\sin^2\theta}\cos(2\theta)\,d\theta$$

Answer:

$$\begin{aligned} d/e \;&=\; constant + \cos\theta \\ &+\left[\frac{1}{4}\left(\frac{e}{L}\right)^2+\frac{1}{16}\left(\frac{e}{L}\right)^4+\frac{15}{512}\left(\frac{e}{L}\right)^6+\ldots\right]\cos 2\theta + \\ &+\left[\frac{-1}{64}\left(\frac{e}{L}\right)^4-\frac{3}{256}\left(\frac{e}{L}\right)^6-\ldots\right]\cos 4\theta + \\ &+\left[\frac{1}{512}\left(\frac{e}{L}\right)^6-\ldots\right]\cos 6\theta - \ldots \end{aligned}$$

$$\boxed{d \cong e\left\{\cos\theta+\left[\tfrac{1}{4}\left(\tfrac{e}{L}\right)+\tfrac{1}{16}\left(\tfrac{e}{L}\right)^3\right]\cos 2\theta-\tfrac{1}{64}\left(\tfrac{e}{L}\right)^3\cos 4\theta\right\}}$$

8.5 Engine Balancing

8.5.1 Primary Balance

A piston engine unavoidably has the mass of each piston moving through the stroke $2e$; to minimize the vibration. It is desirable to have these excitations cancel each other out. For example, in a four-cylinder in-line engine, whenever the outer two cylinders move in one direction, the inner two move in the other: the fundamental sinusoidal motions exactly cancel.

Exercise 8.3 *Show that the fundamental sinusoidal motions cancel and the center-of-gravity is unmoved in a six-cylinder engine. Hint: the first three cranks are offset 120 degrees from each other, the inner two are side-by-side, and the last three are offset 120 degrees from each other in the opposite direction.*

Similar cancellations can be achieved in opposed "boxer" engines, although the necessary offset of the cranks leaves a "rocking-couple" excitation in a two-cylinder boxer.

For only two cylinders, primary balance can be achieved by placing two cylinders in a ninety-degree "V" and connecting both to the same crank so that their motion is ninety degrees out-of-phase. The combined effect of both pistons is a rotating unbalance, which can be cancelled by counterweights on the crankshaft which total the same em as one piston.

Exercise 8.4 *How much counterweight is needed in a seven-cylinder radial engine?*

Even in many-cylinder engines, in which *overall* balance is easily achieved, a V-arrangement is desirable because each pair of pistons can be counterweighted for *local* primary balance to minimize shaft stresses.

If one attempts to balance the motion of a single piston with crank-mounted weights, one discovers that the mass which is added to cancel excitation on the piston-axis, causes excitation at right angles to that axis. To minimize excitation overall, only about 50% of single-cylinder vibration excitation should be balanced-out by means of crank-mounted counterweights. If one follows the center of gravity of the combined system of piston and the 50%-counterbalance, one can show that its path moves in a circle in the opposite direction of crank rotation: the combined effect is the same as a counterweight rotating in the opposite direction! Some large motorcycle singles incorporate the added feature of a counter-rotating balance shaft, with counterweights to cancel this rotary excitation. The overall effect of the weights on the crank and on the counter-rotating shaft is to constitute a "shaker" (Section 8.3) which opposes the primary unbalance of the piston..

We see that engine balance of the fundamental component can be achieved in a number of ways:

- by having four or six cylinders in-line;
- by having four opposed cylinders;
- by having two opposed cylinders with minimal offset;
- by having a 90-degree V-twin with counterweights on the crank;
- by having a single with equal-speed counter-rotating balance shafts.

We have overlooked the complicated effect of the connecting rod; it can be *approximated* by assigning part of its mass to the piston, and part to the crank.

8.5.2 Secondary Balance

The second harmonic in Equation 8.9 is harder to balance out; in a four-cylinder engine, the second-harmonic terms are all in phase, and add rather than cancel; it takes a six-cylinder in-line engine to eliminate them.

Exercise 8.5 *Show that the second-harmonic motions cancel and the center-of-gravity is unmoved in a six-cylinder engine. Recall: cylinders 1 and 6 are paired, as are 2 and 5, and 3 and 4, respectively.*

To reduce vibration, pairs of double-speed balancing shafts have been incorporated in large four-cylinder automotive engines, forming "shakers" (Section 8.3) to oppose the second harmonic of the piston motion.

We see that balancing the second harmonic is harder, but can also be achieved in a number of ways:

- by having six cylinders in-line;
- by having four cylinders and double-speed balance shafts;
- by having two opposed cylinders with minimal offset;
- by having a V-twin or single with double-speed balance shafts.

Exercise 8.6 *What is the unbalance in a V-8 with a flat, 180-degree crankshaft "Up/Down/Down/Up"? Or a 90-degree "East/South/North/West" crankshaft? What is the effect if the pairs of cylinders are slightly offset, i.e., one bank of four is slightly shifted relative of the other bank of four so that the big-ends of the connecting rods can fit side-by-side on the cranks?*

In practical engine design, factors other than vibration are also important; for example, in automotive applications, compactness favors multi-cylinder "V" configurations.

8.5.3 Torsional Balance

We should not overlook torsional excitation around the axis of the crankshaft, which can be felt both at the drive shaft and in the engine mounts. At full throttle, the predominant cause is the force of the power stroke, and the cure is to have many cylinders firing at even intervals.

In addition, the crank-slider mechanism introduces torsional effects which are independent of the power output: accelerating and decelerating the piston requires torque on the crankshaft. Another way of looking at this is that the apparent moment-of-inertia of the crankshaft is a function of crank position, and in order to conserve moment-of-momentum the crankshaft wants to speed up and slow down as it rotates. This effect can be minimized by having either a ninety-degree "V"-configuration or six cylinders in-line; rotating counterbalances do not cancel rotational excitation.

Exercise 8.7 *Find the magnitude of the torsional excitation of a four-cylinder engine in which each piston weighs m and the stroke is $2e$. Reminder: For rotation, the answer should have units of torque, rather than force.*

8.6 Rotating Shafts and Whirling

The deflections of rotating shafts have at least two degrees of freedom; nevertheless, a few special cases can be treated like a one-degree-of-freedom problem.

8.6.1 The "Flat" Shaft

If a shaft is much more flexible in one direction than the other, we can concentrate on looking at the deflection in that direction—a direction which rotates, so that our one-degree-of-freedom system is in rotating coordinates like the mass and spring on a turntable. For a concentrated mass m located mid-span of a flat shaft (where the spring constant is $k = 48EI/L^3$) with an eccentricity e, follows the governing equation

$$\ddot{r} + \left(\frac{kg_c}{m} - \Omega^2\right) \cdot r = \frac{kg_c}{m} \cdot e \tag{8.11}$$

At equilibrium, the acceleration is zero and

$$\overline{r} = \frac{e}{1 - \frac{m}{kg_c} \cdot \Omega^2} \tag{8.12}$$

The equilibrium position diverges $\overline{r} \rightarrow \infty$ when $|\Omega| = \sqrt{kg_c/m}$; the critical speed of the shaft. When $\tilde{r} = 0$, this is the (rotating) deflection, appearing as a whirl synchronous with the rotation of the shaft.

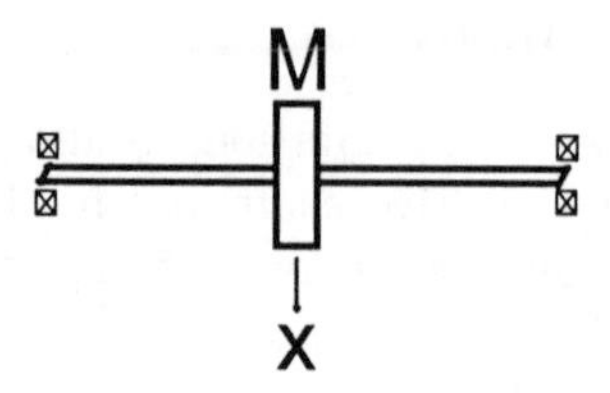

Figure 8.3: Flywheel on Shaft

Substituting $(\overline{r}+\widetilde{r})$ for r, we get

$$\ddot{\widetilde{r}} + \left(\frac{kg_c}{m} - \Omega^2\right) \cdot \widetilde{r} = 0 \tag{8.13}$$

so that, depending on initial conditions, there can be an additional deflection at the frequency

$$2\pi f_n = \omega_n = \sqrt{\frac{kg_c}{m} - \Omega^2} \tag{8.14}$$

To see the meaning of this, we might look at a few special cases: if $\omega_n^2 \gg \Omega^2$, the motion describes a flower-shaped pattern; if $\omega_n^2 \ll \Omega^2$, the motion is a near-circular whirl with slowly varying amplitudes; if $\omega_n^2 \approx \Omega^2$, the motion is a near-synchronous non-circular whirl which advances in the rotational direction if $\omega_n^2 > \Omega^2$, and proceeds in the opposite direction if $\omega_n^2 < \Omega^2$.

8.6.2 The "Circular" Shaft

If a shaft has equal stiffness in all directions (and if internal damping is negligible), it may be treated in fixed coordinate system as if it were not rotating—only the unbalance is rotating—and the two directions of deflection (e.g., vertical and horizontal) are uncoupled. Therefore the governing equations are

$$\begin{aligned} \ddot{x} + \left(\frac{kg_c}{M}\right) \cdot x &= \frac{em}{M} \cdot \Omega^2 \cos \Omega t \\ \ddot{y} + \left(\frac{kg_c}{M}\right) \cdot y &= \frac{em}{M} \cdot \Omega^2 \sin \Omega t \end{aligned} \tag{8.15}$$

and the particular solution is

$$\begin{aligned} x_p &= \frac{em}{M} \cdot \frac{\frac{\Omega^2}{kg_c/M}}{1 - \frac{\Omega^2}{kg_c/M}} \cdot \cos \Omega t \\ y_p &= \frac{em}{M} \cdot \frac{\frac{\Omega^2}{kg_c/M}}{1 - \frac{\Omega^2}{kg_c/M}} \cdot \sin \Omega t \end{aligned} \tag{8.16}$$

which is a circular motion synchronous with the rotation of the shaft, and in-phase with the unbalance for subcritical speeds and out-of-phase for supercritical speeds. We see that this is identical with the solution for other mass excitations of machinery.

Den Hartog[1] has noted that the flexibility of the bearing supports needs to be added in, and is usually different vertically and horizontally, so that the x- and y-equations have different natural frequencies and critical speeds.

Exercise 8.8 *Compare sixty-degree V-6 engines with three cranks, ninety degree V-6 engines with three cranks, and ninety-degree V-6 engines with six cranks arranged for even firing.*

Problem 8.9 *Our discussion on engine balancing has emphasized on four-stroke engines incorporating pairs of alternately firing cylinders. Find the unbalance in a five-cylinder engines with 144 degrees between adjacent cranks.*

Problem 8.10 *Two-stroke engines require fewer cylinders for smooth power output. Find the unbalance in a three-cylinder engine with 120 degrees between cranks.*

Problem 8.11 *Two-stroke cylinders can be combined in opposing pairs, with opposing cranks so that each pair has simultaneous crankcase compression. Find the unbalance in a four-cylinder boxer engine for a drone aircraft. Compare the two possible arrangements: one with the right cylinders shifted relative to the left cylinders, and one for the right cylinders further apart than the left cylinders.*

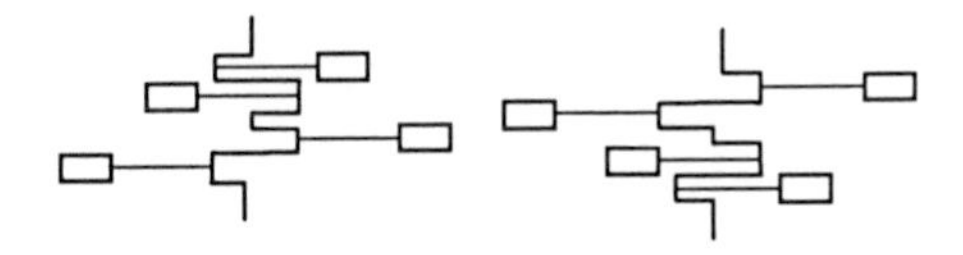

[1] J.P. Den Hartog, *Mechanical Vibrations,* reissued by Dover Publications, New York, 1985, ISBN 0-48664785-4.

Part II

Damped Systems

Chapter 9

DAMPED FREE VIBRATION and logarithmic decrement

In Chapter 4, free vibration without damping was analyzed. We now will add linear damping, define the Standard Form equation with this added term, and study the motion due to Initial Conditions.

9.1 Standard Form

With damping, but in the absence of external excitation, the governing Equation 2.6, which sums up the forces as a function of displacement x, becomes

$$\frac{m}{g_c}\ddot{x} + c\dot{x} + kx = 0 \tag{9.1}$$

As usual, we divide by the coefficient of the first term and obtain

$$\ddot{x} + \left(\frac{cg_c}{m}\right)\dot{x} + \left(\frac{kg_c}{m}\right)x = 0 \tag{9.2}$$

This reduces the number of systems parameters from three to two. For reasons which will soon become evident, it is customary to name them as shown in the Standard Form equation

$$\boxed{\ddot{x} + (2\zeta\omega_n)\dot{x} + \left(\omega_n^2\right)x = 0} \tag{9.3}$$

We can find values for the angular natural frequency ω_n and the parameter ζ (zeta) by comparing Equations 9.2 and 9.3; in this translational-motion example:

$$\omega_n \triangleq \sqrt{\frac{kg_c}{m}} \tag{9.4}$$

$$\zeta \triangleq \frac{cg_c}{2\omega_n m} = \frac{c}{2\sqrt{km/g_c}} \tag{9.5}$$

9.2 Generalization

Other one-dimensional mechanical and electrical problems can be also be brought into the form of Equation 9.3. Depending on the arrangement, the damping-element coefficient c (for translation), c_T (for rotation), or R (for electrical resistance) may appear in the numerator or the denominator. If a translational spring and damper are connected in *series*, the displacements can be summed as a function of the support force F

$$\frac{F}{k} + \int \frac{F}{c} dt + \iint \frac{Fg_c}{m} dtdt + C_1 + C_2 t = 0$$

$$\ddot{F} + \left(\frac{k}{c}\right) \dot{F} + \left(\frac{kg_c}{m}\right) F = 0$$

$$\zeta = \frac{\sqrt{km/g_c}}{2c} \tag{9.6}$$

Torsional springs plus damping by a *parallel* "brake" lead to a torque-summation in terms of angle θ which resembles Equation 9.2

$$\ddot{\theta} + \left(\frac{c_T g_c}{J}\right) \dot{\theta} + \left(\frac{k_T g_c}{J}\right) \theta = 0$$

$$\zeta \triangleq \frac{c_T}{2\sqrt{k_T J/g_c}} \tag{9.7}$$

while the same elements connected in *series* (as in a viscous coupling) are described by an angle-summation in terms of torque T

$$\ddot{T} + \left(\frac{k_T}{c_T}\right) \dot{T} + \left(\frac{k_T g_c}{J}\right) T = 0$$

$$\zeta = \frac{\sqrt{k_T J/g_c}}{2c_T} \tag{9.8}$$

We note that, when dampers are mounted in parallel with springs and coils, respectively, they lead to the largest ζ if they are "stiff" (i.e., when c or c_T have large values); but when they are mounted in series with springs or coils, they are most effective if they are "soft" (i.e., when c or c_T have small values).

By way of contrast, passive electrical elements connected around a loop in series give us a voltage-summation around the circuit in terms of current $\mathfrak{I}$

$$\begin{aligned} L\frac{d\mathfrak{I}}{dt} + R\mathfrak{I} + \int \frac{\mathfrak{I}}{C}dt + \text{constant} &= 0 \\ \ddot{\mathfrak{I}} + \left(\frac{R}{L}\right)\dot{\mathfrak{I}} + \left(\frac{1}{LC}\right)\mathfrak{I} &= 0 \\ \zeta &= \frac{R}{2\sqrt{L/C}} \end{aligned} \tag{9.9}$$

while the same elements in parallel lead to a current-summation as a function of voltage E

$$\begin{aligned} C\frac{dE}{dt} + \frac{1}{R}E + \int \frac{E}{L}dt + \text{constant} &= 0 \\ \ddot{E} + \left(\frac{1}{CR}\right)\dot{E} + \left(\frac{1}{LC}\right)E &= 0 \\ \zeta &= \frac{\sqrt{L/C}}{2R} \end{aligned} \tag{9.10}$$

9.3 Fractional Analysis

In Chapter 4, we found that ω_n is the system time-scale parameter: if we normalize time by defining a dimensionless time-scale $t^* \triangleq \omega_n t$ as we did in Section 4.3, the ω_n disappears from Equation 9.3, leaving only

$$x'' + 2\zeta x' + x = 0 \tag{9.11}$$

As before, we expect to find the product $\omega_n t$ in our solutions. The parameter ζ is a dimensionless measure of the system damping. Since it is the only coefficient which cannot be eliminated by normalizing or rescaling the governing equation, it must be the system parameter which determines the mathematical nature of the solution. It is called a similitude parameter because different systems, large or small, translational or rotational, behave similarly if that parameter is identical.

9.4 Solution

Equation 9.3 is an Ordinary Differential Equation, second order, linear with constant coefficients, and homogeneous. Therefore we will try a solution of the form given in Equation 4.12

$$\begin{aligned} x &= Ae^{st} \\ \dot{x} &= sAe^{st} \\ \ddot{x} &= s^2Ae^{st} \end{aligned} \tag{9.12}$$

where A and s are unknown constants. Substituting this trial solution in Equation 9.3, we obtain the characteristic equation

$$A\left[s^2 + (2\zeta\omega_n)\,s + \left(\omega_n^2\right)\right]e^{st} = 0 \tag{9.13}$$

where the expression in the brackets must equal zero. There are two possible non-trivial solutions for the value of s

$$s_{1,2} = \frac{-2\zeta\omega_n \pm \sqrt{\left(2\zeta\omega_n\right)^2 - 4\omega_n^2}}{2} = \left(-\zeta \pm \sqrt{\zeta^2 - 1}\right)\omega_n \tag{9.14}$$

Therefore s is determined by the system parameters, and our solution from Equation 9.12 becomes

$$x = A_1 e^{s_1 t} + A_2 e^{s_2 t} \tag{9.15}$$

where A_1 and A_2 will be determined from the Initial Conditions. The nature of this solution depends on the value of ζ; there are five possibilities:

CASE I. If $\zeta > 1$, then $\sqrt{\zeta^2 - 1}$ is a real number; also, its absolute magnitude is smaller than ζ so that both s_1 and s_2 are negative real numbers. Therefore the solution for x is the sum of two decaying exponential functions

$$x = A_1 e^{\left(-\zeta + \sqrt{\zeta^2 - 1}\right)\omega_n t} + A_2 e^{\left(-\zeta - \sqrt{\zeta^2 - 1}\right)\omega_n t} \tag{9.16}$$

This solution cannot cross the x-axis more than once: no vibration takes place.

Exercise 9.1 *What are the values of A_1 and A_2 in terms of the Initial Conditions $x_{(0)}$ and $\dot{x}_{(0)}$? When is $x = 0$, and what is the maximum value x can reach?*

CASE II. If $\zeta = 1$, then $s_1 = s_2 = -\zeta$ so that the two solution terms are identical, and we cannot fit them to both of the Initial Conditions that a second-order differential equation requires. This is a repeated-roots situation and we resort to a t-multiplied second term

$$x = A_1 e^{-\omega_n t} + A_2 \omega_n t e^{-\omega_n t} \tag{9.17}$$

This solution, too, can cross the axis at most once (at $\omega_n t = -A_1/A_2$ if A_1 and A_2 have opposite signs). Since $\zeta = 1$ is a borderline case, it is often called "critical damping."

Exercise 9.2 *Show that this solution satisfies the governing equation when $\zeta = 1$, by differentiating the solution twice and plugging back into the equation.*

Exercise 9.3 *What are the values of A_1 and A_2 in terms of the Initial Conditions $x_{(0)}$ and $\dot{x}_{(0)}$? What is the maximum value x can reach?*

CASE III. If $0 < \zeta < 1$, then $\sqrt{\zeta^2 - 1}$ is the imaginary number $i\sqrt{1-\zeta^2}$, and s_1 and s_2 are a conjugate pair of complex numbers

$$\begin{aligned} x &= A_1 e^{\left(-\zeta + i\sqrt{1-\zeta^2}\right)\omega_n t} + A_2 e^{\left(-\zeta - i\sqrt{1-\zeta^2}\right)\omega_n t} \\ &= A_1 e^{-\zeta\omega_n t} e^{i\sqrt{1-\zeta^2}\omega_n t} + A_2 e^{-\zeta\omega_n t} e^{-i\sqrt{1-\zeta^2}\omega_n t} \\ &= e^{-\zeta\omega_n t}\left(A_1 e^{i\sqrt{1-\zeta^2}\omega_n t} + A_2 e^{-i\sqrt{1-\zeta^2}\omega_n t}\right) \end{aligned}$$

Remembering the lessons of Chapter 4, we replace any exponential functions having imaginary exponents by trigonometric functions

$$\boxed{\begin{aligned} x &= e^{-\zeta\omega_n t}\left(A \sin\left(\sqrt{1-\zeta^2}\omega_n t\right) + B\cos\left(\sqrt{1-\zeta^2}\omega_n t\right)\right) \\ &= Ce^{-\zeta\omega_n t}\sin\left(\sqrt{1-\zeta^2}\omega_n t + \phi\right) \end{aligned}} \tag{9.18}$$

This, at last, is a vibrational solution!

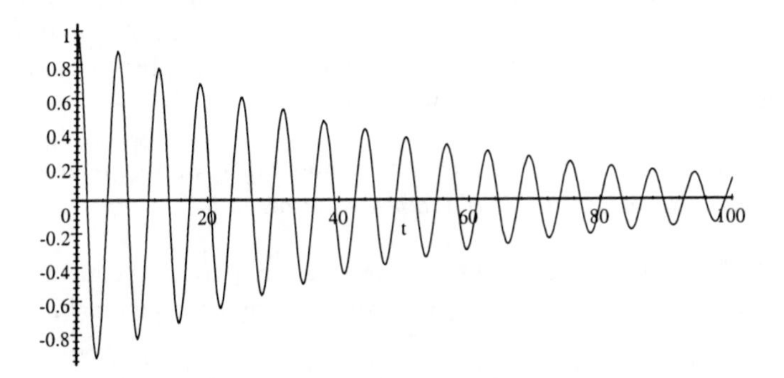

It is the product of a decaying exponential function with an oscillating function: the solution looks like a vibration within an envelope which shrinks with time. The oscillation proceeds at an angular frequency which is *not* at $\omega = \omega_n$, but at the slower damped-response frequency

$$\omega_d = \sqrt{1-\zeta^2}\omega_n$$

In heavily damped systems, this can make a big difference; for example when $\zeta = 0.70 = 70\%$, ω_d is 29% smaller than ω_n. However, in the more common moderately-damped systems, where $\zeta < 0.10 = 10\%$, the value of ω_d is within half a percent of ω_n.

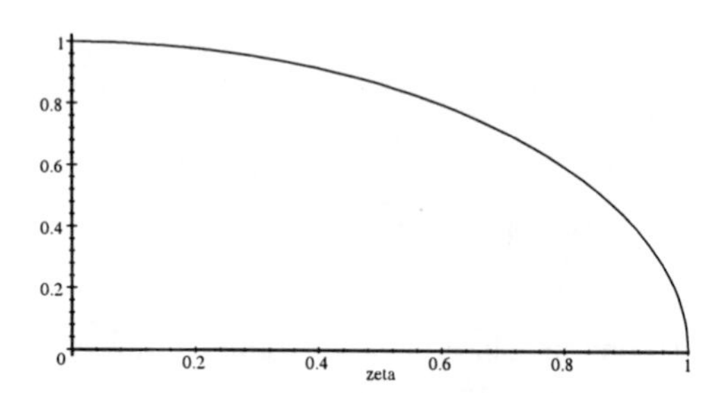

The envelope of the amplitudes is $e^{-\zeta\omega_n t}$, so the oscillation will disappear in time; the larger $\zeta\omega_n$ is, the faster the decay.

Exercise 9.4 *Find C and ϕ in terms of A and B in Equation 9.18.*

CASE IV. If $\zeta = 0$, then $s_{1,2} = \pm i\omega_n$, the same as in Equation 4.14, and we arrive at the solution of Equations 4.19 and 4.22. This solution does not decay with time; we now recognize that an eternally persisting solution is an unrealistic consequence of neglecting damping in our analysis. On the other hand, our Chapter 4 analysis was not wasted: the solution of Equation 9.18 is indistinguishable from the undamped solution if the damping is moderate and the length of time $\omega_n t$ in which we are interested is limited.

Exercise 9.5 *What are the limits on the values of ζ and $\omega_n t$, if we want to limit the error due to using the undamped solution to 5%?*

CASE V. If $\zeta < 0$, then s will have a positive-real part. We conclude that the solution is divergent: systems with negative damping are dynamically unstable.

Exercise 9.6 *Describe the solutions when $-1 < \zeta < 0$; when $\zeta = -1$; and when $\zeta < -1$.*

Exercise 9.7 *Describe the solutions when $\left(\omega_n^2\right)$ is negative, for various positive and negative values of ζ.*

9.5 Initial Conditions

Taking Equation 9.18 and its derivative

$$
\begin{aligned}
x &= e^{-\zeta\omega_n t}\left(A\sin\left(\omega_d t\right) + B\cos\left(\omega_d t\right)\right) \\
\dot{x} &= -\zeta\omega_n e^{-\zeta\omega_n t}\left(A\sin\left(\omega_d t\right) + B\cos\left(\omega_d t\right)\right) \\
&\quad + e^{-\zeta\omega_n t}\left(A\omega_d\cos\left(\omega_d t\right) - B\omega_d\sin\left(\omega_d t\right)\right)
\end{aligned}
$$

and evaluating both of these at $t = 0$, where $\sin(0) = 0$, $\cos(0) = 1$, and $e^{(0)} = 1$

$$\begin{aligned} x_{(0)} &= B \\ \dot{x}_{(0)} &= -\zeta\omega_n (B) + (A\omega_d) = -\zeta\omega_n (B) + \left(A\sqrt{1-\zeta^2}\omega_n\right) \end{aligned}$$

Solving for the coefficients A and B in terms of the Initial Conditions

$$\begin{aligned} B &= x_{(0)} \\ A &= \frac{\dot{x}_{(0)}}{\sqrt{1-\zeta^2}\omega_n} + \left(\frac{\zeta}{\sqrt{1-\zeta^2}}\right) x_{(0)} \end{aligned} \tag{9.19}$$

Exercise 9.8 *Find C and ϕ in Equation 9.18 in terms of $x_{(0)}$ and $\dot{x}_{(0)}$.*

9.6 Logarithmic Decrement

When observing the response of a damped system to initial conditions, it is convenient to observe the reduction in amplitude per cycle, by defining the logarithmic decrement

$$\delta \triangleq \ln\left(\frac{x_n}{x_{n+1}}\right)$$

where x_n and x_{n+1} are two successive maximum positive amplitudes. For linear problems, we can show that δ is a constant which depends only on the system parameter ζ, by looking at the decaying exponential solution for two values of time that are exactly one cycle apart. This occurs when $\omega_d t_2 = \omega_d t_1 + 2\pi$, or $t_2 = t_1 + 2\pi/\omega_d$

$$\delta = \ln \frac{Ce^{-\zeta\omega_n t_1} \sin(\omega_d t_1 + \phi)}{Ce^{-\zeta\omega_n t_1 - 2\pi\zeta\omega_n/\omega_d} \sin(\omega_d t_1 + 2\pi + \phi)}$$

which simplifies to

$$\delta = \frac{2\pi\zeta\omega_n}{\omega_d} = \frac{2\pi\zeta}{\sqrt{1-\zeta^2}} \approx 2\pi\zeta \tag{9.20}$$

δ is a function of ζ; the parameter ω_n cancels out because it governs both the time scale of the decay and the time scale of the frequency. Therefore, if we know ζ we can predict δ; conversely, if we measure δ we can obtain ζ

$$\zeta = \frac{(\delta/2\pi)}{\sqrt{1+(\delta/2\pi)^2}} \approx \frac{\delta}{2\pi} \tag{9.21}$$

If the damping is very small, the difference between two successive amplitudes is tiny and difficult to measure; it is easier to compare amplitudes that are ten or twenty cycles apart. Since the ratios of successive amplitudes stay constant in linear systems, we can write for m cycles

$$\begin{aligned}
\delta &= \frac{1}{m}\left(\ln\left(\frac{x_n}{x_{n+1}}\right) + \ln\left(\frac{x_{n+1}}{x_{n+2}}\right) + \ldots + \ln\left(\frac{x_{n+m-1}}{x_{n+m}}\right)\right) \\
&= \frac{1}{m}\ln\left(\left(\frac{x_n}{x_{n+1}}\right) \times \left(\frac{x_{n+1}}{x_{n+2}}\right) \times \ldots \times \left(\frac{x_{n+m-1}}{x_{n+m}}\right)\right) \\
&= \frac{1}{m}\ln\left(\frac{x_n}{x_{n+m}}\right)
\end{aligned} \tag{9.22}$$

If the damping is very large, the second positive amplitude is tiny and difficult to measure; it is easier to measure the "overshoot," which is defined as the ratio of the maximum negative amplitude to the initial maximum positive amplitude

$$\beta \triangleq \left|\frac{x_{n+\frac{1}{2}}}{x_n}\right| \tag{9.23}$$

which is related to the logarithmic decrement

$$\delta \triangleq \ln\left(\frac{x_n}{x_{n+1}}\right) = \ln\left(\left|\frac{x_n}{x_{n+\frac{1}{2}}}\right| \times \left|\frac{x_{n+\frac{1}{2}}}{x_{n+1}}\right|\right) = \ln\left(\frac{1}{\beta} \times \frac{1}{\beta}\right) = 2\ln\frac{1}{\beta} \tag{9.24}$$

For example, an overshoot of $\beta = 0.05 = 5\%$ indicates a logarithmic decrement of $\delta = 6.0$ and a dimensionless damping ratio $\zeta = 0.69 = 69\%$.

Exercise 9.9 *In gauges and electrical instruments, it is desirable to introduce the right amount of damping to have the indicator settle as quickly as possible on the new value after a step change in the measured variable. Too little damping will let the indicator oscillate a long time before settling on the final value. Too much damping will make the indicator sluggish and slow to approach the final value. What is the ideal amount of damping to make the indicator overcome the first* 95% *of a step change as quickly as possible?*

9.7 Summary

We have learned the response of a damped system to Initial Conditions, and found three ways of describing damping:

1. in engineering design and specification, by the dimensional damping coefficient c;

2. for the purpose of mathematical analysis, by the dimensionless damping ratio ζ; and

3. in relation to experimental observation, by the logarithmic decrement δ.

Through the Equations given in this chapter, you should be able to convert between the coefficient c, the ratio ζ, and the logarithmic decrement δ.

Problem 9.10 *A simple system composed of a mass of* 0.50 *kg, a spring of* 0.10 *N/m, and an unknown damper is tested in free vibration. It is observed to oscillate with a period of about* 14 *seconds, and the amplitude decreases by one-half every ten cycles. What is the damping coefficient in N·s/m? Caution: For this small amount of damping,* ω_n *and* ω_d *are almost the same, because* $\sqrt{1-\zeta^2}$ *is almost exactly unity. Therefore, trying to deduce damping from the measured frequency is hopelessly inaccurate. Approximate solution:*

$$\begin{aligned} \delta &\approx 0.07 \\ \zeta &\approx 0.01 \\ c &\approx 0.005\ N \cdot s/m \end{aligned}$$

Problem 9.11 *A welded structure is claimed to have a damping ratio of only* $\frac{1}{2}\%$*. To what extent will vibrations die out in* 20 *cycles?*

Problem 9.12 *Compute the logarithmic decrement* δ *and the overshoot* β *for a dimensionless damping ratio* $\delta = 1/\sqrt{2}$.

Problem 9.13 *Specify the coefficient c for a damper which will limit to* 10% *the overshoot of a* 660*-pound mass on a* 66*-pound-per-inch spring.*

Problem 9.14 *Specify the damping coefficient c required to limit to* 15% *the overshoot of system consisting of a* 300*-kg mass on a* 12*-kN/m spring.*

Problem 9.15 *Devise a test procedure for determining the low-load drag of a bicycle's front roller bearing by adding a modest weight to the rim, letting it oscillate around the axle as a pendulum, and observing the decay of the amplitude of oscillation. Estimate the errors. Discuss what parameters could be varied for systematic experimentation.*

Chapter 10

FORMULATION OF DAMPING TERMS and hereditary damping

In Chapters 2 and 3, the formulation of undamped problems was demonstrated. We now need to add the damping term. Chapter 9 analyzed the effect of linear damping to show us that we can do this by observation of the decay of free vibration. For design problems, we also need to be able to formulate damped problems from the description of the system. If system layout were always as simple as Figure 2.1, we could insert the damping coefficient c as we did in Chapter 9. In more complex one-degree-of-freedom systems, with several coupled coordinates, we need to express the damping in terms of the chosen main coordinate.

10.1 Effective Damping

If we have a system of levers and gears with many masses, moments of inertia, and springs, but only one damper (translational or rotational), an effective strategy is to choose the coordinate x or θ associated with the damper as the reference coordinate, and obtain equivalent mass M_{equiv} and equivalent spring constant K_{equiv} in terms of that coordinate, as outlined in Section 5.3. The damping term can then be simply written in.

However, if there are several translational dampers on a lever, it is necessary to transfer the effect of at least some of the dampers to another location on the lever (or to the angular rotation of the lever). Contrary to untrained intuition, it is the *square* of the lever-ratio (or the *square* of the distance to the pivot) which must be applied—long levers not only increase the velocity and, therefore, the force at the damper, they also increase the effect of that force. In the following section we will look at a procedure for breaking complex problems down into simple steps.

10.2 Power Loss

The power dissipated in each linear damper, in terms of its coordinate x (or θ), is

$$-\frac{dE}{dt} = -F_{\text{damp}} \cdot \frac{dx}{dt} = \left(c\,\dot{x}\right) \cdot \dot{x} = c\,\dot{x}^2 \quad \text{or } c_T\,\dot{\theta}^2 \tag{10.1}$$

If we have many dampers, we can write the total power dissipation as the sum of all of them, each in its own coordinate system

$$-\frac{dE}{dt} = c_1\,\dot{x}_1^2 + c_2\,\dot{x}_2^2 + \ldots \tag{10.2}$$

The next step is to use the kinematic relationships between the dampers' respective coordinate, to transform this sum to an expression in terms of one reference coordinate, just as we did in Section 5.2. In the resulting quadratic form, we call whatever combined constant shows up: C_{equiv} relative to that reference coordinate

$$-\frac{dE}{dt} = [C_{\text{equiv}}]\,\dot{x}_{ref}^2 \tag{10.3}$$

noting that there is no additional factor of 1/2 as there was in Equations 5.3 and 5.4 for equivalent mass and spring constant.

We can incorporate the equivalent damping into the Conservation of Energy Equation 5.2, recalling that we have defined dE/dt as a *loss* of energy from the system

$$\frac{d}{dt}(T+U) = \frac{dE}{dt} \tag{10.4}$$

which in the case of linear element laws (i.e., quadratic energy expressions) leads to

$$\begin{aligned}
\frac{d}{dt}\left(\frac{1}{2}\left[\frac{M_{\text{equiv}}}{g_c}\right]\dot{x}^2 + \frac{1}{2}[K_{\text{equiv}}]\,x^2\right) &= -[C_{\text{equiv}}]\,\dot{x}^2 \\
\left[\frac{M_{\text{equiv}}}{g_c}\right]\dot{x}\ddot{x} + [K_{\text{equiv}}]\,x\,\dot{x} &= -[C_{\text{equiv}}]\,\dot{x}^2
\end{aligned}$$

By dividing all terms by $\dot{x}$, we obtain the desired equation

$$\left[\frac{M_{\text{equiv}}}{g_c}\right]\ddot{x} + [C_{\text{equiv}}]\,\dot{x} + [K_{\text{equiv}}]\,x = 0 \tag{10.5}$$

The actual value of C_{equiv} depends on the reference coordinate, so it is important to use the same reference coordinate for M_{equiv}, K_{equiv}, and C_{equiv}.

Example: Let us consider small motions of an unsymmetrical "seesaw" lever, with a lumped m_1, k_1, and c_1 attached ℓ_1 from the center, and m_2, k_2, and c_2 attached ℓ_2 from the center. For small angles, the energy expressions are in terms of upward coordinates x_1 and x_2

$$\begin{aligned} T &= \frac{1}{2}\frac{m_1}{g_c}\dot{x}_1^2 + \frac{1}{2}\frac{m_2}{g_c}\dot{x}_2^2 \\ U &= \frac{1}{2}k_1x_1^2 + \frac{1}{2}k_1x_1^2 + \frac{m_1 g}{g_c}x_1 + \frac{m_2 g}{g_c}x_2 \\ -\dot{E} &= c_1\dot{x}_1^2 + c_2\dot{x}_2^2 \end{aligned}$$

and the kinematic relationship is

$$\frac{x_1}{\ell_1} = -\frac{x_2}{\ell_2} \approx \theta$$

which we can use to simplify the energy expressions

$$\begin{aligned} T &= \frac{1}{2}\left[\frac{m_1}{g_c} + \frac{\ell_2^2}{\ell_1^2}\frac{m_2}{g_c}\right]\dot{x}_1^2 = \frac{1}{2}\left[\frac{\ell_1^2}{\ell_2^2}\frac{m_1}{g_c} + \frac{m_2}{g_c}\right]\dot{x}_2^2 \approx \frac{1}{2}\left[\frac{m_1\ell_1^2}{g_c} + \frac{m_2\ell_2^2}{g_c}\right]\dot{\theta}^2 \\ U &= \frac{1}{2}\left[k_1 + \frac{\ell_2^2}{\ell_1^2}k_2\right]x_1^2 + \left(\frac{m_1 g}{g_c} - \frac{\ell_2}{\ell_1}\frac{m_2 g}{g_c}\right)x_1 \\ &= \frac{1}{2}\left[\frac{\ell_1^2}{\ell_2^2}k_1 + k_2\right]x_2^2 + \left(-\frac{\ell_1}{\ell_2}\frac{m_1 g}{g_c} + \frac{m_2 g}{g_c}\right)x_2 \\ &\approx \frac{1}{2}\left[k_1\ell_1^2 + k_2\ell_2^2\right]\theta^2 + \left(\frac{m_1 g\ell_1}{g_c} - \frac{m_2 g\ell_2}{g_c}\right)\theta \\ -\dot{E} &= \left[c_1 + \frac{\ell_2^2}{\ell_1^2}c_2\right]\dot{x}_1^2 = \left[\frac{\ell_1^2}{\ell_2^2}c_1 + c_2\right]\dot{x}_2^2 \approx \left[c_1\ell_1^2 + c_2\ell_2^2\right]\dot{\theta}^2 \end{aligned}$$

The governing Equation 10.4 therefore is

$$\begin{aligned} \left[\frac{m_1}{g_c} + \frac{\ell_2^2}{\ell_1^2}\frac{m_2}{g_c}\right]\ddot{x}_1 + \left[c_1 + \frac{\ell_2^2}{\ell_1^2}c_2\right]\dot{x}_1 + \left[k_1 + \frac{\ell_2^2}{\ell_1^2}k_2\right]x_1 &= \frac{-m_1 g}{g_c} + \frac{\ell_2}{\ell_1}\frac{m_2 g}{g_c} \\ \left[\frac{\ell_1^2}{\ell_2^2}\frac{m_1}{g_c} + \frac{m_2}{g_c}\right]\ddot{x}_2 + \left[\frac{\ell_1^2}{\ell_2^2}c_1 + c_2\right]\dot{x}_2 + \left[\frac{\ell_1^2}{\ell_2^2}k_1 + k_2\right]x_2 &= \frac{\ell_1}{\ell_2}\frac{m_1 g}{g_c} - \frac{m_2 g}{g_c} \\ \left[\frac{m_1\ell_1^2}{g_c} + \frac{m_2\ell_2^2}{g_c}\right]\ddot{\theta} + \left[c_1\ell_1^2 + c_2\ell_2^2\right]\dot{\theta} + \left[k_1\ell_1^2 + k_2\ell_2^2\right]\theta &\approx \frac{-m_1 g\ell_1}{g_c} + \frac{m_2 g\ell_2}{g_c} \end{aligned}$$

where the quantities in the square brackets are the equivalent components with respect to the associated coordinate.

Exercise 10.1 *Show that the natural frequency ω_n and the damping ratio ζ are identical in these three equations, even though each is written in terms of a different coordinate.*

We can also incorporate the equivalent damping into Lagrange's equation 5.8

$$\frac{d}{dt}\left(\frac{\partial T}{\partial v}\right) - \frac{\partial T}{\partial x} + \frac{dU}{dx} = Q \tag{10.6}$$

where Q is the generalized force

$$Q = -\frac{dE}{dt} \cdot \frac{dt}{dx} = \frac{-1}{v}\frac{dE}{dt} \tag{10.7}$$

which in the case of linear damping elements leads to

$$Q = -\left[C_{\text{equiv}}\right] \dot{x} \tag{10.8}$$

10.3 Linearization

Although damping elements are usually non-linear, we would like to represent them with linear mathematical expressions. One way to do this is to observe the logarithmic decrement of decaying free vibrations, as defined in the preceding chapter; it will generally not be a constant, but a function of the amplitude x_{max}. Therefore the calculated damping constant will also be a function of the amplitude

$$\frac{m}{g_c} \ddot{x} + c_{(x_{\text{max}})} \dot{x} + kx = 0$$

This is a highly unsatisfactory representation not only because the damping coefficient is a function of something other than the contemporaneous values of x and its derivatives, but also because it is obtained by observing the decay of free vibrations, and therefore is applicable to only at the frequency of free vibrations.

A slightly better approach is to obtain the energy dissipated per cycle, from element laws

$$W_{\text{cycle}} = \oint f_{\text{damp}} \cdot dx$$

or from the response to periodic excitation (Chapter 11, Section 11.5). For a linear element, the energy dissipated in one cycle (assuming sinusoidal motion) would be

$$\begin{aligned} W_{\text{cycle}} &= \oint c\, \dot{x}\, dx = \oint c\, \dot{x}^2\, dt = \int_0^{2\pi} c\omega^2 x_{\text{max}}^2 \cos^2(\omega t)\, \frac{d(\omega t)}{\omega} \\ &= \pi c \omega x_{\text{max}}^2 \end{aligned}$$

so that the energy dissipated per cycle is proportional to c, ω, and the square of the amplitude x_{max}. Since the energy stored within the system is also proportional to the square of the maximum amplitude, we see that a linear damper removes the same percentage of the remaining energy during each cycle. If we know the energy *actually* dissipated, we can calculate back to an equivalent linear damping coefficient

$$c_{\text{equiv}} = \frac{W_{\text{cycle}}}{\pi \omega x_{\max}}$$

which will generally be a function of $x_{\max}$ and ω

$$\frac{m}{g_c} \ddot{x} + c_{(\omega, x_{\max})} \dot{x} + kx = 0$$

This is still an unsatisfactory representation, because frequency-domain quantities appear within the time-domain equation. Non-linear analysis (Part V) is a better approach for studying non-linear systems.

10.4 Hereditary Damping

Structural or materials damping is observed in the form of a hysteretic delay in the action of a spring: we don't get back all of the energy we try to store in it. The simplest conceptual model is description of the stiffness by a complex number $(1 + i\lambda)\, k$ where the real part is the traditional spring constant k and the imaginary part λk incorporates the "structural damping factor" λ. This representation is only valid if the motion is sinusoidal; which is approximately true if the damping is very small. In that instance, structural damping is equivalent to a viscous damping coefficient of $c_{\text{equiv}} = \lambda k / \omega$, which is another example of a representation which is only good for tracking the energy absorbed per cycle, and not for writing a valid differential equation.

10.5 Summary

We have learned three ways of obtaining damping terms:

1. in Chapter 2, from force-velocity relationship, e.g., $F_{\text{damper}} = -c\, \dot{x}$;
2. in Chapter 10, from energy dissipation $-dE/dt$; and
3. in Chapter 8, from experimental observation of the log decrement δ.

We note that damping is usually non-linear, and we have attempted linearization of the damping. The resulting equivalent damping is only good for estimating the energy absorbed per cycle, and does not lead to a valid differential equation for the damped system.

Problem 10.2 *Find a differential equation to describe the vertical bouncing of a car, if the suspension at each corner consists of a shock absorber inside a coil spring, connecting the frame of the car with the half-way point of the lower A-arm.*

Problem 10.3 *Find a differential equation to describe the vertical bouncing of a car, if the suspension at each corner consists of a coil-over-shock element angled at 45 degrees to connect the frame of the car with the tip of the lower A-arm.*

Problem 10.4 *Find a differential equation to describe the vertical bouncing of a car, if the suspension at each corner consists of a coil-over-shock strut angled at* 15 *degrees to connect the frame of the car with the hub carrier.*

Problem 10.5 *Find a differential equation to describe the lateral rocking of a car, if it is suspended by parallel leaf springs attached a distance* L_k *apart on beam axles, and additional damping is provided by shock absorbers angled* 30 *degrees from the vertical and attached to the solid axles at a larger separation* L_c.

Chapter 11

PERIODIC EXCITATION OF DAMPED SYSTEMS and forces at the base

We will now move on to systems that contain all of the types of terms that showed up in the development of the governing equations of vibration in Chapter 2: mass, damping, spring, and excitation terms.

11.1 Governing Equations

We would like to solve Equation 2.6

$$\frac{m}{g_c}\ddot{x} + c\dot{x} + kx = F(t) \tag{11.1}$$

We will let $F(t)$ be a sinusoidal function of arbitrary amplitude F_o and arbitrary angular frequency ω_{ex}

$$\frac{m}{g_c}\ddot{x} + c\dot{x} + kx = F_o \sin(\omega_{ex} t) \tag{11.2}$$

As usual, we normalize to make the first coefficient unity

$$\ddot{x} + \left(\frac{cg_c}{m}\right)\dot{x} + \left(\frac{kg_c}{m}\right)x = \left(\frac{F_o}{k}\right)\left(\frac{kg_c}{m}\right)\sin(\omega_{ex} t) \tag{11.3}$$

and defining ω_n and ζ as before, write this in Standard Form

$$\boxed{\ddot{x} + (2\zeta\omega_n)\dot{x} + \left(\omega_n^2\right)x = \left(\tfrac{F_o}{k}\right)\left(\omega_n^2\right)\sin(\omega_{ex} t)} \tag{11.4}$$

This is the form of the equation we will solve; later, when we have the solution, we can substitute the appropriate expressions for the Standard-Form parameters ω_n and ζ, and the pseudo-static-deflection (F_o/k).

11.2 Method of Solution

As we saw in Section 7.2, the solution $x(t)$ of Equation 11.4 is composed of the particular integral x_p and the complementary function x_h

$$x = x_p + x_h \tag{11.5}$$

where x_h is identical to the homogeneous solution of Equation 9.18

$$x = e^{-\zeta\omega_n t}\left(A\sin\left(\sqrt{1-\zeta^2}\omega_n t\right) + B\cos\left(\sqrt{1-\zeta^2}\omega_n t\right)\right) \tag{11.6}$$

but we cannot evaluate A and B from the Initial Conditions yet, because the particular-solution terms x_p may have some effect.

In order for the particular-solution terms x_p to balance off against the right-hand forcing terms, they must have a similar form to the forcing terms, containing the excitation parameter ω_{ex}; we will try

$$\begin{aligned} x_p &= C\sin(\omega_{ex}t) + D\cos(\omega_{ex}t) \qquad (11.7) \\ \dot{x}_p &= C\omega_{ex}\cos(\omega_{ex}t) - D\omega_{ex}\sin(\omega_{ex}t) \\ \ddot{x}_p &= -C\omega_{ex}^2\sin(\omega_{ex}t) - D\omega_{ex}^2\cos(\omega_{ex}t) \end{aligned}$$

Plugging these into Equation 11.4, we get

$$\begin{array}{l} -D\omega_{ex}^2\cos(\omega_{ex}t) + 2\zeta\omega_n C\omega_{ex}\cos(\omega_{ex}t) + \omega_n^2 D\cos(\omega_{ex}t) + \\ -C\omega_{ex}^2\sin(\omega_{ex}t) - 2\zeta\omega_n D\omega_{ex}\sin(\omega_{ex}t) + \omega_n^2 C\sin(\omega_{ex}t) = \frac{F_o}{k}\omega_n^2\sin(\omega_{ex}t) \end{array}$$

which we can sort into coefficients of cosines which must cancel out against each other, and coefficients of sines which must cancel out against the excitation

$$\begin{aligned} -D\omega_{ex}^2 + (2\zeta\omega_n)C\omega_{ex} + \left(\omega_n^2\right)D &= 0 \\ -C\omega_{ex}^2 - (2\zeta\omega_n)D\omega_{ex} + \left(\omega_n^2\right)C &= \left(\frac{F_o}{k}\right)\left(\omega_n^2\right) \end{aligned}$$

or

$$\begin{aligned} (2\zeta\omega_n\omega_{ex})C + \left(\omega_n^2 - \omega_{ex}^2\right)D &= 0 \\ \left(\omega_n^2 - \omega_{ex}^2\right)C - (2\zeta\omega_n\omega_{ex})D &= \left(\frac{F_o}{k}\right)\left(\omega_n^2\right) \end{aligned}$$

We can solve these two equations for the two unknowns C and D by Cramer's rule

$$C = \frac{-\left(\frac{F_o}{k}\right)\left(\omega_n^2\right)\left(\omega_n^2 - \omega_{ex}^2\right)}{-\left(\omega_n^2 - \omega_{ex}^2\right)^2 - (2\zeta\omega_n\omega_{ex})^2} = \frac{\left(\frac{F_o}{k}\right)\left(1 - \frac{\omega_{ex}^2}{\omega_n^2}\right)}{\left(1 - \frac{\omega_{ex}^2}{\omega_n^2}\right)^2 + \left(2\zeta\frac{\omega_{ex}}{\omega_n}\right)^2} \tag{11.8}$$

$$D = \frac{+\left(\frac{F_o}{k}\right)\left(\omega_n^2\right)(2\zeta\omega_n\omega_{ex})}{-\left(\omega_n^2 - \omega_{ex}^2\right)^2 - (2\zeta\omega_n\omega_{ex})^2} = \frac{-\left(\frac{F_o}{k}\right)\left(2\zeta\frac{\omega_{ex}}{\omega_n}\right)}{\left(1 - \frac{\omega_{ex}^2}{\omega_n^2}\right)^2 + \left(2\zeta\frac{\omega_{ex}}{\omega_n}\right)^2} \tag{11.9}$$

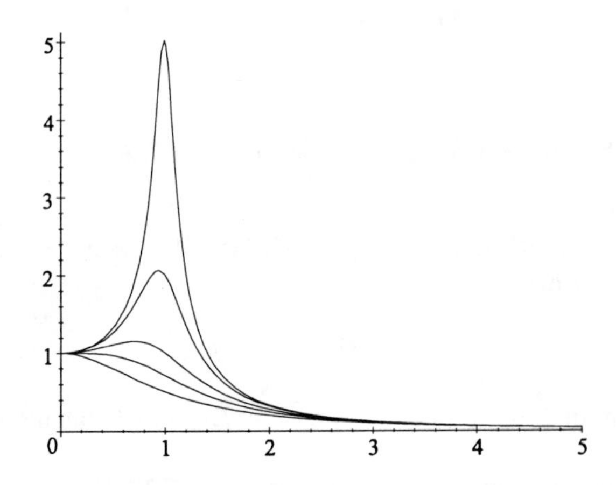

Figure 11.1: Amplitude Reseponse to Force Excitation

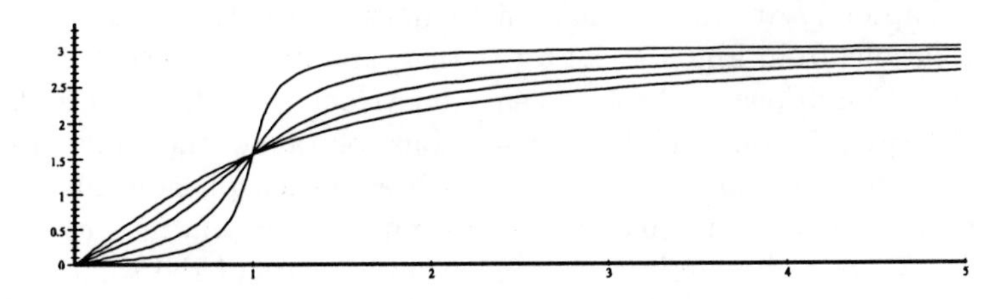

Figure 11.2: Phase Response to Force Excitation

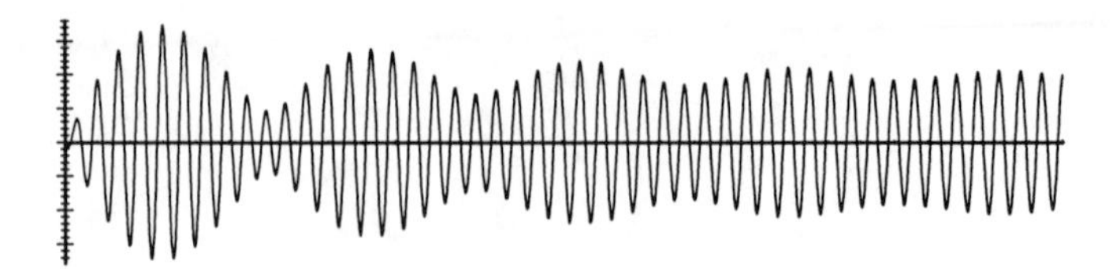

Figure 11.3: Complete Solution near Resonance

which we can insert into Equation 11.7. The result is easier to interpret if we convert it to the form

$$x_p = X\sin(\omega_{ex}t - \phi) \tag{11.10}$$

where the maximum amplitude is $X = \sqrt{C^2 + D^2}$, which calculates out to

$$\frac{X}{(F_o/k)} = \frac{\omega_n^2}{\sqrt{(\omega_n^2 - \omega_{ex}^2)^2 + (2\zeta\omega_n\omega_{ex})^2}} = \frac{1}{\sqrt{\left(1 - \frac{\omega_{ex}^2}{\omega_n^2}\right)^2 + \left(2\zeta\frac{\omega_{ex}}{\omega_n}\right)^2}} \tag{11.11}$$

as plotted in Figure 11.1, and the phase lag of the response is $\phi = \arctan(D/C)$, leading to

$$\tan\phi = \frac{2\zeta\omega_n\omega_{ex}}{\omega_n^2 - \omega_{ex}^2} = \frac{2\zeta\frac{\omega_{ex}}{\omega_n}}{1 - \frac{\omega_{ex}^2}{\omega_n^2}} \tag{11.12}$$

as plotted in Figure 11.2. For $\zeta = 0$, these reduce to the undamped relationships $Xk/F_o = 1/\left(1 - \omega_n^2/\omega_{ex}^2\right)$ and $\phi = 0$ or π, respectively.

As in Chapter 7, we can now add this particular solution and the complementary solutions from Equation 11.6 together, and fit the constants A and B to the Initial Conditions of displacement and velocity. Unlike the undamped system of Chapter 7, however, the homogeneous solution of the damped system decays. As in Section 7.5, it is possible to observe such phenomena as mixed frequencies and beats for a while (see, for example, a *zero-initial-state* solution in Figure 11.3), but ultimately the complementary part of the solution disappears. Hence the particular solution given by Equations 11.10 and 11.11 is also the **steady-state solution** of this system.

11.3 Interpretation of Response Curves

From these equations, evaluated at various values of ζ, we obtain a family of steady-state response curves with the following features:

- At left, for $\omega_{ex}/\omega_n \ll 1$, the response is dominated by the deflection of the **spring**, and X approaches F_o/k; phase angle ϕ is near zero.

Exercise 11.1 *Show that the response curves have horizontal tangents at the vertical coordinate.*

- To the right, for $\omega_{ex}/\omega_n \gg 1$, the response is dominated by acceleration of the **mass**; X approaches $F_o g_c/m\omega_{ex}^2$; phase lag ϕ approaches π radians or 180 degrees of angle.

Exercise 11.2 *Show that a log-log plot of the response is asymptotic to a line sloping at negative 2 as $\omega_{ex}/\omega_n \Rightarrow \infty$.*

- In the middle, for $\omega_{ex}/\omega_n \approx 1$, the response is dominated by the velocity at the **damper**; for $\omega_{ex}/\omega_n = 1$

$$\frac{X}{(F_o/k)} = \frac{1}{2\zeta} \tag{11.13}$$

or $X = F_o/c\omega_{ex}$, with a phase lag ϕ of $\pi/2$ radians or 90 degrees at that point.

Exercise 11.3 *Show that the actual maximum of X is located slightly to the left of resonance, at*

$$\omega_{ex}/\omega_n = \sqrt{1 - 2\zeta^2}$$

and is slightly higher than $1/(2\zeta)$, with a peak value equal to

$$\frac{X}{(F_o/k)} = \frac{1}{2\zeta\sqrt{1-\zeta^2}}$$

Hint: to find the location of the maximum of the curve for X, we should make the derivative of the whole expression equal to zero—but the derivative of the square of the inverse of X is a much simpler expression, and has its extrema at the same values of the abscissa.

- For heavy damping, $\zeta \geq 1/\sqrt{2} = 70.7\%$, the response $Xk/F_o \leq 1$ for all values of the frequency ratio ω_{ex}/ω_n.

11.4 Sharpness of Resonance

We have seen that the peak response is approximately $Xk/F_o = 1/(2\zeta)$. We can also see that, as we reduce damping, the width of the amplitude response curve becomes narrower *relative to its height.* In electrical engineering, this is quantified by identifying the frequencies of the "half-power" points, above and below resonance, where the amplitude X is $1/\sqrt{2}$ of the maximum resonant response. The difference $\Delta\omega$ between these two frequencies is used to define

$$Q \triangleq \frac{\omega}{\Delta\omega} \cong \frac{1}{2\zeta} \tag{11.14}$$

The relationship of Q to ζ is usually demonstrated by setting the response in Equation 11.11 equal to 50% of the power—70.7% of the amplitude—relative to resonance, i.e.,

$$X_{\text{halfpower}} k/F_o = \frac{1}{\sqrt{2}} \cdot \frac{1}{2\zeta}$$

and solving for the half-power frequencies

$$\frac{\omega_{1,2}}{\omega_n} = \sqrt{1 - 2\zeta^2 \mp 2\zeta\sqrt{1 - 2\zeta^2}}$$

Since Q is normally only employed when damping is low, we can use the binomial expansion and leave off higher-order terms

$$\frac{\omega_{1,2}}{\omega_n} \cong \sqrt{1 \mp 2\zeta} \approx 1 \mp \zeta$$

leading to $\Delta\omega/\omega_n \cong 2\zeta$.

Exercise 11.4 *Solve for the location of the "half-power" points and the value of Q relative to the actual peak response, by setting*

$$\frac{X_{halfpower}}{(F_o/k)} = \frac{1}{\sqrt{2}} \cdot \frac{1}{2\zeta\sqrt{1-\zeta^2}} = \frac{1}{\sqrt{\left(1 - \frac{\omega_{1,2}^2}{\omega_n^2}\right)^2 + \left(2\zeta\frac{\omega_{1,2}}{\omega_n}\right)^2}}$$

and show that

$$\begin{aligned}\frac{\omega_{1,2}}{\omega_n} &= \sqrt{1 - 2\zeta^2 \mp 2\zeta\sqrt{1 - \zeta^2}} \\ \frac{\Delta\omega}{\omega_n} &= (2\zeta + \zeta^3 - \ldots)\sqrt{1 - \zeta^2}\end{aligned}$$

11.5 Power

At steady-state, the rate at which energy is absorbed in the damper when the motion is $x_{ss} = X\sin(\omega_{ex}t - \phi)$, can be obtained from the force×distance integral

$$\begin{aligned} W_{\text{cycle}} &= \oint F_{\text{damp}} \cdot dx = \oint c\, \dot{x}_{ss}\, dx = \oint c\, \dot{x}_{ss}^{2}\, dt \\ &= \int_{-\phi}^{2\pi-\phi} c\omega_{ex}^2 X^2 \cos^2(\omega_{ex} t - \phi) \frac{d(\omega_{ex} t)}{\omega_{ex}} \qquad (11.15) \\ &= \pi c \omega_{ex} X^2 \\ \left(\frac{dW}{dt}\right)_{\text{avg}} &= W_{\text{cycle}} \cdot \frac{\omega_{ex}}{2\pi} = \frac{c\omega_{ex}^2 X^2}{2} = \frac{\zeta \omega_{ex}^2 X^2 k}{\omega_n} \end{aligned}$$

where X is given by Equation 11.11.

Exercise 11.5 *We could instead have determined the power provided by the excitation force* $F(t) = F_o \sin(\omega_{ex} t)$

$$\begin{aligned} W_{cycle} &= \oint F_{(t)} \cdot dx = \oint F_{(t)} \cdot \frac{dx_{ss}}{dt} dt \\ &= \oint F_o \omega_{ex} \cos(\omega_{ex} t) X \omega_{ex} \cos(\omega_{ex} t - \phi) \end{aligned}$$

inserting ϕ *from Equation 11.12. Show that this leads to the same result.*

Exercise 11.6 *Compare the power with the average energy stored in the system,* $T + U$.

11.6 Force Transmissibility

If we are interested in the amount of force that is transmitted to the foundation, we need to add up the force transmitted by the spring, which is in phase with the motion x

$$F_{\text{spring}} = kx = kX \sin(\omega_{ex} t - \phi) = \frac{F_o \sin(\omega_{ex} t - \phi)}{\sqrt{\left(1 - \frac{\omega_{ex}^2}{\omega_n^2}\right)^2 + \left(2\zeta \frac{\omega_{ex}}{\omega_n}\right)^2}}$$

and the force transmitted by the damper, which is ninety degrees out-of-phase

$$F_{\text{damp}} = c\, \dot{x} = cX\omega_{ex} \cos(\omega_{ex} t - \phi) = \frac{F_o \cdot 2\zeta \frac{\omega_{ex}}{\omega_n} \cos(\omega_{ex} t - \phi)}{\sqrt{\left(1 - \frac{\omega_{ex}^2}{\omega_n^2}\right)^2 + \left(2\zeta \frac{\omega_{ex}}{\omega_n}\right)^2}}$$

The sum of these forces can be written as

$$F_{base} = F_B \sin(\omega_{ex} t - \psi) \qquad (11.16)$$

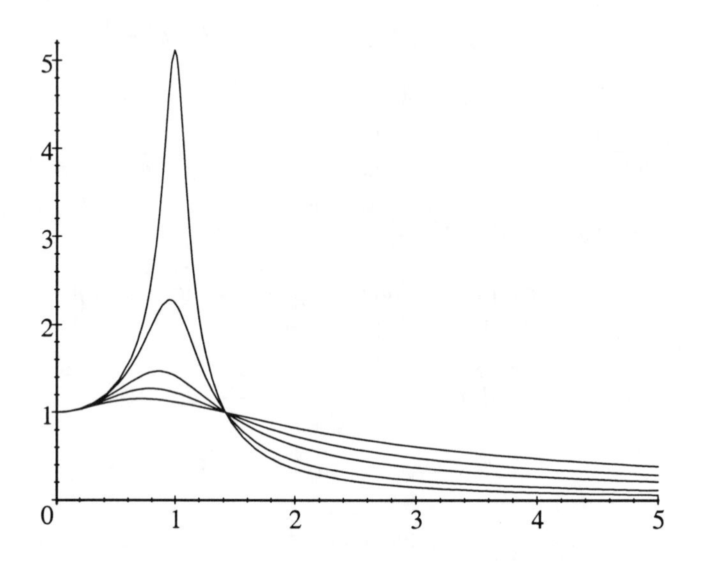

Figure 11.4: Transmissibility of Excitation Force to Base

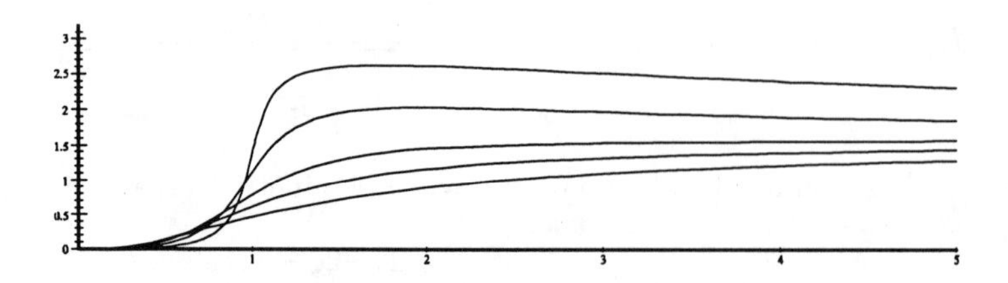

Figure 11.5: Phase of Force Transmitted to Base

where

$$\frac{F_B}{F_o} = \frac{\sqrt{1 + \left(2\zeta\frac{\omega_{ex}}{\omega_n}\right)^2}}{\sqrt{\left(1 - \frac{\omega_{ex}^2}{\omega_n^2}\right)^2 + \left(2\zeta\frac{\omega_{ex}}{\omega_n}\right)^2}} \tag{11.17}$$

as plotted in Figure 11.4, and

$$\psi = \phi + \arctan\left(2\zeta\frac{\omega_{ex}}{\omega_n}\right) = \arctan\left(\frac{2\zeta\left(\frac{\omega_{ex}}{\omega_n}\right)^3}{1 - \left(\frac{\omega_{ex}}{\omega_n}\right)^2 + \left(2\zeta\frac{\omega_{ex}}{\omega_n}\right)^2}\right) \tag{11.18}$$

as plotted in Figure 11.5. In the absence of damping, these expressions reduce to the relationships $F_B/F_o = \left|1/\left(1 - \omega_n^2/\omega_{ex}^2\right)\right|$ and $\phi = 0$ or π.

Exercise 11.7 *Derive these relationships. Hint: it is easier to do this in the* $C\sin\omega_{ex}t + D\cos\omega_{ex}t$ *formulation, converting to* $F_B\sin(\omega_{ex}t - \psi)$ *at the end.*

Exercise 11.8 *Show that the maximum transmissibility occurs at*

$$\frac{\omega_{ex}}{\omega_n} = \frac{\sqrt{\sqrt{1+8\zeta^2} - 1}}{2\zeta} \approx 1$$

and determine its magnitude.

Exercise 11.9 *Show that at high excitation frequencies the transmissibility is asymptotic to* $2\zeta\omega_n/\omega_{ex}$*. How does this appear on a log-log plot?*

11.7 Isolation of Force from the Base

This family of curves differs from the previous one in that all the curves cross at $\omega_{ex}/\omega_n = \sqrt{2}$; that means that, if we are to reduce the transmission of the exciting force to the base, we must make $\omega_{ex}/\omega_n > \sqrt{2}$ and damping as small as possible. For $\zeta \ll 1$, transmissibility in the region above resonance is often written as

$$\frac{F_B}{F_o} \approx \frac{1}{\left|\frac{\omega_{ex}^2}{\omega_n^2} - 1\right|}$$

However, damping is never negligible at high excitation frequencies, where

$$\frac{F_B}{F_o} \Rightarrow \frac{2\zeta}{\omega_{ex}/\omega_n}$$

so it is safer to use the entire transmissibility Equation 11.17.

Because of reciprocity, the transmissibility of force to the base is identical to the transmissibility of base motion $y(t) = Y_o\sin\omega_{ex}t$ to mass motion x. This is explored in Chapter 12.

Problem 11.10 *A harmonic force with an amplitude of* 0.3 *lbf and a frequency of* 10 *Hz acts on a* 2.0 *lbm mass attached to a spring of stiffness* 15 *lbf/inch. What is the amplitude of the steady-state response, if the damping is* 0.67 *lbf-sec/ft? What is the amplitude of the force transmitted to the base? Would you increase or decrease the damping to reduce the amplitude?*

Problem 11.11 *A harmonic force with an amplitude of* 0.3 *lbf and a frequency of* 10 *Hz acts on a* 1.0 *lbm mass attached to a spring of stiffness* 15 *lbf/inch. What is the amplitude of the steady-state response, if the damping is* 0.67 *lbf-sec/ft? What is the amplitude of the force transmitted to the base? Would you increase or decrease the damping to reduce the force transmitted to the base?*

Problem 11.12 *Use Lagrange's equation to obtain the governing equation of a pendulum with a laterally (horizontally) oscillating pivot* $y(t) = Y_o \sin(\omega_{ex} t)$. *Linearize for small oscillations. Solution:*

$$\begin{aligned} U &= \frac{mgL}{g_c}(1 - \cos\theta) \\ T &= \frac{1}{2} \cdot \frac{m}{g_c}\left\{\left[Y_o\omega_{ex}\cos(\omega_{ex}t) + L\,\dot{\theta}\cos\theta\right]^2 + \left[L\,\dot{\theta}\sin\theta\right]^2\right\} \\ \ddot{\theta} + \frac{g}{L}\sin\theta &= \frac{Y_o\left(\omega_{ex}^2\sin(\omega_{ex}t)\cos\theta\right)}{L} \end{aligned}$$

Problem 11.13 *How would a vertically oscillating pivot* $z(t) = Z_o \sin(\omega_{ex} t)$ *be different? Solution:*

$$\begin{aligned} U &= \frac{mg}{g_c}\left[L(1 - \cos\theta) + Z_o\sin(\omega_{ex}t)\right] \\ T &= \frac{1}{2} \cdot \frac{m}{g_c}\left\{\left[Z_o\omega_{ex}\cos(\omega_{ex}t) + L\,\dot{\theta}\sin\theta\right]^2 + \left[L\,\dot{\theta}\cos\theta\right]^2\right\} \\ 0 &= \ddot{\theta} + \frac{g - Z_o\omega_{ex}^2\sin(\omega_{ex}t)}{L}\sin\theta \end{aligned}$$

Problem 11.14 *Using a mathematical program like* Maple, *plot the zero-initial-state solution (like Figure 11.3) for the frequency ratio* $r \triangleq f_{ex}/f_n = \omega_{ex}/\omega_n$ *listed with the first letter of your last name in the table below:*

ini.	r	*ini.*	r	*ini.*	r	*ini.*	r	*ini.*	r
A	0.1	*F*	0.6	*L*	0.99	*Q*	1.2	*V*	2.5
B	0.2	*G*	0.7	*M*	1.00	*R*	1.3	*W*	3.0
C	0.3	*H*	0.8	*N*	1.01	*S*	1.5	*X*	4.0
D	0.4	*I,J*	0.9	*O*	1.05	*T*	1.8	*Y*	5.0
E	0.5	*K*	0.95	*P*	1.10	*U*	2.0	*Z*	10.0

Problem 11.15 *A mass-spring-damper system is excited by a calibrated force generator with adjustable* F_o *and frequency* f_{ex}*, and the steady-state response amplitude* X *measured. Design an experiment to determine the resonant frequency, the damping ratio, the spring constant, the mass, and the damping coefficient.*

Chapter 12

BASE EXCITATION and dynamic instrumentation

We closed Chapter 11 by looking at the transmissibility of excitation forces to the base. We will now look at base excitation, and the transmissibility of base motion to a spring-isolated mass.

12.1 Governing Equations

We would like to solve the equation

$$\frac{m}{g_c}\ddot{x} + c\left(\dot{x} - \dot{y}\right) + k\left(x - y\right) = 0 \tag{12.1}$$

Where $y(t)$ is a periodic function such as $y = Y_o \sin\left(\omega_{ex} t\right)$, so that

$$\frac{m}{g_c}\ddot{x} + c\dot{x} + kx = kY_o \sin(\omega_{ex} t) + cY_o\omega_{ex}\cos(\omega_{ex} t) \tag{12.2}$$

As usual, we normalize to make the first coefficient unity

$$\ddot{x} + \left(\frac{cg_c}{m}\right)\dot{x} + \left(\frac{kg_c}{m}\right)x = \left(\frac{kg_c}{m}\right)Y_o \sin(\omega_{ex} t) + \left(\frac{cg_c}{m}\right)Y_o\omega_{ex}\cos(\omega_{ex} t) \tag{12.3}$$

and defining ω_n as before, write this in Standard Form

$$\boxed{\ddot{x} + \left(2\zeta\omega_n\right)\dot{x} + \left(\omega_n^2\right)x = \left(\omega_n^2\right)Y_o \sin(\omega_{ex} t) + \left(2\zeta\omega_n\right)Y_o\omega_{ex}\cos(\omega_{ex} t)} \tag{12.4}$$

This is the form of the equation we will solve; later, when we have the solution, we can substitute the appropriate expressions for the Standard-Form parameters ω_n, ζ, and Y_o.

12.2 Solution

The transmissibility of base motion $y(t) = Y_o \sin \omega_{ex} t$ to mass motion x follows the same laws as the transmissibility of force to the base (Equation 11.17). By reciprocity, the expression for force ratio F_b/F_o also holds for X/Y_o, and

$$x = X \sin(\omega_{ex} t - \psi) \tag{12.5}$$

where

$$\frac{X}{Y_o} = \frac{\sqrt{1 + \left(2\zeta \frac{\omega_{ex}}{\omega_n}\right)^2}}{\sqrt{\left(1 - \frac{\omega_{ex}^2}{\omega_n^2}\right)^2 + \left(2\zeta \frac{\omega_{ex}}{\omega_n}\right)^2}} \tag{12.6}$$

as plotted in Figure 11.4; and the same expression for ψ also applies

$$\psi = \arctan\left(\frac{2\zeta\left(\frac{\omega_{ex}}{\omega_n}\right)^3}{1 - \left(\frac{\omega_{ex}}{\omega_n}\right)^2 + \left(2\zeta \frac{\omega_{ex}}{\omega_n}\right)^2}\right) \tag{12.7}$$

as plotted in Figure 11.5.

12.3 Vibration Isolation

Whether we are interested in reducing the transmission of machinery forces to the floor, or reducing the transmission of floor vibration to a delicate mechanism, the principle is the same: good isolation is achieved when $\omega_{ex}/\omega_n \gg \sqrt{2}$ and $\zeta \ll 1$

$$\boxed{\left|\frac{X}{Y_o}\right| = \left|\frac{F_b}{F_o}\right| \approx \frac{1}{\frac{\omega_{ex}^2}{\omega_n^2} - 1}} \tag{12.8}$$

In practice this means aiming for low natural frequency ω_n; since Equation 4.29 tells us that $\omega_n^2 = g/\overline{x}$, this requires specifying large static deflection $\overline{x}$ in spring supports with negligible damping

$$\left|\frac{X}{Y_o}\right| = \left|\frac{F_b}{F_o}\right| \approx \frac{1}{\frac{\overline{x}}{g}\omega_{ex}^2 - 1} \tag{12.9}$$

This is a viable strategy when we can be sure that there is no possibility of any excitation near resonance persisting for a significant length of time, lest the low damping would permit the build-up of large amplitudes.

In the case of automotive suspension, a wide range of frequencies is possible, from high-speed road chatter, to long waves and deep culverts. A typical strategy is to choose a natural frequency $f_n = 1$ to 2 Hertz—natural frequencies much below 1 Hz. are sick-making for many passengers—and ζ about 50 to 70% of critical damping: we can pick out the corresponding response curves from Figure 11.5

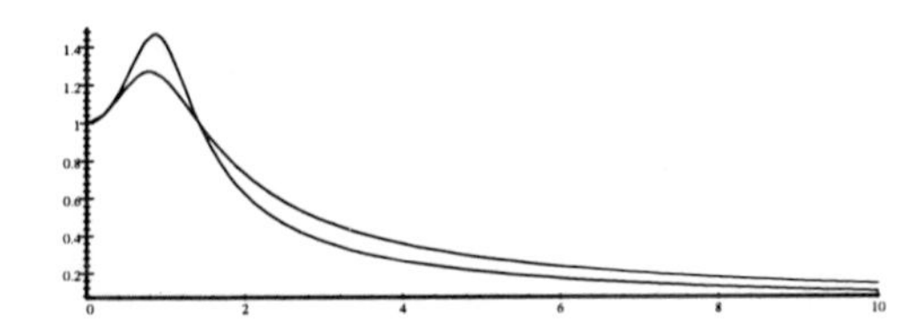

This large damping avoids having much of a resonant peak, and still gives some isolation at high excitation frequencies.

An adaptive strategy is a variable damper, which is set to less damping when there is no persistent low-frequency input from the road, but can be adjusted towards more damping when large resonant oscillations are encountered. Many classic automobiles had this type of adjustment, controlled by the driver; in some modern luxury cars, an automatic system detects the effects of road conditions and trims the shock-absorbers accordingly.

12.4 Dynamic Instrumentation

We can detect the motion of an object by fastening a small spring-mass system to it, and observing the deflection of the spring when the object moves. We can apply Equation 12.1 to this system: the motion of the object is the base motion $y(t)$, and the motion of the small mass in our instrument is x. However, we prefer to express the governing equation in terms of the observed spring deflection

$$z = x - y(t) \tag{12.10}$$

which changes Equation 12.1 to

$$\frac{m}{g_c}\left(\ddot{z} + \ddot{y}\right) + c\,\dot{z} + kz = 0 \tag{12.11}$$

where $y = Y_o \sin(\omega_{ex} t)$ so that

$$\begin{aligned}
\frac{m}{g_c}\ddot{z} + c\,\dot{z} + kz &= \frac{m}{g_c} Y_o \omega_{ex}^2 \sin(\omega_{ex} t) \\
\ddot{z} + \left(\frac{c g_c}{m}\right)\dot{z} + \left(\frac{k g_c}{m}\right) z &= Y_o \omega_{ex}^2 \sin(\omega_{ex} t) \\
\ddot{z} + (2\zeta\omega_n)\,\dot{z} + \left(\omega_n^2\right) z &= Y_o \omega_{ex}^2 \sin(\omega_{ex} t) \qquad (12.12)
\end{aligned}$$

which has the solution

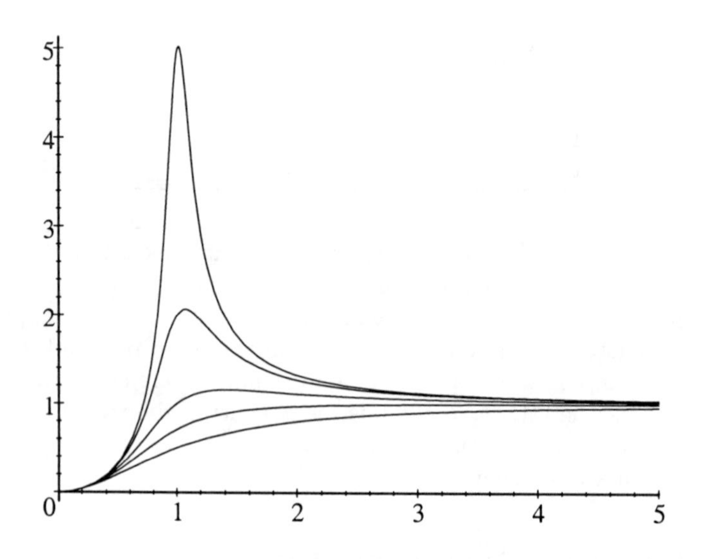

Figure 12.1: Relative-Motion Due to Base Excitation

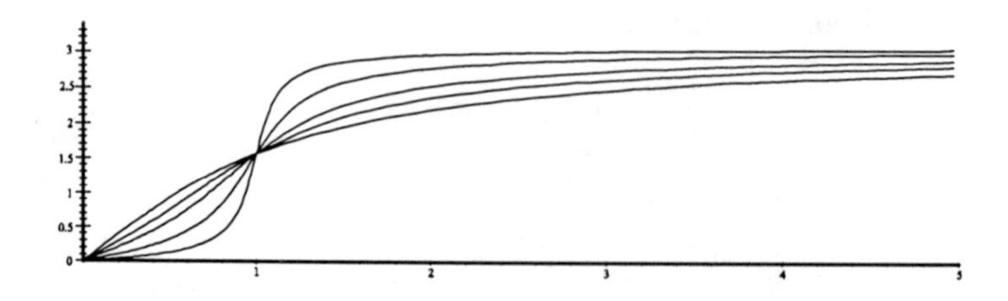

Figure 12.2: Relative-Motion Phase Relative to Base Excitation

$$z_{ss} = Z \sin(\omega_{ex} t - \phi) \tag{12.13}$$

where, by comparison of Equation 12.12 with Equation 11.4

$$\frac{Z}{Y_o} = \frac{\left(\frac{\omega_{ex}}{\omega_n}\right)^2}{\sqrt{\left(1 - \frac{\omega_{ex}^2}{\omega_n^2}\right)^2 + \left(2\zeta\frac{\omega_{ex}}{\omega_n}\right)^2}} \tag{12.14}$$

as plotted in Figure 12.1, and the phase is identical to that of Equation 11.12

$$\tan\phi = \frac{2\zeta\frac{\omega_{ex}}{\omega_n}}{1 - \frac{\omega_{ex}^2}{\omega_n^2}} \tag{12.15}$$

as plotted in Figure 12.2. In order to make the spring-mass-based instrument easy to use, the response Z should be as uniform as possible. There are two approaches to this: the Seismograph and the Accelerometer.

12.4.1 Seismographs

If we make the natural frequency of the "seismic" mass in our instrument very low, much lower than any expected excitation frequencies, our response will be approximately

$$\frac{Z}{Y_o} \cong 1 \text{ for } \frac{\omega_{ex}}{\omega_n} \gg 1 \tag{12.16}$$

Seismographs for the detection of earth motion are constructed this way. Traditionally, masses are mounted on pendulums with nearly vertical pivot-axes, yielding very low natural frequencies, for detecting horizontal ground motion. Vertical ground motion can be detected by mounting a mass on a pendulum with a horizontal axis, and preloading it with a soft torsional spring until the pendulum extends horizontally from the pivot, yielding a very low natural frequency in the vertical direction. The relative motion between the earth and the seismic mass can be recorded mechanically or electronically.

If we want to extend the useful range downward into the resonance region, we can introduce viscous damping $\zeta \cong 1/\sqrt{2}$, so that the resonant peak disappears (but the phase-shift region widens) and the response stays as flat as possible. Very low frequency motions, however, are not effectively detected.

Furthermore, the displacement-oriented approach is not practical when large-amplitude low-frequency excitation is present: if we mounted a seismograph in an automobile, it would deflect off-scale as soon as we picked up speed, or turned, or started up a hill.

12.4.2 Accelerometers

If we make the natural frequency of the spring-mass system high compared to the excitation frequencies we want to detect, we find it useful to write the response of Z relative to the ground acceleration $\ddot{Y}_o = -\omega_{ex}^2 Y_o$

$$\frac{Z\omega_n^2}{\ddot{Y}_o} = \frac{-1}{\sqrt{\left(1-\frac{\omega_{ex}^2}{\omega_n^2}\right)^2 + \left(2\zeta\frac{\omega_{ex}}{\omega_n}\right)^2}} \tag{12.17}$$

$$= -1 \text{ for } \frac{\omega_{ex}}{\omega_n} \ll 1 \tag{12.18}$$

Therefore, we can detect ground acceleration if we can measure Z or the force in the instrument's spring, if we also know the instrument's ω_n, and if the frequencies of interest stay well below the resonance frequency of the instrument. We can build such an instrument by mounting a small mass on a piezoelectric crystal, which acts as a spring and generates an electric charge when it is deflected. A charge amplifier converts the crystal's output to voltages that can be logged by analog or digital instruments.

If we want to extend the useful range upwards towards resonance we can introduce viscous damping $\zeta \cong 1/\sqrt{2}$, so that the resonant peak disappears (but the phase-shift region widens) and the response stays as flat as possible. Very high frequency accelerations, however, are not effectively detected.

The design and application of accelerometers is a trade-off among three specifications:

1. The sensor should be as *small and light* as possible, so that attaching it to the object of interest has little influence on the motion to be measured—but making it light means that sensitivity, i.e., the output for a given input, is limited.

2. The sensor should have a *high natural frequency*, so that high frequency accelerations can be accurately measured—but making it stiff means that sensitivity is limited.

3. The sensor should have *high sensitivity*, so that small accelerations can be detected—but a high output energy requires a large mass and a soft spring, opposing the previous two requirements.

Sensors are available in a wide range of size, frequency, and sensitivity, which must be matched to the experiment.

12.4.3 Laser Vibrometers

If a vibrating object can be observed from a solid base nearby, the motion of its surface can be detected by optical means. A laser interferometer reflects a beam from the (non-specular) surface, and mixes the reflected light with a reference beam split off from the common source laser. If the distance to the surface changes, the combined beams go through augmentation and cancellation whenever the path from the instrument to the surface and back changes by a wave-length of the laser light. Therefore, the magnitude (but not the direction) of normal *displacement* of the surface can be obtained in half-wave increments by counting "fringes."

Laser velocimeters work similarly, but detect *velocity* by measuring the beat frequency of the combined reflected and reference beams. Their operation can be explained either from interferometric principles (but measuring the rate rather than the number of fringe passages) or from Doppler analysis (heterodyning of Doppler-shifted light frequency with the original-source light frequency).

Since only the magnitude of the velocity is detected, sophisticated laser velocimeters phase-shift the reference beam at a very high frequency by means of a Bragg cell, superimposing a known apparent vibration which has a known direction of motion. When that apparent vibration is subtracted out at the end of signal processing, both the magnitude and the direction of the actual motion are obtained.

Crossing-beam geometries have also been developed for translational surface motion and for fluid-suspended-particle anemometry.

Problem 12.1 *Show that the amplitude of the forces experienced by the moving base itself, as a result of base-displacement excitation is*

$$\frac{F_B}{kY_o} = \frac{\frac{\omega_{ex}^2}{\omega_n^2}\sqrt{1+\left(2\zeta\frac{\omega_{ex}}{\omega_n}\right)^2}}{\sqrt{\left(1-\frac{\omega_{ex}^2}{\omega_n^2}\right)^2+\left(2\zeta\frac{\omega_{ex}}{\omega_n}\right)^2}}$$

Hint: you can either sum up the forces in the spring and the damper due to relative displacement and velocity z and $\dot{z}$ from Equation 12.14; or else you can determine the forces needed to obtain the acceleration $\ddot{x}$ in Equation 12.6.

Problem 12.2 *An turntable for old-fashioned "high-fidelity" records is to be mounted in such a way that* 20*-Hertz bass vibration from the supporting shelf is reduced at least ten-fold. What is the minimum static deflection needed in the mounting?*

Problem 12.3 *A delicate electronic instrument is mounted shipboard, where the propulsive machinery produces a* 0.5 *mm double-amplitude vibration at* 100 *Hz, and wave motion produces low-frequency accelerations of* 2 *g. Specify a support system which will limit the acceleration to a maximum of* 5 *g. What is the necessary static deflection?*

Chapter 13

UNBALANCE EXCITATION OF DAMPED SYSTEMS and forces at the base

In many cases, the source of the excitation is not an externally imposed force, but an unbalanced mass. We can convert the effect of that mass to an equivalent force. We will show this on an undamped case first.

13.1 Governing Equation

If a spring- and damper-supported machine of mass M includes a mass excitation $em \sin \omega_{ex} t$, we can add the damping term to Equation 8.1 to yield

$$\begin{aligned}
-kx - c\,\dot{x} &= \frac{(M-m)}{g_c}\,\ddot{x} + \frac{m}{g_c}\left(\ddot{x} - e \cdot \omega_{ex}^2 \cdot \sin \omega_{ex} t\right) \\
\frac{M}{g_c}\,\ddot{x} + c\,\dot{x} + kx &= \left(\frac{em\omega_{ex}^2}{g_c}\right) \sin \omega_{ex} t \\
\ddot{x} + \left(\frac{cg_c}{M}\right)\dot{x} + \left(\frac{kg_c}{M}\right)x &= \left(\frac{em\omega_{ex}^2}{M}\right) \sin \omega_{ex} t \\
\ddot{x} + (2\zeta\omega_n)\,\dot{x} + \left(\omega_n^2\right)x &= \left(\frac{em}{M}\right)\left(\frac{\omega_{ex}^2}{\omega_n^2}\right)\omega_n^2 \sin \omega_{ex} t \qquad (13.1)
\end{aligned}$$

which has the same form as Equation 11.4 if we replace F_o/k in that equation with

$$\left(\frac{F_{equiv}}{k}\right) = \left(\frac{em}{M}\right)\left(\frac{\omega_{ex}^2}{\omega_n^2}\right) \tag{13.2}$$

so that the solution is

$$x_{ss} = X\sin(\omega_{ex}t - \phi) \tag{13.3}$$

where

$$\frac{XM}{em} = \frac{\left(\frac{\omega_{ex}}{\omega_n}\right)^2}{\sqrt{\left(1-\frac{\omega_{ex}^2}{\omega_n^2}\right)^2 + \left(2\zeta\frac{\omega_{ex}}{\omega_n}\right)^2}} \tag{13.4}$$

$$\tan\phi = \frac{2\zeta\frac{\omega_{ex}}{\omega_n}}{1-\frac{\omega_{ex}^2}{\omega_n^2}} \tag{13.5}$$

identical to the expressions plotted in Figures 12.1 and 12.2. This response has the following characteristics:

- For $\omega_{ex}/\omega_n \ll 1$, X approaches zero, because the equivalent exciting force goes towards zero for small ω_{ex}.
- For $\omega_{ex}/\omega_n \gg 1$, XM/em approaches unity—negative, since the phase lag at that point is π radians or 180 degrees—which means that $XM = -em$, so that the center of gravity of the combined masses stays stationary.
- At resonance, when $\omega_{ex}/\omega_n = 1$, $XM/em - 1/2\zeta$ and $\phi = \pi$ radians or 90 degrees.

Exercise 13.1 *Show that the actual maximum response is located slightly to the right of resonance, at*

$$\frac{\omega_{ex}}{\omega_n} = \frac{1}{\sqrt{1-2\zeta^2}}$$

and is slightly higher than $1/(2\zeta)$, *with a peak value of*

$$\frac{XM}{em} = \frac{1}{2\zeta\sqrt{1-\zeta^2}}$$

Hint: this is more easily done after multiplying both numerator and denominator of the equation by $(\omega_n/\omega_{ex})^2$ *so that it is expressed as a function of the inverse of the traditional abscissa*

$$\frac{XM}{em} = \frac{1}{\sqrt{\left(\frac{\omega_n^2}{\omega_{ex}^2}-1\right)^2 + \left(2\zeta\frac{\omega_n}{\omega_{ex}}\right)^2}} \tag{13.6}$$

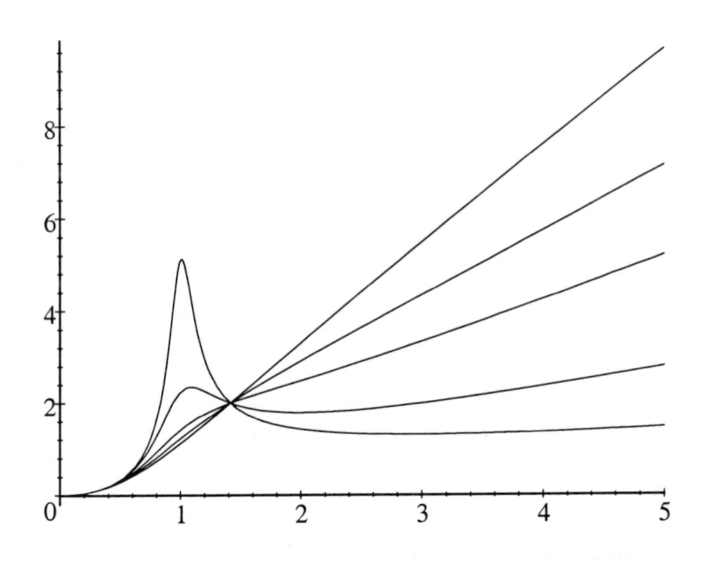

Figure 13.1: Excitation Forces at Base due to Mass Unbalance

13.2 Forces at the Base

The transmission of forces to the base includes both the spring and the damper forces. Adding them up, we obtain

$$F_{\text{base}} = F_B \sin(\omega_{ex}t - \varphi) \tag{13.7}$$

where

$$\frac{F_B}{em\omega_n^2} = \frac{\frac{\omega_{ex}^2}{\omega_n^2}\sqrt{1+\left(2\zeta\frac{\omega_{ex}}{\omega_n}\right)^2}}{\sqrt{\left(1-\frac{\omega_{ex}^2}{\omega_n^2}\right)^2+\left(2\zeta\frac{\omega_{ex}}{\omega_n}\right)^2}} \tag{13.8}$$

as plotted in Figure 13.1, and the phase is the same as in Equation 11.18

$$\tan\psi = \frac{2\zeta\left(\frac{\omega_{ex}}{\omega_n}\right)^3}{1-\left(\frac{\omega_{ex}}{\omega_n}\right)^2+\left(2\zeta\frac{\omega_{ex}}{\omega_n}\right)^2} \tag{13.9}$$

as plotted in Figure 13.2.

The curves all pass through $F_B/em\omega_n^2 = 2$ at $\omega_{ex}/\omega_n = \sqrt{2}$. Because the equivalent excitation force of a particular unbalance em keeps increasing with

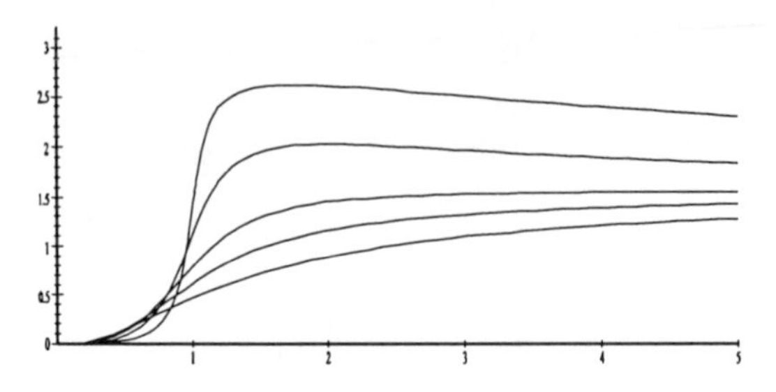

Figure 13.2: Phase of Force at Base

increasing ω_{ex}, the force at the base levels out at $F_B/em\omega_n^2 = 1$ for $\omega_{ex}/\omega_n \gg \sqrt{2}$ if the damping is negligible; but any damping at all will ultimately cause the force transmitted to the base to increase with increasing excitation speed.

13.3 Isolation of Mass Excitation

For the purpose of isolating vibration at the base, we are interested in how isolating springs will perform compared with bolting the machine directly to the base, without springs, in which case the transmitted force would be the entire equivalent force

$$F_{\text{equiv}} = em\omega_{ex}^2 \tag{13.10}$$

Comparing the force at the base (Equation 13.8) with this reference case, we in obtain the familiar transmissibility Equation 11.17

$$\frac{F_B}{F_{equiv}} = \frac{\sqrt{1 + \left(2\zeta\frac{\omega_{ex}}{\omega_n}\right)^2}}{\sqrt{\left(1 - \frac{\omega_{ex}^2}{\omega_n^2}\right)^2 + \left(2\zeta\frac{\omega_{ex}}{\omega_n}\right)^2}} \approx \frac{1}{\left|\frac{\omega_{ex}^2}{\omega_n^2} - 1\right|}$$

as plotted in Figure 11.4, so that the same strategies for isolation are employed.

13.4 Overview of Periodic Excitation

Reviewing the periodic-excitation chapters, we note that there are four different plots that recur as we look at nine cases of displacement, relative-motion, or base-force response to excitation by force, unbalance, or base motion.

Type I: Amplitude response to force excitation, and accelerometer response, follow Figure 11.2 curves that start at unity, peak near resonance, and descend towards zero

$$\frac{Xk}{F} = \frac{-\omega_n^2 Z}{\ddot{Y}_o} = \frac{1}{\sqrt{\left(1-\frac{\omega_{ex}^2}{\omega_n^2}\right)^2 + \left(2\zeta\frac{\omega_{ex}}{\omega_n}\right)^2}} \approx \frac{1}{\left|1-\frac{\omega_{ex}^2}{\omega_n^2}\right|}$$

$$\tan\phi = \frac{2\zeta\frac{\omega_{ex}}{\omega_n}}{1-\frac{\omega_{ex}^2}{\omega_n^2}}$$

Type II: Transmissibility of force excitation to the base, of base motion to mass motion, and of equivalent-force excitation to the base all follow Figure 11.4 curves which start at zero, peak near resonance, intersect at $\left(\sqrt{2}, 1\right)$, and ultimately decline

$$\frac{F_B}{F_o} = \frac{X}{Y_o} = \frac{F_B}{em\omega_{ex}^2} = \frac{\sqrt{1+\left(2\zeta\frac{\omega_{ex}}{\omega_n}\right)^2}}{\sqrt{\left(1-\frac{\omega_{ex}^2}{\omega_n^2}\right)^2 + \left(2\zeta\frac{\omega_{ex}}{\omega_n}\right)^2}} \approx \frac{1}{\left|1-\frac{\omega_{ex}^2}{\omega_n^2}\right|}$$

$$\tan\phi = \frac{2\zeta\left(\frac{\omega_{ex}}{\omega_n}\right)^3}{1-\left(\frac{\omega_{ex}}{\omega_n}\right)^2 + \left(2\zeta\frac{\omega_{ex}}{\omega_n}\right)^2}$$

Type III: Relative-motion response to base excitation. and amplitude response of mass-unbalance excitation, obey Figure 12.1 curves which start at zero, peak near resonance, and level out at unity

$$\frac{Z}{Y_o} = \frac{XM}{em} = \frac{\left(\frac{\omega_{ex}}{\omega_n}\right)^2}{\sqrt{\left(1-\frac{\omega_{ex}^2}{\omega_n^2}\right)^2 + \left(2\zeta\frac{\omega_{ex}}{\omega_n}\right)^2}} \approx \frac{\frac{\omega_{ex}^2}{\omega_n^2}}{\left|1-\frac{\omega_{ex}^2}{\omega_n^2}\right|}$$

$$\tan\phi = \frac{2\zeta\frac{\omega_{ex}}{\omega_n}}{1-\frac{\omega_{ex}^2}{\omega_n^2}}$$

Type IV: Base Force due to mass-unbalance excitation (as well as the force at a moving base) is the Figure 13.1 set of curves which start and zero, peak near resonance, intersect at $\left(\sqrt{2}, 2\right)$, and ultimately increase indefinitely

$$\frac{F_B}{em\omega_n^2} = \frac{F_B}{kY_o} = \frac{\frac{\omega_{ex}^2}{\omega_n^2}\sqrt{1+\left(2\zeta\frac{\omega_{ex}}{\omega_n}\right)^2}}{\sqrt{\left(1-\frac{\omega_{ex}^2}{\omega_n^2}\right)^2 + \left(2\zeta\frac{\omega_{ex}}{\omega_n}\right)^2}} \approx \frac{\frac{\omega_{ex}^2}{\omega_n^2}}{\left|1-\frac{\omega_{ex}^2}{\omega_n^2}\right|}$$

$$\tan\psi = \frac{2\zeta\left(\frac{\omega_{ex}}{\omega_n}\right)^3}{1-\left(\frac{\omega_{ex}}{\omega_n}\right)^2 + \left(2\zeta\frac{\omega_{ex}}{\omega_n}\right)^2}$$

13.5 Application

In Chapter 7, we demonstrated that arbitrary periodic excitation can be decomposed into a Fourier series of harmonic terms. The relationships above can be used to obtain the amplitude and phase of the response of the system at each component frequency. In these linear systems, the complete response is obtained by superposing each component of response. The calculation effort is a little higher with damping, since the phase-angles of the components of the response can have any value from zero to π radians.

Problem 13.2 *A piston engine is to be supported by elastic mounts to minimize vibration-transmission at operating speeds up to* 6000 *r.p.m. Idle speed is* 600 *r.p.m. Specify the static deflection and damping ratio of the mounts.*

Chapter 14

TRANSIENTS by convolution

So far we have studied Initial Condition solutions and periodic excitation. We will now look at the response of linear systems to transient excitations. Because we are dealing with linear systems, the zero-initial-state solutions to different input components can be added to each other (and to the Initial Condition solution) to get the total solution.

14.1 Impulse Response

The zero-initial-state response to an short, sharp blow can be obtained by idealizing the blow as a Dirac delta function: an infinitesimally brief force of infinitely large amplitude. The product of force and time, obtained from $\hat{I} = \int F(t)dt$ across this singularity, has a non-zero, non-infinite value. For the Dirac delta function this value is unity, but we can multiply the function by the factor $\hat{I}$ to get the size of blow we need.

From Newton's law, this blow on a mass causes a change in velocity $mdv/g_c = Fdt$. If the velocity $v_{(0-)}$ before the impulse is zero, then the velocity immediately after the impulse is $v_{(0+)} = \hat{I}g_c/m$. After the excitation is over, the motion will proceed like the homogeneous-equation solution of Equation 9.18 with an initial velocity $\hat{I}g_c/m$ and zero initial displacement:

$$x = \frac{\hat{I}g_c}{m\omega_d} e^{-\zeta\omega_n t} \sin(\omega_d t) \tag{14.1}$$

$$= \frac{\hat{I}\omega_n}{k\sqrt{1-\zeta^2}} e^{-\zeta\omega_n t} \sin\left(\sqrt{1-\zeta^2}\,\omega_n t\right) \tag{14.2}$$

which we can express as the product of an excitation function $\hat{I}$ with a system response function $h_{imp}(t)$:

$$x = \hat{I} \cdot h_{imp}(t) \tag{14.3}$$

where

$$\boxed{h_{imp}(t) = \frac{\omega_n}{k\sqrt{1-\zeta^2}} e^{-\zeta\omega_n t} \sin\left(\sqrt{1-\zeta^2}\omega_n t\right) \approx \frac{\omega_n}{k} \sin\left(\omega_n t\right)} \tag{14.4}$$

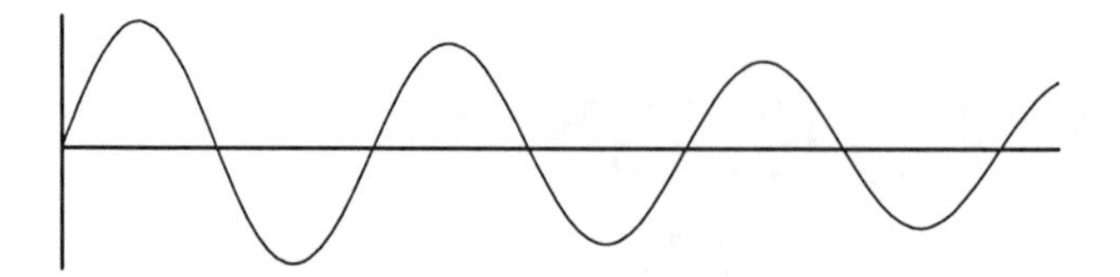

We should ask two questions: First, how short must the pulse be in order for this analysis to be appropriate? The answer is that the mass should have no significant motion *while* the impulse takes place, compared to the later motion described by this equation. Therefore, the time duration Δt of the blow should be much smaller than the quarter-period of the system, or $\omega_n \Delta t \ll \pi/2$.

Secondly, when can we use the simple undamped solution at right rather than the full damped solution? To use the simple expression, we must limit ourselves both to moderate damping $\zeta \ll 1$ and also to a limited time frame $\omega_n t < 2\pi$. The logarithmic decrement δ gives us an indication of the difference between damped and undamped solutions during one period of free oscillation: at 1.5% of critical damping, the damped solution decays about 10%. Since many structural systems are very lightly damped, and since we are interested mostly in the maximum amplitudes, which occur soon (about a quarter-cycle) after the impulse, the undamped solution is often adequate.

We can generalize this solution to impulses occurring at some time t_1 other than zero,

$$\begin{aligned} h_{imp}(t - t_1) &= \frac{\omega_n}{k\sqrt{1-\zeta^2}} e^{-\zeta\omega_n(t-t_1)} \sin\left(\sqrt{1-\zeta^2}\omega_n (t - t_1)\right) \\ &\approx \frac{\omega_n}{k} \sin\left(\omega_n t - \omega_n t_1\right) \end{aligned} \tag{14.5}$$

and we can superpose many different impulses

$$x = \sum_n \hat{I}_n \cdot h_{imp}(t - t_n) \tag{14.6}$$

Therefore, the solution to a series of impulses is always the summation of the responses of all *previous* impulses—in other words, the form of the solution changes and acquires an additional term each time an impulse passes by.

Example The undamped response to a "double hammer," a pair of equal impulses separated by time interval t_1, is

$$\begin{aligned} x &= 0 \text{ for } t \leq 0 \\ x &= \frac{\hat{I}\omega_n}{k} \sin(\omega_n t) \text{ for } 0 \leq t \leq t_1 \\ x &= \frac{\hat{I}\omega_n}{k} (\sin(\omega_n t) + \sin(\omega_n t - \omega_n t_1)) \text{ for } t \geq t_1 \end{aligned}$$

Note that the final part of the solution can be rearranged, using the trigonometric formula for $\sin(A-B)$, into a steady oscillation, the amplitude of which is a function of the excitation-function time interval t_1 relative to the system time scale ω_n

$$x = \frac{\hat{I}\omega_n}{k} ((1 + \cos\omega_n t_1) \sin\omega_n t - (\sin\omega_n t_1) \cos\omega_n t) \text{ for } t \geq t_1$$

which approximates $x = 2\left(\hat{I}\omega_n/k\right) \sin\omega_n t$ for $\omega_n t_1 \ll 1$.

Exercise 14.1 *Calculate the undamped response to a "doublet," an opposed pair of equal impulses separated by a time interval t_1. Show that the final part of the solution is*

$$x = \frac{\hat{I}\omega_n}{k} ((1 - \cos\omega_n t_1) \sin\omega_n t - (\sin\omega_n t_1) \cos\omega_n t) \text{ for } t \geq t_1)$$

which approximates zero for $\omega_n t_1 \ll 1$.

14.2 Step Response

The response to a step of height F_o imposed at $t = 0$ can be described as an Initial Condition response due to a steady force which causes a shift in equilibrium:

$$\frac{m}{g_c} \ddot{x} + c\,\dot{x} + kx = F_o$$

or, in Standard Form,

$$\ddot{x} + (2\zeta\omega_n)\,\dot{x} + \left(\omega_n^2\right) x = \left(\frac{F_o}{k}\right)\left(\omega_n^2\right)$$

where $x(0) = 0$ and $\dot{x} = 0$; this results in the solution

$$x = \frac{F_o}{k} e^{-\zeta\omega_n t} \left(1 - \cos(\omega_d t) - \frac{\zeta}{\sqrt{1-\zeta^2}} \sin(\omega_d t)\right) \approx \frac{F_o}{k}(1 - \cos(\omega_n t)) \tag{14.7}$$

Exercise 14.2 *Derive this solution, or check it by differentiating and plugging back into the equation.*

We can express this as the product of a step input F_o with a unit-step response function $h_{stp}(t)$:

$$x = F_o \cdot h_{stp}(t) \tag{14.8}$$

where

$$h_{stp}(t) = \frac{1}{k} e^{-\zeta\omega_n t} \left(1 - \cos(\omega_d t) - \frac{\zeta}{\sqrt{1-\zeta^2}} \sin(\omega_d t)\right) \approx \frac{1}{k}(1 - \cos(\omega_n t)) \tag{14.9}$$

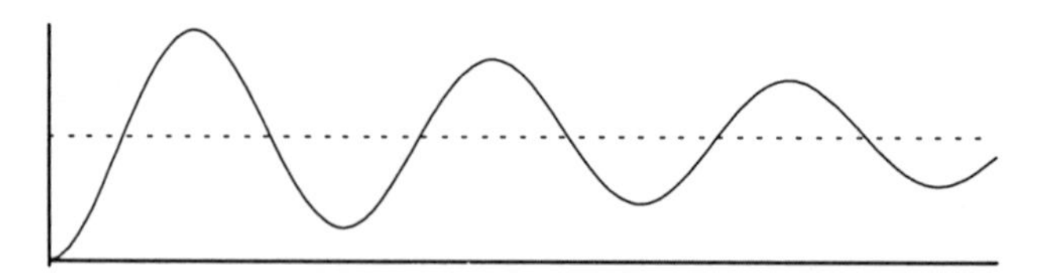

and generalize it for a succession of steps F_n superimposed on each other at different times t_n:

$$x = \sum_n F_n \cdot h_{stp}(t - t_n) \tag{14.10}$$

where

$$\begin{aligned} h_{stp}(t - t_n) &= \frac{e^{-\zeta\omega_n (t-t_n)}}{k} \left(1 - \cos\omega_d (t - t_n) - \frac{\zeta}{\sqrt{1-\zeta^2}} \sin\omega_d (t - t_n)\right) \\ &\approx \frac{1}{k}(1 - \cos(\omega_n (t - t_n))) \end{aligned} \tag{14.11}$$

Therefore we can represent a forcing function as a series of steps, and superpose the solutions to these steps to obtain the total response.

Example The undamped response to a square pulse of length $t_1$is the response of a positive step plus the response of a negative step which cancels out the first step after time t_1

$$\begin{aligned} x &= 0 \text{ for } t \leq 0 \\ x &= \frac{F_o}{k}(1 - \cos(\omega_n t)) \text{ for } 0 \leq t \leq t_1 \\ x &= \frac{F_o}{k}(1 - \cos(\omega_n t) - 1 + \cos(\omega_n t - \omega_n t_1)) \text{ for } t \geq t_1 \end{aligned}$$

which can be simplified to

$$x = \frac{F_o}{k}((\sin\omega_n t_1)\sin\omega_n t + (-1 + \cos\omega_n t_1)\cos\omega_n t) \text{ for } t \geq t_1$$

Exercise 14.3 *Show that, if* $\omega_n t_1 \ll 1$, *this is equivalent to the impulse response. Hint: let* $F_o \cdot t_1 = \hat{I}$.

14.3 Ramp Response

Although we can simulate arbitrarily shaped pulses as a series of small steps, we can do it more efficiently if we obtain the response of a ramp function of slope $\overset{\circ}{F}$, by solving the differential equation

$$\frac{m}{g_c} \ddot{x} + c \dot{x} + kx = \overset{\circ}{F} t$$

or, in Standard Form,

$$\ddot{x} + (2\zeta\omega_n) \dot{x} + \left(\omega_n^2\right) x = \left(\frac{\overset{\circ}{F}}{k}\right) \left(\omega_n^2\right) t$$

which has the approximate solution

$$\begin{aligned} x &= \frac{\overset{\circ}{F}}{k}\left(t - \frac{e^{-\zeta\omega_n t}}{\omega_n}\left(2\zeta\left(1 - \cos\omega_d t\right) + \frac{1 - 2\zeta^2}{\sqrt{1-\zeta^2}} \sin\omega_d t\right)\right) \\ &\approx \frac{\overset{\circ}{F}}{k}\left(t - \frac{\sin\omega_n t}{\omega_n}\right) \end{aligned} \tag{14.12}$$

Exercise 14.4 *Verify the solution for a ramp input, by differentiating it twice and plugging back into the differential equation.*

This can be expressed as the product of a ramp excitation $\overset{\circ}{F}$ with a ramp response $h_{rmp}(t)$

$$x = \overset{\circ}{F}_o \cdot h_{rmp}(t) \tag{14.13}$$

where

$$\begin{aligned} h_{rmp}(t) &= \frac{1}{k}\left(t - \frac{e^{-\zeta\omega_n t}}{\omega_n}\left(2\zeta\left(1 - \cos\omega_d t\right) + \frac{1 - 2\zeta^2}{\sqrt{1-\zeta^2}} \sin\omega_d t\right)\right) \\ &\approx \frac{1}{k}\left(t - \frac{\sin\omega_n t}{\omega_n}\right) \end{aligned} \tag{14.14}$$

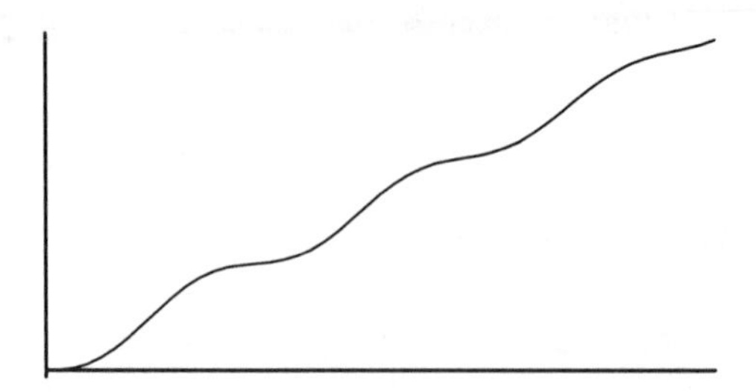

or, for several ramps (i.e., changes in slope) occurring at various times t_n

$$x = \sum_n \overset{\circ}{F}_n \cdot h_{rmp}(t - t_n) \tag{14.15}$$

where

$$h_{rmp}(t - t_n) \approx \frac{1}{k}\left(t - t_n - \frac{\sin\left(\omega_n\left(t - t_n\right)\right)}{\omega_n}\right) \tag{14.16}$$

Therefore, we can take an arbitrary transient excitation by representing the forcing function as a series of straight line segments, and adding-in a ramp at every corner to represent the change in slope.

The undamped solution for a triangular pulse of length $2t_1$ and height $\overset{\circ}{F}_o \cdot t_1$ is

$$\begin{aligned}
x &= 0 \text{ for } t \le 0 \\
x &= \frac{\overset{\circ}{F}_o}{k}\left(t - \frac{\sin\omega_n t}{\omega_n}\right) \text{ for } 0 \le t \le t_1 \\
x &= \frac{\overset{\circ}{F}_o}{k}\left(t - \frac{\sin\omega_n t}{\omega_n}\right) - 2\frac{\overset{\circ}{F}_o}{k}\left(t - t_1 - \frac{\sin\left(\omega_n\left(t - t_1\right)\right)}{\omega_n}\right) \text{ for } t_1 \le t \le 2t_1 \\
x &= \left\{ \begin{array}{c} \frac{\overset{\circ}{F}_o}{k}\left(-\frac{\sin\omega_n t}{\omega_n}\right) - 2\frac{\overset{\circ}{F}_o}{k}\left(-t_1 - \frac{\sin(\omega_n(t-t_1))}{\omega_n}\right) \\ +\frac{\overset{\circ}{F}_o}{k}\left(-2t_1 - \frac{\sin(\omega_n(t-2t_1))}{\omega_n}\right) \end{array} \right\} \text{ for } t \ge 2t_1
\end{aligned}$$

Exercise 14.5 *Show that the final solution can be put into the form*

$$x = \frac{\overset{\circ}{F}_o}{k\omega_n}\left(\left(-1 + 2\cos\omega_n t_1 - \cos 2\omega_n t_1\right)\sin\omega_n t + \left(-2\sin\omega_n t_1 + \sin 2\omega_n t_1\right)\cos\omega_n t\right)$$

and that this approximates the impulse solution when $2\omega_n t_1 \ll 1$. *Hint: let* $F_o \cdot t_1 = \hat{I}$.*)*

Exercise 14.6 *Obtain the solution for a step of height* F_o *with a rise time* t_1. *Compare with the step-response of the previous section.*

14.4 Convolution Integration

By superposing discrete impulses, steps, and/or ramps, we can represent forcing functions in a rough way. We could improve on this by representing the forcing function $F(t)$ as infinite number of infinitesimal impulses $\hat{I} = F(t)dt$, substituting integration over time for the summation:

$$x(t) = \int_{-\infty}^{t} F(\xi) \cdot h_{imp}(t-\xi)d\xi \tag{14.17}$$

Note that we are careful to replace the different impulse times t_n with the dummy variable ξ; since time also occurs elsewhere within the kernel and in the upper limit of the integration, we must not be sloppy about distinguishing between the dummy variable and the time t in the answer $x(t)$. The lower limit of $-\infty$ can be replaced with zero if we are careful to start the time scale early enough, at a point where we have a zero-initial-state condition.

This "superposition integral" or "convolution integral" takes the function F point-by-point for each occurrence time ξ, up to the final time t. It obtains each point's response, which is the magnitude of the solution that develops in the intervening time between ξ and the final t. It then adds up all the responses due to all the points before time t. Therefore, the integral sweeps through all possible occurrence times from zero to t, and backwards through all possible response intervals from t to zero. When it is more convenient, we can just as well run through all combinations of excitation and response in the opposite direction:

$$x(t) = \int_{-\infty}^{t} F(t-\xi) \cdot h_{imp}(\xi)d\xi \tag{14.18}$$

Example The response to a square pulse, $F(t) = F_o$ for $0 \prec t \prec t_1$ and zero everywhere else, can be obtained by writing

$$\begin{aligned}
x &= 0 \text{ for } t \leq 0 \\
x &= \int_0^t F_o \cdot h_{imp}(\xi)d\xi \approx F_o \int_0^t \frac{\omega_n}{k} \sin(\omega_n \xi)(\xi)d\xi \\
&\approx \frac{F_o}{k}(1-\cos(\omega_n t)) \text{ for } 0 \leq t \leq t_1 \\
x &= \int_0^{t_1} F_o \cdot h_{imp}(\xi)d\xi + \int_{t_1}^{t} F_o \cdot h_{imp}(\xi)d\xi \\
&\approx \frac{F_o}{k}(-\cos(\omega_n t) + \cos(\omega_n t - \omega_n t_1)) \text{ for } t \geq t_1
\end{aligned}$$

which is identical to the result we obtained above in another way.

Exercise 14.7 *Using the superposition integral, verify the response to a unit step.*

Exercise 14.8 *Using the superposition integral, verify the response to ramp of unit slope.*

Instead of representing the function as a "bar-graph" of many impulses, Duhamel represented it as a series of steps, taking the slope of the exiting function $\dot{F} \equiv dF/dt$ as a measure of the magnitude of the steps, an multiplying it with the unit-step response:

$$x(t) = \int_{-\infty}^{t} \dot{F}(\xi) \cdot h_{stp}(t-\xi)d\xi \tag{14.19}$$

Exercise 14.9 *Use Duhamel's integral to verify the response to a ramp of unit slope.*

If it is more convenient, we can also use change in direction or curvature of the exciting function $\ddot{F} \equiv d^2F/dt^2$ and the ramp response:

$$x(t) = \int_{-\infty}^{t} \ddot{F}(\xi) \cdot h_{rmp}(t-\xi)d\xi \tag{14.20}$$

14.5 Base Excitation

These methods can be adapted to base excitation by replacing x by $z = (x-y)$ and F by $-\ddot{y}$.

Example: Obtain an expression for the response $x(t)$ of an undamped system due to a ramp input at the base

$$\begin{aligned} y &= 0 \text{ for } t \leq 0 \\ y &= \dot{Y}_o \cdot t \text{ for } t \geq 0 \\ \dot{y} &= \dot{Y}_o \text{ for } t \geq 0 \end{aligned}$$

where $\dot{Y}_o$ is a constant.

Solution: The governing equation for a base-excited system is

$$\frac{m}{g_c}\ddot{x} + c\left(\dot{x} - \dot{y}\right) + k(x-y) = 0$$

which normalizes to

$$\ddot{x} + 2\zeta\omega_n \dot{x} + \omega_n^2 x = 2\zeta\omega_n \dot{y} + \omega_n^2 y$$

In the absence of damping this becomes

$$\ddot{x} + \omega_n^2 x = \omega_n^2 y$$

Plugging in our excitation

$$\ddot{x} + \omega_n^2 x = \omega_n^2 \dot{Y}_o t$$

The complementary solution has the form

$$x_h = A \sin \omega_n t + B \cos \omega_n t$$

and the particular solution has the form

$$\begin{aligned} x_p &= C + Dt \\ \dot{x}_p &= D \\ \ddot{x}_p &= 0 \end{aligned}$$

Which we can plug into the differential equation to give

$$0 + \omega_n^2 (C + Dt) = \omega_n^2 \dot{Y}_o t$$

which tells us that $C = 0$ and $D = \dot{Y}_o$. Therefore, the complete solution is

$$\begin{aligned} x &= \dot{Y}_o t + A \sin \omega_n t + B \cos \omega_n t \\ \dot{x} &= \dot{Y}_o + A\omega_n \cos \omega_n t - B\omega_n \sin \omega_n t \end{aligned}$$

for the initial conditions

$$\begin{aligned} x(0) &= B \\ \dot{x}(0) &= \dot{Y}_o + A\omega_n \end{aligned}$$

Therefore, the zero-initial-state solution has $B = 0$ and $A = -\dot{Y}_o / \omega_n$

$$x = \dot{Y}_o t - \frac{\dot{Y}_o}{\omega_n} \sin \omega_n t$$

Example: Obtain an expression for the response $x(t)$ of a damped system due to a ramp input at the base

$$\begin{aligned} y &= 0 \text{ for } t \leq 0 \\ y &= \dot{Y}_o \cdot t \text{ for } t \geq 0 \\ \dot{y} &= \dot{Y}_o \text{ for } t \geq 0 \end{aligned}$$

where $\dot{Y}_o$ is a constant.

Solution: The governing equation for a base-excited system is

$$\frac{m}{g_c} \ddot{x} + c\left(\dot{x} - \dot{y}\right) + k(x - y) = 0$$

which normalizes to

$$\ddot{x} + 2\zeta\omega_n \dot{x} + \omega_n^2 x = 2\zeta\omega_n \dot{y} + \omega_n^2 y$$

Plugging in our excitation

$$\ddot{x} + 2\zeta\omega_n \dot{x} + \omega_n^2 x = 2\zeta\omega_n \dot{Y}_o + \omega_n^2 \dot{Y}_o t$$

The complementary solution has the form

$$x_h = e^{-\zeta\omega_n t}\left(A \sin\sqrt{1-\zeta^2}\omega_n t + B\cos\sqrt{1-\zeta^2}\omega_n t\right)$$

and the particular solution has the form

$$\begin{aligned} x_p &= C + Dt \\ \dot{x}_p &= D \\ \ddot{x}_p &= 0 \end{aligned}$$

Which we can plug into the differential equation to give

$$0 + 2\zeta\omega_n D + \omega_n^2 (C + Dt) = \omega_n^2 \dot{Y}_o t$$

which tells us that $D = \dot{Y}_o$and $C = \frac{-2\zeta}{\omega_n} D = \frac{-2\zeta}{\omega_n} \dot{Y}_o$. Therefore, the complete solution is

$$\begin{aligned} x &= \frac{-2\zeta}{\omega_n}\dot{Y}_o + \dot{Y}_o t + e^{-\zeta\omega_n t}\left(A \sin\sqrt{1-\zeta^2}\omega_n t + B\cos\sqrt{1-\zeta^2}\omega_n t\right) \\ \dot{x} &= \dot{Y}_o - \zeta\omega_n e^{-\zeta\omega_n t}\left(A \sin\sqrt{1-\zeta^2}\omega_n t + B\cos\sqrt{1-\zeta^2}\omega_n t\right) \\ &\quad + e^{-\zeta\omega_n t}\left(A\sqrt{1-\zeta^2}\omega_n \cos\sqrt{1-\zeta^2}\omega_n t - B\sqrt{1-\zeta^2}\omega_n \sin\sqrt{1-\zeta^2}\omega_n t\right) \end{aligned}$$

for the initial conditions,

$$\begin{aligned} x(0) &= \frac{-2\zeta}{\omega_n}\dot{Y}_o + B \\ \dot{x}(0) &= \dot{Y}_o - \zeta\omega_n B + e^{-\zeta\omega_n t} A\sqrt{1-\zeta^2}\omega_n \end{aligned}$$

Therefore, the zero-initial-state solution has

$$\begin{aligned} B &= \frac{2\zeta}{\omega_n}\dot{Y}_o \\ A &= \frac{-\left(1-2\zeta^2\right)}{e^{-\zeta\omega_n t}\sqrt{1-\zeta^2}\omega_n}\dot{Y}_o \end{aligned}$$

and the complete solution is

$$x = \frac{\dot{Y}_o}{\omega_n}\left(\omega_n t - e^{-\zeta\omega_n t}\frac{-\left(1-2\zeta^2\right)}{e^{-\zeta\omega_n t}\sqrt{1-\zeta^2}}\sin\sqrt{1-\zeta^2}\omega_n t - 2\zeta\left(1-e^{-\zeta\omega_n t}\cos\sqrt{1-\zeta^2}\omega_n t\right)\right)$$

which, for the undamped case, simplifies to

$$x = \frac{\dot{Y}_o}{\omega_n}\left(\omega_n t - \sin\omega_n t\right)$$

Problem 14.10 *Obtain the response to an extended "doublet" consisting of a positive step at time zero, twice as large a negative step at time t_1, and another positive step at time $2t_1$.*

Problem 14.11 *Obtain the solution for a triangular pulse with a rise time t_1, and a symmetrical fall-off time t_1. Hint: you will need three ramps: one to start the rise, a double-strength negative one at the peak F_o when $t = t_1$ to stop the rise and start the fall, and one to stop the fall when $t = 2t_1$. Can you simplify the solution?*

Problem 14.12 *Obtain the solution for a nearly square pulse with a rise time t_1, a mean duration of nt_1, and a symmetrical fall-off time. Hint: you will need four ramps: one to start the rise, one to stop at the plateau F_o when $t = t_1$, one to start the decay when $t = nt_1$, and one to stop the fall when $t = (nt_1 + t_1)$*

Problem 14.13 *Using the superposition integral, obtain the response to a pulse that is shaped like one-half of a sine wave.*

Chapter 15

SHOCK SPECTRA and similitude

From the previous chapter, we can compute the response of a system of known natural frequency and damping to a transient input of a particular shape. We observe that the amplitude of the response scales to the amplitude of the input, but that it depends in a complex way on the relative time-scales of the input and the system $\omega_n t_1$, and on the damping ratio ζ. If we pick the maximum positive or negative amplitude $x_{\max}$ as the most interesting feature of the response, we can plot it for $\zeta = 0$ or any other value of ζ against the abscissa $\omega_n t_1$. We will illustrate this with several examples.

15.1 Three Examples

15.1.1 The Impulse Doublet

If we have two equal and opposite impulses $\hat{I}$, separated by the time interval t_1, the response of an undamped system is

$$x = 0 \text{ for } t \leq 0 \tag{15.1}$$

$$x = \frac{\hat{I}\omega_n}{k} \sin(\omega_n t) \text{ for } 0 \leq t \leq t_1 \tag{15.2}$$

$$x = \frac{\hat{I}\omega_n}{k} (\sin(\omega_n t) - \sin(\omega_n t - \omega_n t_1)) \text{ for } t \geq t_1 \tag{15.3}$$

We have two possibilities for when the maximum amplitude occurs: It could occur as part of the *early response,* before t_1, or it could occur as part of the *late response*, after t_1. We should examine both and see which is larger.

The *early response,* during $0 \leq t \leq t_1$, is whatever value x reaches before the window of time for this solution ends. If $\omega_n t_1$ happens to be less than $\pi/2$ or

90 degrees, that is the last value that the response reaches; but if $\omega_n t_1$ happens to be more than $\pi/2$ or 90 degrees, then $\sin(\omega_n t_1)$ will have reached a value of unity:

$$x_{\max} = \frac{\hat{I}\omega_n}{k}\sin(\omega_n t_1) \text{ for } \omega_n t_1 \leq \frac{\pi}{2} \tag{15.4}$$

$$x_{\max} = \frac{\hat{I}\omega_n}{k} \text{ for } \omega_n t_1 \geq \frac{\pi}{2} \tag{15.5}$$

Note that time t does not appear in these expressions for $x_{\max}$; we can now plot the early-response $x_{\max}k/\hat{I}\omega_n$ against $\omega_n t_1$.

The *late response* is the maximum value of x in the final solution

$$\begin{aligned} x &= \frac{\hat{I}\omega_n}{k}\left(\sin(\omega_n t) - \sin(\omega_n t - \omega_n t_1)\right) \text{ for } t \geq t_1 \\ &= \frac{\hat{I}\omega_n}{k}\left((1-\cos\omega_n t_1)\sin\omega_n t - (\sin\omega_n t_1)\cos\omega_n t\right) \\ &= \frac{\hat{I}\omega_n}{k}\sqrt{(1-\cos\omega_n t_1)^2 + (\sin\omega_n t_1)^2}\sin(\omega_n t + \theta) \end{aligned}$$

Therefore, the maximum late response is

$$\begin{aligned} x_{\max} &= \frac{\hat{I}\omega_n}{k}\sqrt{(1-\cos\omega_n t_1)^2 + (\sin\omega_n t_1)^2} \\ &= \frac{\hat{I}\omega_n}{k}\sqrt{1 - 2\cos\omega_n t_1 + \cos^2\omega_n t_1 + \sin^2\omega_n t_1} \\ &= \frac{\hat{I}\omega_n}{k}\sqrt{2 - 2\cos\omega_n t_1} \\ &= \frac{\hat{I}\omega_n}{k}2\sqrt{(1-\cos\omega_n t_1)/2} \\ &= \frac{\hat{I}\omega_n}{k}2\sin\frac{\omega_n t_1}{2} \end{aligned} \tag{15.6}$$

We can now plot the late-response $x_{\max}k/\hat{I}\omega_n$ against $\omega_n t_1$. Of the two curves, early and late response, whichever one is higher is the one that represents the true $x_{\max}$.

This example demonstrates several principles:

- Although some of these plots may have the superficial appearance of oscillations, $x_{\max}$ is *not* a function of elapsed time t, but a function of the time *scale* of the excitation t_1.

- Fractional Analysis of undamped systems tells us that the independent variable has the form $\omega_n t_1$; in other words, it is the relationship of the time scale of the excitation t_1 to the period of the system $\tau_n = 2\pi/\omega_n$ that determines the maximum response.

- The abscissa $\omega_n t_1 = 2\pi t_1/\tau_n$ of our plots of transient spectra is inverse to the abscissa of our periodic-excitation spectra, where we use the ratio of the frequency of excitation to the natural frequency of the system ω_{ex}/ω_n.
- Because of superposition principles, the dependent variable can be expressed in the dimensionless form $x_{\max}k/\hat{I}\omega_n = x_{\max}m/\hat{I}g_c$.

Exercise 15.1 *Plot the maximum response of an undamped system to a "double hammer," page 139, as a function of the time interval between the two blows. For what values of $\omega_n t_1$ is the $x_{\max}$ of a double hammer within 5% of a single impulse of $2\hat{I}$?*

15.1.2 The Square Pulse

If we have a square pulse, page 140, of height F_o and length t_1, the response of an undamped system is the superposition of a positive step at $t = 0$ and an opposite step at $t = t_1$

$$x = 0 \text{ for } t \leq 0 \tag{15.7}$$

$$x = \frac{F_o}{k}\left(1 - \cos\left(\omega_n t\right)\right) \text{ for } 0 \leq t \leq t_1 \tag{15.8}$$

$$x = \frac{F_o}{k}\left(1 - \cos\left(\omega_n t\right) - 1 + \cos\left(\omega_n t - \omega_n t_1\right)\right) \text{ for } t \geq t_1 \tag{15.9}$$

We have two possibilities for when the maximum amplitude occurs: It could occur as part of the *early response*, before t_1, or it could occur as part of the *late response*, after t_1. We should examine both and see which is larger.

The *early response*, during $0 \leq t \leq t_1$, is whatever value x reaches before the window of time for this solution ends. If $\omega_n t_1$ happens to be less than π or 180 degrees, that is the last value that the response reaches; but if $\omega_n t_1$ happens to be more than π or 180 degrees, then $1 - \cos\omega_n t_1$ will have reached a value of 2:

$$x_{\max} = \frac{F_o}{k}\left(1 - \cos\omega_n t_1\right) \text{ for } \omega_n t_1 \leq \pi \tag{15.10}$$

$$x_{\max} = 2\frac{F_o}{k} \text{ for } \omega_n t_1 \geq \pi \tag{15.11}$$

Note that time t does not appear in these expressions for $x_{\max}$; we can now plot the early-response $x_{\max}k/F_o$ against $\omega_n t_1$.

The *late response* is the maximum value of x in the final solution

$$\begin{aligned} x &= \frac{F_o}{k}\left(-\cos\left(\omega_n t\right) + \cos\left(\omega_n t - \omega_n t_1\right)\right) \text{ for } t \geq t_1 \\ &= \frac{F_o}{k}\left(\left(\sin\omega_n t_1\right)^2 + \left(\cos\omega_n t_1 - 1\right)^2\right) \\ &= \frac{F_o}{k}\left(\left(\sin\omega_n t_1\right)\sin\omega_n t + \left(\cos\omega_n t_1 - 1\right)\cos\omega_n t\right) \\ &= \frac{F_o}{k}\sqrt{\left(\sin\omega_n t_1\right)^2 + \left(\cos\omega_n t_1 - 1\right)^2}\sin\left(\omega_n t + \theta\right) \end{aligned}$$

Therefore, the maximum late-response of an undamped system is

$$\begin{aligned}
x_{\max} &= \frac{F_o}{k}\sqrt{(\sin\omega_n t_1)^2 + (\cos\omega_n t_1 - 1)^2} \\
&= \frac{F_o}{k}\sqrt{\sin^2\omega_n t_1 + \cos^2\omega_n t_1 - 2\cos\omega_n t_1 + 1} \\
&= \frac{F_o}{k}\sqrt{2 - 2\cos\omega_n t_1} \\
&= \frac{F_o}{k}2\sqrt{(1 - \cos\omega_n t_1)/2} \\
&= \frac{F_o}{k}2\sin\frac{\omega_n t_1}{2} \qquad (15.12)
\end{aligned}$$

We can now plot the late-response $x_{\max}k/F_o$ against $\omega_n t_1$. Of the two curves, early and late response, whichever one is higher is the one that represents the true $x_{\max}$; in this case, the early response dominates the spectrum.

Again, the independent variable is $\omega_n t_1$; the dimensionless dependent variable is $x_{\max}k/F_o$.

Exercise 15.2 *We can compare a square step with an impulse if we take $\hat{I} = F_o t_1$ so that we replace F_o/k with $\hat{I}/t_1 k$. For what values of $\omega_n t_1$ is the $x_{\max}$ of a square pulse within 5% of a singular impulse with the same magnitude $\hat{I}$?*

Exercise 15.3 *What is the maximum response due to a pair of opposite square pulses, consisting of a step F_o at $t = 0$, a step $-2F_o$ at t_1, and a final step F_o at t_1? Hint: There are three possibilities for the maximum: it might occur early, before t_1; it might occur mid-life, between t_1 and $2t_1$; or it might occur late, after $2t_1$. Compare the result with the impulse doublet in Section 15.1.1.*

15.1.3 The Ramped Step

If we have a ramped step of height $F_o$and rise time t_1 the response of an undamped system is the superposition of a ramp at $t = 0$ and an opposite ramp at $t = t_1$:

$$x = 0 \text{ for } t \leq 0 \qquad (15.13)$$

$$x = \frac{F_o}{kt_1}\left(t - \frac{1}{\omega_n}\sin\omega_n t\right) \text{ for } 0 \leq t \leq t_1 \qquad (15.14)$$

$$x = \frac{F_o}{kt_1}\left(t - \frac{1}{\omega_n}\sin\omega_n t - (t - t_1) + \frac{1}{\omega_n}\sin(\omega_n t - \omega_n t_1)\right) \text{ for } t \geq t_1 \qquad (15.15)$$

We have two possibilities for when the maximum amplitude occurs: It could occur as part of the *early response,* before t_1, or it could occur as part of the *late response*, after t_1. We should examine both and see which is larger.

The *early response*, during $0 \leq t \leq t_1$, is whatever value x reaches before the window of time for this solution ends. Since the solution rises monotonically (or, at worst, stays level), that is the last value that the response reaches:

$$x_{\max} = \frac{F_o}{k}\left(\frac{\omega_n t_1 - \sin\omega_n t_1}{\omega_n t_1}\right) \tag{15.16}$$

Note that time t does not appear in this expressions for $x_{\max}$; we can now plot the early-response $x_{\max}k/F_o$ against $\omega_n t_1$.

The *late response* is the maximum value of x in the final solution

$$\begin{aligned}
x &= \frac{F_o}{k\omega_n t_1}\left(-\sin(\omega_n t) + \omega_n t_1 + \sin(\omega_n t - \omega_n t_1)\right) \\
&= \frac{F_o}{k}\left(1 + \frac{\sin(\omega_n t - \omega_n t_1) - \sin(\omega_n t)}{\omega_n t_1}\right) \\
&= \frac{F_o}{k}\left(1 + \frac{(\cos\omega_n t_1 - 1)\sin(\omega_n t) - (\sin\omega_n t_1)\cos\omega_n t}{\omega_n t_1}\right) \\
&= \frac{F_o}{k}\left(1 + \frac{\sqrt{(\cos\omega_n t_1 - 1)^2 + (\sin\omega_n t_1)^2}}{\omega_n t_1}\sin(\omega_n t + \theta)\right)
\end{aligned}$$

Therefore, the maximum late-response value of the undamped system is

$$\begin{aligned}
x_{\max} &= \frac{F_o}{k}\left(1 + \frac{\sqrt{(\cos\omega_n t_1 - 1)^2 + (\sin\omega_n t_1)^2}}{\omega_n t_1}\right) \\
&= \frac{F_o}{k}\left(1 + \frac{\sqrt{\cos^2\omega_n t_1 - 2\cos\omega_n t_1 + 1 + \sin^2\omega_n t_1}}{\omega_n t_1}\right) \\
&= \frac{F_o}{k}\left(1 + \frac{\sqrt{1 - \cos^2\omega_n t_1}}{\omega_n t_1}\right) \\
&= \frac{F_o}{k}\left(1 + \frac{2\sqrt{(2 - 2\cos^2\omega_n t_1)/2}}{\omega_n t_1}\right) \\
&= \frac{F_o}{k}\left(1 + \frac{2}{\omega_n t_1}\sin\frac{\omega_n t_1}{2}\right)
\end{aligned} \tag{15.17}$$

We can now plot the late-response $x_{\max}k/F_o$ against $\omega_n t_1$. Of the two curves, early and late response, whichever one is higher is the one that represents the true $x_{\max}$.

Exercise 15.4 *For what values of $\omega_n t_1$ is the $x_{\max}$ of a ramped step within 5% of a square step of the same magnitude F_o?*

15.2 Similitude

15.2.1 Background

There are considerable advantages to setting problems up in terms of dimensionless numbers. Fractional analysis (Sections 4.3 and 9.3) gave us the dimensionless time coordinate $\omega_n t$ and the dimensionless damping ratio ζ. Superposition in linear systems (Chapters 11 to 13) has lead us to dimensionless output/input ratios like $x/(F_o/k)$ (for force excitation) or x/Y_o (for base excitation) or $x/(em/M)$ (for mass excitation).

If we did not recognize these dimensionless ratios, we would believe that an initial velocity v_o has a response which is a function of five parameters, and a maximum response which is a function of four parameters

$$x_h = \mathfrak{F}(t, m, c, k, v_o) = v_o e^{\frac{-cg_c}{2m}t} \sin\left(\frac{t\sqrt{4kmg_c - c^2 g_c^2}}{2m}\right) \quad (15.18)$$

$$x_{\max} = \mathfrak{F}(m, c, k, v_o) \quad (15.19)$$

but recognizing the dimensionless groups, we can write functions of three fewer parameters

$$\pi_{1a} = \mathfrak{F}(\pi_{2a}, \pi_3) = e^{-\pi_3 \pi_{2a}} \sin\left(\pi_{2a}\sqrt{1-\pi_3^2}\right) \quad (15.20)$$

$$\pi_{1b} = \mathfrak{F}(\pi_3) \quad (15.21)$$

where

$$\pi_{1a} \triangleq \frac{x_h}{v_o}\sqrt{\frac{m}{kg_c}} \quad (15.22)$$

$$\pi_{1b} \triangleq \frac{x_{\max}}{v_o}\sqrt{\frac{m}{kg_c}} \quad (15.23)$$

$$\pi_{2a} \equiv \omega_n t \triangleq t\sqrt{\frac{kg_c}{m}} \quad (15.24)$$

$$\pi_3 \equiv \zeta \triangleq \frac{c}{2\sqrt{km/g_c}} \quad (15.25)$$

If we found it difficult to obtain $x_{\max}$ mathematically, we could obtain it experimentally, inducing initial velocities in a model and measuring the maximum deflection. Were we to depend on the raw Equation 15.19 as our guide, we might conduct experiments for ten values of m, times ten values of c, times ten values of k, times ten values of v_o, a total of 10,000 tests, to get reliable information about $x_{\max}$. With luck, we might recognize that the results could plot into a single curve, if we picked the right coordinates. But using the dimensionless Equation 15.21, we immediately recognize that it should be sufficient to vary only m, or only c, or only k, in order to investigate ten values of $\pi_3 = \zeta$, for a total of only 10 tests, to obtain the plot of π_{1b} versus π_3.

The underlying reason for this simplification is that, from the governing Equation 9.11 of the system, we recognized that there is "similitude" between all systems that have the same value of the dimensionless parameter ζ. This concept of similitude was developed in the study of fluid dynamics, but is useful in all branches of engineering.[1]

Exercise 15.5 *Find a discussion of the Buckingham pi-theorem in a fluid-dynamics text, and determine why our examples all have three fewer dimensionless parameters than dimensional parameters.*

15.2.2 Response Spectra

The response to a single impulse, a single step, or a single ramp, like the response to initial conditions, is particularly simple because the excitation is a singularity. As a result, our normalized maximum response to these excitations is a function only of ζ.

Exercise 15.6 *Show that the maximum response to an initial velocity v_o is*

$$\frac{x_{\max}\omega_n}{v_o} = e^{\left(\frac{-\zeta}{\sqrt{1-\zeta^2}} \arccos \zeta\right)}$$

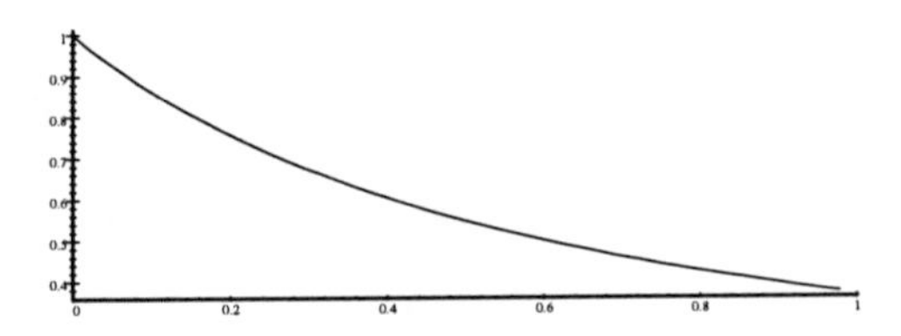

Is this expression valid for $\zeta = 0$? For $\zeta = 1$? For $\zeta > 1$?

Exercise 15.7 *Find the maximum of the response to a single impulse, Equation 14.1.*

Exercise 15.8 *Find the maximum of the response to a single step, Equation 14.7.*

Any extended excitation, such as the steady-state sinusoidal excitation of Chapter 11, has a time-scale, such as the period $\tau_{ex} = 2\pi/\omega_{ex}$. The relationship of that excitation to the time-scale of the system affects the response; therefore,

[1] Stephen J. Kline, *Similitude and Approximation Theory,* reissued 1986, Springer-Verlag, Berlin and Heidelberg, ISBN 0-38716518-5.

we must include a parameter π_{2b} such as τ_{ex}/τ_n, or its inverse, ω_{ex}/ω_n, and we find relationships involving this additional parameter:

$$\begin{aligned}\frac{x_{ss}k}{F_{eq}} &= \mathfrak{F}\left(\omega_n t, \frac{\omega_{ex}}{\omega_n}, \varsigma\right) \\ \frac{X_{\max}k}{F_{eq}} &= \mathfrak{F}\left(\frac{\omega_{ex}}{\omega_n}, \varsigma\right)\end{aligned}$$

and we plotted the maximum $X/(F_o/k)$ or X/Y_o or $X/(em/M)$, against the abscissa ω_{ex}/ω_n, with ζ as the family-of-curves independent variable. If we were particularly interested in some other kind of periodic excitation, such as square or saw-tooth waves, we could develop similar $X_{\max}$ plots for them.

Extended transients also have a characteristic time scale, represented in our examples—doublets, pulses, and ramped steps—by the interval t_1. Therefore we introduced the time-constant ratio $2\pi t_1/\tau_n = \omega_n t_1$, and expressed the response as

$$\begin{aligned}\left(\frac{x_{\max}k}{F_o}\right) \text{ or } \left(\frac{x_{\max}}{Y_o}\right) &= \mathfrak{F}(\omega_n t, \omega_n t_1, \varsigma) \\ \left|\frac{x_{\max}k}{F_o}\right| \text{ or } \left|\frac{x_{\max}}{Y_o}\right| &= \mathfrak{F}(\omega_n t_1, \varsigma)\end{aligned}$$

and plot the ordinate $x_{\max}k/F_o$ against the abscissa $\omega_n t_1$, with ζ as the second independent variable.

15.2.3 Generalization

Starting from our examples—doublets and pulses—we can generalize that any particular shape of transient force function applied to a undamped or damped system, leads to a universally applicable response *for that particular shape of excitation function*

$$\left|\frac{xk}{F_{\max}}\right| = \mathfrak{F}(\omega_n t, \omega_n t_{ex}, \varsigma)$$

where the force and time scales are given by $F_{\max}$, the magnitude of the excitation, and t_{ex}, the duration of the excitation.

Continuing with this thought, we can generalize that any particular shape of transient force function applied to a undamped or damped system, leads to a universally applicable *maximum* response for that particular shape of excitation function, or "shock,"

$$\left|\frac{x_{\max}k}{F_{\max}}\right| = \mathfrak{F}(\omega_n t_{ex}, \varsigma)$$

which can be graphed once and for all as a single plot (for an undamped system) or one family of curves (for damped systems).

There are many particular shapes of excitation functions that can become of interest—a typical earthquake, the thrust from a typical rocket "burn," the impact from a particular tool—and, if we can obtain the spectral response for each shape, we can then apply it to other situations involving the same shape of shock.

15.3 Closure

Given the effort needed to obtain solutions for complex transients, we have introduced similitude in order to apply any solutions we obtain to as wide a range of problems as possible. In Chapter 16, we will explore additional methods for obtaining solutions.

Problem 15.9 *A large mass M is initially at rest, supported by a spring k. A small mass m falls on it from a height h and bounces off elastically. What is the maximum deflection of the large mass from its initial equilibrium position?*

Problem 15.10 *Repeat the problem for two masses, released time t_1 apart.*

Problem 15.11 *A large mass M is initially at rest, supported by a spring k. A small mass m falls on it from a height h and sticks. What is the maximum deflection of the large mass from its initial position?*

Problem 15.12 *A large mass M is initially at rest, supported by a spring k an a damper c. A small mass m falls on it from a height h and bounces off elastically. What is the maximum deflection of the large mass from its initial equilibrium position?*

Problem 15.13 *Find the response of an undamped system ω_n to a half-sine pulse of duration π/ω_{ex} for at least two different values of ω_n/ω_{ex} and examine the response to find the maximum excursion $X_{\max}$.*

Chapter 16

TRANSIENTS by simulation

The superposition methods of Chapter 14, whether using the summation of responses of several non-infinitesimal impulses, steps, and ramps, or a convolution integral of an excitation function, require extensive mathematical effort. Let us examine some of the alternatives.

16.1 Scale Models

If we want to determine the response of a building to the lateral ground-motion of earthquakes, we can build a model of the building, mount it on a movable plate, and shake it laterally with a cam-driven mechanism. We can cut the cam to have the same motion as some well-documented earthquake—suitable seismographic records have been available since the Long Beach earthquake of 1933—in order to determine the structure's resistance to a "typical" earthquake.

The model can be of any size and stiffness, as long as its period has the same relationship to the duration of the excitation, as the period of the prototype has to the duration of the earthquake. In other words, if the model has a higher natural frequency than the prototype, we need to run the excitation cam proportionately faster in order to keep the value of a parameter $\pi_{2b} \triangleq \omega_n t_{ex}$ in the model the same as in the prototype.

If we want to investigate the effect of making the structure either stiffer or else more compliant, we need not build another model; all we need to is run the excitation cam slower or faster, because the value of ω_n does not matter, only the dimensionless number $\omega_n t_{ex}$.

The amplitude of the cam's motion, relative to the earthquake's amplitude, must be known, but need not be adjustable: If the system is fundamentally linear, we interpret the output in terms of the output/input ratio $\pi_1 \triangleq x/y_{\text{base}}$

$$\frac{x_{\text{prototype}}}{y_{\text{earthquake}}} = \frac{x_{\text{model}}}{y_{\text{cam}}}$$

If the damping in the prototype is believed to have an influence, the we must adjust the corresponding similarity parameter $\zeta_{\text{model}} = \zeta_{\text{prototype}}$. If the damping c, like the stiffness k, is inherent in the elastic walls of the model structure, the mass m of the roof can be changed in order to match the dimensionless damping ζ to that of the prototype. This will also change the natural frequency ω_n in the model, but the similitude match of $\omega_n t_{ex}$ with the prototype can be restored by changing the cam speed and therefore t_{ex}.

It is important to differentiate between *similitude* and *extrapolation*. If the governing equation has been correctly identified, and if all the pertinent dimensionless parameters have been matched, similitude is achieved and the use of a model does not entail any approximations. On the other hand, if all the models have different values of ζ than the prototype, and the response at the ζ of the prototype is predicted by interpolation or extrapolation, then we cannot be certain that our results are applicable to the prototype, especially if we extrapolate very far from the measured values of ζ.

16.2 Analogs

We can avoid cutting a mechanical cam for every earthquake we want to try on our structure, by substituting an electromagnetic shaker, which is driven by a moving electrical coil in a magnetic field, similar to the voice coil in a speaker system (but much more robust). Instead of changing cams, we only need to change the analog or digital input actuating the shaker.

We can even avoid building a mechanical model of the structure: in Section 9.2 we noted that electrical circuits can have the same governing equation as our mechanical systems. Therefore we can simulate the response to any given shock by applying an electrical input to an appropriate passive circuit, and recording the response. To apply the analogy, we compare the appropriate output/input ratios or model and prototype, and make sure we match the time scaling and the independent parameter ζ.

However, if we have identified the analogy by comparing governing equations, we can also go directly to a mathematical model on a computer.

16.3 Numerical Simulation

To solve for transients by means of digital computers, it is possible to carry out convolution integration numerically. However, it is more efficient to use a numerical simulation method which directly solves the governing equation

$$\ddot{x} + (2\zeta\omega_n)\,\dot{x} + \left(\omega_n^2\right)x = \left(\frac{F(t)}{k}\right)\left(\omega_n^2\right) \tag{16.1}$$

similar to Equation 11.4 but not restricted to a periodic function. We will demonstrate a Finite-Difference approach.

16.3.1 Central-Difference Method

Instead of describing functions in a continuous way, digital computers treat them as lists of numbers representing the amplitude at discrete time intervals. This constitutes an approximate representation; on the one hand, the function is known only at specific times; on the other hand, it is known only to the accuracy of the number of digits recorded. This is not an insurmountable limitation, because the function can be represented at time intervals Δt as short as needed, and with as many significant digits as needed for accuracy.

If we number the input and output data points with subscripts, designated as F_n and x_n, we can substitute the following discrete expressions for the functions $F(t)$ and x near the point n

$$\begin{aligned} F &\cong F_n \\ x &\cong x_n \\ \dot{x} &\cong \frac{x_{n+1} - x_{n-1}}{2\Delta t} \\ \ddot{x} &\cong \frac{\frac{x_{n+1}-x_n}{\Delta t} - \frac{x_n - x_{n-1}}{\Delta t}}{\Delta t} \end{aligned} \tag{16.2}$$

Substituting this into our governing Equation 16.1 for one-degree-of-freedom systems

$$\left(\frac{x_{n+1} - 2x_n + x_{n-1}}{(\Delta t)^2}\right) + 2\zeta\omega_n\left(\frac{x_{n+1} - x_{n-1}}{2\Delta t}\right) + \omega_n^2\,(x_n) \cong \left(\frac{F_n}{k}\right)\omega_n^2$$

which can be solved for the last output data point x_{n+1}

$$\begin{aligned} x_{n+1} &\cong \frac{2 - (\omega_n\Delta t)^2}{1 + \zeta\,(\omega_n\Delta t)}x_n + \frac{-\left(1 - \zeta\,(\omega_n\Delta t)\right)}{1 + \zeta\,(\omega_n\Delta t)}x_{n-1} + \frac{(\omega_n\Delta t)^2}{\left(1 + \zeta\,(\omega_n\Delta t)\right)}\frac{F_n}{k} \\ &\approx \left(2 - (\omega_n\Delta t)^2\right)x_n + (-1)x_{n-1} + \left(\frac{(\omega_n\Delta t)^2}{k}\right)F_n \end{aligned} \tag{16.3}$$

This is a simple algorithm of the form

$$x_{new} = (A)\,x_{old} + (B)\,x_{older} + (C)\,F_{old} \tag{16.4}$$

where A, B, and C are constants that can be determined from ζ and ω_n, once Δt has been chosen, by comparing Equations 16.3 and 16.4. If the time intervals Δt are constant, we can program the algorithm on a computer by taking the following steps:

1. Calculate the constants A, B, and C in the algorithm from the system parameters and the time interval Δt, by comparing the damped or undamped Equation 16.3 with Equation 16.4.

2. Enter two beginning values x_{old} and x_{older} which fit the Initial Conditions; set both values to zero if you want to start at zero initial displacement and velocity.

3. Start the procedure:

   - Read-in an F_{old}
   - Obtain $x_{new} = A \cdot x_{old} + B \cdot x_{older} + C \cdot F_{old}$
   - Print-out x_{new}
   - Replace $x_{older} = x_{old}$
   - Replace $x_{old} = x_{new}$
   - Return to start of procedure

   This can be implemented using a choice of recursive procedures and/or arrays of variables. Also, procedures for counting the numbers of repetitions and for tracking the time variable (the number of repetitions times Δt) should be added to the program.

4. Stop the procedure after recording at least one full cycle of zero input.

16.3.2 Stability

The stability of this procedure depends on the choice of Δt. If $\omega_n \Delta t > \sqrt{2}$, or if $\zeta\,(\omega_n \Delta t) > 1$, the numerical model diverges even though the physical prototype does not. We can demonstrate this by carrying out a few repetitions of the calculation. This kind of instability is common in explicit numerical simulations of vibration problems.

Exercise 16.1 *Starting with the Initial Conditions $x_{older} = 0$ and $x_{old} = 1/\omega_n \Delta t$, letting damping $\zeta = 0$, and keeping all excitation $F = 0$, compute on a hand calculator the vibration for four choices of Δt: $(\omega_n \Delta t) = 2.0$, 1.4142, 1.0, and 0.1, and plot them. Continue each calculation until the output value x crosses the axis at least three times. For each choice of Δt, does the period correspond to the expected value of $(\omega_n t_n) = 2\pi$ for free vibrations? Do successive amplitudes stay constant as expected for undamped vibrations?*

16.3.3 Accuracy

The accuracy also depends on the choice of single or double accuracy, and on the choice of Δt; to obtain good precision, we should make $\omega_n \Delta t \ll \sqrt{2}$, and $\zeta(\omega_n \Delta t) \ll 1$.

Exercise 16.2 *Write a simple program to compute the response of a system with* $\zeta = 0.10$ *to a unit impulse. Start with the Initial Conditions* $x_{older} = 0$ *and* $x_{old} = 0$*, and keep all excitation* $F = 0$*, except for* $F = 1/\Delta t$ *on the first time-step. Try four choices of* Δt*:* $(\omega_n \Delta t) = 0.3,\ 0.1,$ *and* 0.03*, and plot the results for at least two complete cycles. How close is the observed logarithmic decrement to the expected value? How close is the observed period to the expected value?*

Exercise 16.3 *Estimate the order-of-magnitude of the error of this procedure.*

16.3.4 Initial Conditions

Because any error in the first step stays with us throughout the calculation, it is a good idea to pay special attention to the first two values of x. If time is supposed to start with $t = 0$ the first x_{old}, then that first $x_{old} = x_{(0)}$, but a fictitious preceding x_{older} must be generated to give us the right slope $\dot{x}_{(0)}$ at $t = 0$. the simplest approach is to write $x_{older} \cong x_{old} - \dot{x}_{(0)}\,\Delta t$, even though that imposes the correct slope about one-half time step Δt too early, and proceed to compute a rough **prediction** of x_{new} from the fundamental algorithm above. At this point, we can have second thoughts, recalculate $x_{older} \cong x_{new} - 2\,\dot{x}_{(0)}\,\Delta t$, which gets the correct slope $\dot{x}_{(0)}$ closer to $t = 0$, and repeat the computation of x_{new} from the basic algorithm. We can repeat this **correction** procedure until the values no longer change when we recalculate them and we are satisfied that these first few points are as accurate as they need to be.

This **iteration** procedure is not efficient, but gives us a simple way of converting the Initial Conditions to the best possible starting values for a given choice of Δt. For the bulk of the calculations, it is computationally more efficient to stick to our simple algorithm, choosing a sufficiently small Δt to achieve the needed precision.

16.3.5 Other Finite-Difference Methods

We have written a simple central-difference algorithm in order to demonstrate the concepts of Finite-Difference simulation, recursive procedures, iteration, stability, and accuracy. There are more sophisticated numerical-simulation procedures, which often have computational advantages, available in application programs like *Matlab* and *Maple* (Section 1.5).

16.4 Analog Computers

It is also possible to construct components which add, integrate, or differentiate either mechanical rotation or else electrical voltage, and connect them to simulate differential equations. In the early 1940s, the Institute of Technology at Darmstadt (Germany) operated several very precise mechanical integrators, geared together with differentials, for that purpose. However, electrical systems are much easier to construct.

Exercise 16.4 *Prepare a survey of mechanical methods for obtaining integrals.*

Unlike the passive circuits mentioned in Section 16.2, electrical analog computers employ active circuits in the form of amplifiers which increase the input voltage by a very large ratio A, and also reverse its polarity for convenience in adding negative-feedback loops. Such operational amplifiers can be used to construct a variety of components:

- With a feed-back resistor and one or more input resistors, they reverse the sign of an electrical input, apply a constant scaling factor, and/or add several inputs together.
- With a feed-back capacitor and one or more input resistors, they reverse the sign and integrate the input(s).
- With a feed-back coil and an input resistor (or a feed-back resistor and an input capacitor), they reverse the sign and differentiate the input(s).

This is enough to model a linear differential equation (non-linear elements require more complex circuitry). The required hook-up of the components is similar to a signal-flow chart. The desired input is applied by a voltage generator, and the output voltages recorded or observed on an oscilloscope.

In general, analog computers use integrators rather than differentiators:

Differentiators emphasize higher frequencies (by the multiplier ω), increasing the high-frequency random noise present in all electrical circuits.

Integrators tend to reduce high-frequency noise. However, they are subject to low-frequency drift, because even small input bias will gradually accumulate a significant output.

For transient simulations, drift is controlled by limiting the duration of operation; for steady-state simulations, it is controlled by filtering-out D.C.

Analog computers depend on precision components: high accuracy requires expensive components. As a result, they have been largely replaced by mass-produced digital computers, except for educational demonstrations. To replicate the signal-flow hook-up, digital-computer programs are available which permit the combination of subprograms for integration and addition in a similarly transparent way.

In general, digital integration is preferred to digital differentiation:

Differentiation causes the round-off errors in the input data to show up as spurious spikes in the output.

Integration tends to smooth out spikes. However, the round-off errors in the input can lead to low-frequency drift.

For transient analysis, low-frequency drift presents less of a problem than do high-frequency spikes.

Problem 16.5 *Write a simple program to find the response of a damped system to a single impulse, for several values of* ζ. *Identify the maximum excursion for each value of damping, and plot* $x_{\max}$ *versus* ζ. *You may do this by hand, or use a programming language like Basic, Fortran, Pascal, C, or a spreadsheet (but not a packaged differential-equation solver). Calculate at least two full cycles of output.*

Problem 16.6 *Write a simple program to find the response of an undamped system to a square pulse, for several different values of* $\omega_n t_1$. *Identify the maximum excursion for each pulse duration (or for each value of natural frequency), and plot* $x_{\max}$ *versus* $\omega_n t_1$.

Problem 16.7 *Write a simple program to find the response of a damped system to a square pulse, for one value of* ζ. *Identify the maximum excursion for each value of* $\omega_n t_1$, *and* $x_{\max}$ *versus* $\omega_n t_1$.

Problem 16.8 *Plot the maximum response of an undamped system to a "double hammer," page 139, as a function of the time interval between the two blows.*

Problem 16.9 *What is the maximum response of an undamped system to a pair of opposite square pulses, consisting of a step* F_o *at* $t = 0$, *a step* $-2F_o$ *at* t_1, *and a final step* F_o *at* t_1?

Problem 16.10 *Write a simple program to find the response of an undamped system* ω_n *to a half-sine pulse of duration* π/ω_{ex}. *Run it for at least two different values of* ω_n/ω_{ex} *and examine the response to find the maximum excursion* $X_{\max}$. *Enter the resulting (two) points on a plot of* $X_{\max}/(F_{\max}/k)$ *versus* ω_n/ω_{ex} *on an inch scale. Repeat for a damped system* $\zeta = 0.25$. *You may do this by hand, or use a programming language like Basic, Fortran, Pascal, C, or a spreadsheet (but not a packaged differential-equation solver).*

Problem 16.11 *You just wrote a simple program, using the central-difference method, to find the response of an undamped system to a half-sine pulse of force. If the pulse duration is* $t_1 = 0.45$ *sec, and the system parameters are* $m = 5.7$ *kg and* $k = 8107$ *N/m, what is the maximum usable time-step* Δt?

Problem 16.12 *Use the differential-equation solver in an application program (Section 1.5) to find the response of a damped system to a single impulse, for several values of* ζ. *Identify the maximum excursion for each value of damping, and plot* $x_{\max}$ *versus* ζ.

Problem 16.13 *Use the differential-equation solver in an application program to find the response of an undamped system to a square pulse, for several different values of* $\omega_n t_1$. *Identify the maximum excursion for each pulse duration (or for each value of natural frequency), and plot* $x_{\max}$ *versus* $\omega_n t_1$.

Problem 16.14 *Use the differential-equation solver in an application program to find the response of a damped system to a square pulse, for one value of* ζ. *Identify the maximum excursion for each value of* $\omega_n t_1$, *and* $x_{\max}$ *versus* $\omega_n t_1$.

Problem 16.15 *Use the differential-equation solver in an application program to find the maximum response of an undamped system to a "double hammer," page 139, as a function of the time interval between the two blows.*

Problem 16.16 *Use the differential-equation solver in an application program to find the maximum response of an undamped system to a pair of opposite square pulses, consisting of a step* F_o *at* $t = 0$, *a step* $-2F_o$ *at* t_1, *and a final step* F_o *at* t_1?

Problem 16.17 *Use the differential-equation solver in an application program to find the response of an undamped system* ω_n *to a half-sine pulse of duration* π/ω_{ex}. *Run it for at least two different values of* ω_n/ω_{ex} *and examine the response to find the maximum excursion* $X_{\max}$. *Enter the resulting (two) points on a plot of* $X_{\max}/(F_{\max}/k)$ *versus* ω_n/ω_{ex} *on an inch scale. Repeat for a damped system* $\zeta = 0.25$.

Chapter 17

TRANSIENTS by integral transforms

Transients in linear systems are routinely solved by transform methods. Laplace transforms are formulated around the expected exponential-form solutions of linear systems with constant coefficients.

17.1 Laplace Transforms

The Laplace transform of a function f can be defined by the integral

$$\mathcal{L}\left(f_{(t)}\right) \triangleq \int_0^\infty f_{(t)} e^{-st} dt = \mathcal{F}(s) \tag{17.1}$$

While f is a function of time, t is the dummy variable in the integral and disappears, so that the Laplace transform is a function of the complex variable s. From our experience with exponential and trigonometric solutions, we anticipate that s is related to the frequency domain and is equivalent to $i\omega$.

The Laplace transform has been especially formulated for transient analysis: the lower limit of the integration is zero, because it is assumed that there is no excitation or response before $t = 0$. Therefore, a constant unit value $f_{(t)} = 1$ actually represents a unit step at $t = 0$; carrying out the integration, we see that the corresponding transform is $\mathcal{F}(s) = 1/s$. (Excitation with a later start, at $t = t_1$, can be obtained by multiplying the transform by $e^{-t_1 s}$.)

Laplace transforms avoid the need for convolution integration, because the product of the transforms of two functions is the transform of the convolution integral of the two functions

$$\mathcal{L}\left(\int_0^t F(\xi) \cdot h_{imp}(t-\xi) d\xi\right) = \mathcal{L}\left(F(t)\right) \times \mathcal{L}\left(h_{imp}(t)\right) = \mathcal{F}(s)\,\mathcal{H}(s)$$

The transforms for commonly occurring functions can be evaluated once-and-for-all and tabulated; for example

$f(t)$	$\mathcal{F}(s)$
$\delta_{(0+)}$	1
1	$\frac{1}{s}$
t	$\frac{1}{s^2}$
$e^{-\alpha t}$	$\frac{1}{s+\alpha}$
$\sin \omega t$	$\frac{\omega}{s^2+\omega^2}$
$\cos \omega t$	$\frac{s}{s^2+\omega^2}$
$e^{-\alpha t} \sin \omega t$	$\frac{\omega}{(s+\alpha)^2+\omega^2}$

Mathematical handbooks list additional function pairs. We can also handle linear combinations of tabulated functions: multiplying any of these $f_{(t)}$ by a constant factor leads to $\mathcal{F}(s)$ being multiplied by the same constant factor; and adding any two $f_{(t)}$ leads to the corresponding two $\mathcal{F}(s)$ being added.

Processing derivative like df/dt through the integral gives us additional relationships

$$\begin{aligned} \mathcal{L}\left(\frac{df}{dt}\right) &= s\mathcal{F}(s) - f_{(t=0)} \qquad (17.2) \\ \mathcal{L}\left(\frac{d^2f}{dt^2}\right) &= s^2\mathcal{F}(s) - sf_{(t=0)} - \left[\frac{df}{dt}\right]_{t=0} \end{aligned}$$

Example: These relationships allow us to transform our Standard-Form equations from the time-domain

$$\frac{d^2x}{dt^2} + 2\zeta\omega_n\frac{dx}{dt} + \omega_n^2 x_{(t)} = \left(\frac{F(t)}{k}\right)\omega_n^2$$

to the frequency domain

$$s^2\mathcal{X}_{(s)} - sx_{(0)} - \dot{x}_{(0)} + 2\zeta\omega_n\left(s\mathcal{X}_{(s)} - x_{(0)}\right) + \omega_n^2\mathcal{X}_{(s)} = \left(\frac{\mathcal{F}(s)}{k}\right)\omega_n^2$$

which can be solved for $\mathcal{X}_{(s)}$

$$\begin{aligned} \mathcal{X}_{(s)} &= \left[\left(\frac{\omega_n}{k}\right)\frac{\omega_n}{s^2 + (2\zeta\omega_n)s + (\omega_n^2)}\right]\mathcal{F}(s) + \frac{(s + 2\zeta\omega_n)x_{(0)} + \dot{x}_{(0)}}{s^2 + (2\zeta\omega_n)s + (\omega_n^2)} \\ &\approx \left[\left(\frac{\omega_n}{k}\right)\frac{\omega_n}{s^2 + \omega_n^2}\right]\mathcal{F}(s) + \frac{sx_{(t=0)} + \dot{x}_{(0)}}{s^2 + \omega_n^2} \qquad (17.3) \end{aligned}$$

The expression in the big bracket is the "transfer function" $\mathcal{H}(s)$ from the input $\mathcal{F}(s)$ to the output $\mathcal{X}_{(s)}$, and the right-most term is the initial-condition solution. For the zero-initial-state solution we can write

$$\begin{aligned} \mathcal{X}_{\text{zis}} &= \mathcal{H}(s)\,\mathcal{F}(s) \qquad (17.4) \\ \text{where } \mathcal{H}(s) &= \left[\left(\frac{\omega_n}{k}\right)\frac{\omega_n}{s^2+(2\zeta\omega_n)\,s+(\omega_n^2)}\right] \approx \left[\left(\frac{\omega_n}{k}\right)\frac{\omega_n}{s^2+\omega_n^2}\right] \end{aligned}$$

which we recognize as the transform of the impulse response function h_{imp} from Equation 14.4.

17.1.1 Procedure

The analysis of transients in linear systems can be now be broken down into routine steps:

1. Using the rules of Laplace transforms, such as Equation 17.2, we transform the system equation from the time domain into the frequency domain. For zero-initial-state solutions, $x_{(0)}$ and $\dot{x}_{(0)}$ are zero, and the left-hand side of the governing Equation 2.6 for one-degree-of-freedom systems reduces to $(s^2+2\zeta\omega_n s+\omega_n^2)\mathcal{L}(x)$. Within the parentheses, we recognize the characteristic Equation 9.13 of the system.

2. Using Laplace transform tables, which cover many common functions, we transform the excitation function $f(t)$ on the right-hand side of the governing equation to $\mathcal{F}(s)$.

3. The resulting algebraic equation is solved for the transform of the response transform of the response $\mathcal{L}(x)$. We recognize that $\omega_n^2/(s^2+2\zeta\omega_n s+\omega_n^2)k$ is the Laplace transform of h_{imp}. Therefore, convolution integration has been replaced by $\mathcal{L}(x) = \mathcal{L}(F) \times \mathcal{L}(h_{imp})$.

4. The resulting transform is generally in the form of a ratio of polynomials of s. If the numerator is of equal (or greater) order than the denominator, it is necessary to perform a long division to expand the transform into impulses (and other singularities), plus a proper fraction having a lower-order polynomial in the numerator.

5. In order to obtain expressions which can be found in Laplace transform tables, it is necessary to factor the denominator, and expand the proper fraction into a summation of partial fractions. (In theoretical problems, there may be exactly repeated roots, which require special treatment.) Computer application programs like *Matlab* can assist in this process.

6. For the inverse transformation back into the time domain, we look for each partial fraction in the Laplace transform tables.

Exercise 17.1 *Obtain the homogeneous solution for undamped one-degree-of-freedom systems by means of Laplace transforms:*

$$\mathcal{L}(x_h) = \frac{\dot{x}_{(0)}}{\omega_n} \cdot \frac{\omega_n}{s^2+\omega_n^2} + x_{(0)} \cdot \frac{s}{s^2+\omega_n^2}$$

Compare with Equations 4.19 to 4.21.

Exercise 17.2 *Obtain the homogeneous solutions for damped one-degree-of-freedom systems by means of Laplace transforms:*

$$\begin{aligned}
\mathcal{L}(x) &= \frac{x_{(0)} \cdot s + \dot{x}_{(0)} + 2\zeta\omega_n x_{(0)}}{s^2 + 2\zeta\omega_n s + \omega_n^2} = \frac{x_{(0)} \cdot s + \dot{x}_{(0)} + 2\zeta\omega_n x_{(0)}}{(s+\zeta\omega_n)^2 + (1-\zeta^2)\,\omega_n^2} \\
&= \frac{\dot{x}_{(0)} + \zeta\omega_n x_{(0)}}{\sqrt{1-\zeta^2}\omega_n} \cdot \frac{\sqrt{1-\zeta^2}\omega_n}{(s+\zeta\omega_n)^2 + (1-\zeta^2)\,\omega_n^2} \\
&\quad + x_{(0)} \cdot \frac{s+\zeta\omega_n}{(s+\zeta\omega_n)^2 + (1-\zeta^2)\,\omega_n^2}
\end{aligned}$$

Compare with Equations 9.18 and 9.19.

Exercise 17.3 *Obtain the response to a unit impulse by means of Laplace transforms:*

$$\begin{aligned}
\mathcal{L}(x_{zis}) &= \frac{\omega_n}{k} \cdot \frac{\omega_n}{s^2 + 2\zeta\omega_n s + \omega_n^2} \mathcal{F}(s) \\
&= \frac{\omega_n}{k\sqrt{1-\zeta^2}} \cdot \frac{\sqrt{1-\zeta^2}\omega_n}{(s+\zeta\omega_n)^2 + (1-\zeta^2)\,\omega_n^2} \mathcal{F}(s) \approx \frac{\omega_n}{k} \cdot \frac{\omega_n}{s^2+\omega_n^2} \mathcal{F}(s)
\end{aligned}$$

Compare with Equation 14.4.

Exercise 17.4 *Obtain the zero-initial-state solution to an harmonic force excitation by means of Laplace transforms. Identify the steady-state response terms.*

$$\begin{aligned}
\mathcal{L}(x_{zis}) &= \frac{F_o\omega_{ex}}{s^2+\omega_{ex}^2} \cdot \frac{\omega_n}{k} \cdot \frac{\omega_n}{s^2 + 2\zeta\omega_n s + \omega_n^2} \\
&= \frac{As+B}{s^2+\omega_{ex}^2} + \frac{Cs+D}{(s+\zeta\omega_n)^2 + (1-\zeta^2)\,\omega_n^2} \\
\mathcal{L}(x_{ss}) &= \frac{As+B}{s^2+\omega_{ex}^2}
\end{aligned}$$

where A, B, C, and D are obtained from partial fraction expansion; e.g., for negligible damping

$$\begin{aligned}
B &= \frac{F_o\omega_{ex}}{k} \cdot \frac{\omega_n^2}{\omega_n^2+\omega_{ex}^2} \\
\mathcal{L}(x_{ss}) &= \frac{F_o}{k} \cdot \left(\frac{\omega_n^2}{\omega_n^2+\omega_{ex}^2}\right) \cdot \frac{\omega_{ex}}{s^2+\omega_{ex}^2}
\end{aligned}$$

Compare with Equations 7.6, 7.7, and 11.11.

17.1.2 Overview of Transfer Functions

Transfer functions for a one-degree-of-freedom system are, for force input,

$$\begin{aligned}\mathcal{L}(x) &= \frac{\omega_n^2}{s^2+2\zeta\omega_n s+\omega_n^2}\cdot\frac{\mathcal{L}(F_{ex})}{k}\\ \mathcal{L}(F_{base}) &= \frac{2\zeta\omega_n s+\omega_n^2}{s^2+2\zeta\omega_n s+\omega_n^2}\cdot\mathcal{L}(F_{ex})\end{aligned}$$

Transfer functions for base motion input

$$\begin{aligned}\mathcal{L}(x) &= \frac{2\zeta\omega_n s+\omega_n^2}{s^2+2\zeta\omega_n s+\omega_n^2}\cdot\mathcal{L}(y_{base})\\ \mathcal{L}(z_{rel}) &= \frac{-s^2}{s^2+2\zeta\omega_n s+\omega_n^2}\cdot\mathcal{L}(y_{base})\\ \mathcal{L}(z_{rel}) &= \frac{-1}{s^2+2\zeta\omega_n s+\omega_n^2}\cdot\mathcal{L}\left(\ddot{y}_{base}\right)\\ \mathcal{L}(F_{base}) &= \frac{\left(2\zeta\omega_n s+\omega_n^2\right)s^2}{\left(s^2+2\zeta\omega_n s+\omega_n^2\right)\omega_n^2}\cdot k\mathcal{L}(y_{base})\end{aligned}$$

Transfer functions for mass excitation by an inner mass m moving a distance e_{rel} within the total mass M

$$\begin{aligned}\mathcal{L}(x) &= \frac{-s^2}{s^2+2\zeta\omega_n s+\omega_n^2}\cdot\frac{m\mathcal{L}(e_{rel})}{M}\\ \mathcal{L}(z_{rel}) &= \frac{\omega_n^2}{s^2+2\zeta\omega_n s+\omega_n^2}\cdot\frac{m\mathcal{L}(e_{rel})}{M}\\ \mathcal{L}(F_{\text{base}}) &= \frac{\left(2\zeta\omega_n s+\omega_n^2\right)s^2}{\left(s^2+2\zeta\omega_n s+\omega_n^2\right)\omega_n^2}\cdot\frac{km}{M}\mathcal{L}(e_{rel})\\ \mathcal{L}(F_{\text{base}}) &= \frac{2\zeta\omega_n s+\omega_n^2}{s^2+2\zeta\omega_n s+\omega_n^2}\cdot\mathcal{L}(F_{\text{equiv}})\end{aligned}$$

Exercise 17.5 *Compare these expressions with the periodic-excitation steady-state responses in Section 13.4.*

17.1.3 s-Plane Analysis

We can learn a great deal about a system even without completing the inverse transformation. The crucial step in the inverse transformation is the factoring of the denominator, which in our problem is the characteristic-equation expression

$$\begin{aligned}
s^2 + 2\zeta\omega_n + \omega_n^2 &= (s+\zeta\omega_n)^2 + \left(1-\zeta^2\right)\omega_n^2 \\
&= \left(s+\zeta\omega_n - i\sqrt{1-\zeta^2}\omega_n\right)\left(s+\zeta\omega_n + i\sqrt{1-\zeta^2}\omega_n\right) \\
&= (s-s_1)(s-s_1) \\
\text{where } s_{1,2} &= -\zeta\omega_n \pm i\sqrt{1-\zeta^2}\omega_n
\end{aligned} \tag{17.5}$$

The roots s_1 and s_2 of the characteristic equation are the **poles** of the function in the s-domain; at those values of s, the denominator goes to zero and the ratio of polynomials $\mathcal{L}(x)$ spikes to infinity. Recalling Section 9.4, we see five cases for the values of $s_{1,2}$:

I. If $\zeta > 1$: two separate negative real numbers;

II. If $\zeta = 1$: repeated roots at $s_{1,2} = -\omega_n$;

III. If $0 < \zeta < 1$: a conjugate pair of complex numbers, with the negative real part $-\zeta\omega_n$ representing the decaying envelope of the oscillation, and the complex parts $\pm i\sqrt{1-\zeta^2}\omega_n$ representing the oscillation at the damped frequency;

IV. If $\zeta = 0$: a pair of imaginary numbers $\pm i\omega_n$, representing the oscillation at the natural frequency;

V. If $\zeta < 0$: numbers with a positive real part, representing an unstable, divergent solution.

The roots of the polynomial in the numerator are **zeros** and represent values of s where the function $\mathcal{L}(x)$ goes to zero, potentially canceling out poles.

The general method for converting a Laplace transform back into the time domain is the contour integral of the Inverse Laplace Transform

$$\mathcal{L}^{-1}\left[\mathcal{F}(s)\right] \triangleq \frac{1}{2\pi i}\int_{c-i\infty}^{c+i\infty} \mathcal{F}(s)\, e^{st} = f_{(t)}$$

which is obviously depends on the location of the poles and zeros in the s-plane. Much of the stability analysis in linear control theory is done by studying the singularities in the s-plane. Poles in the negative-real half of the plane indicate decaying solutions; in the positive-real half, divergent solutions.

When even a simple system is multiplied by a complicated input, the output may be laborious to factor into partial fractions. However, the initial and final responses may be obtained directly from the s-multiplied transform

$$\lim_{t\to 0} f(t) = \lim_{s\to\infty} s\mathcal{F}(s) \tag{17.6}$$

$$\lim_{t\to\infty} f(t) = \lim_{s\to 0} s\mathcal{F}(s) \tag{17.7}$$

17.2 Zeta Transform

Systems governed by difference equations, such as electrical ladder-networks, can be studied by z-transforms. Finite-difference mathematical methods, and instrumentation systems which use samples taken at non-infinitesimal intervals, are also governed by finite-difference equations (Eqn. 16.3). We would like to compare them with the underlying differential equations.

The z-transform replaces the integral of Equation 17.1 with the summation

$$\mathcal{L}(f_n) \cong \sum_{n=0}^{\infty} f_{(n\Delta t)} e^{-sn\Delta t} = \mathcal{F}(n\Delta t)$$

where Δt is the sampling interval and n is the counter of the samples of the function f_n. Substituting

$$z \triangleq e^{s\Delta t} \tag{17.8}$$

we obtain

$$\mathcal{Z}(f_n) \triangleq \sum_{n=0}^{\infty} f_{(n\Delta t)} z^{-n} = \mathcal{F}(z) \tag{17.9}$$

For convenience, we rescale time by a factor of $1/\Delta t$; in our normalized time coordinate, $\Delta t = 1$ and

$$\mathcal{Z}(f_n) \triangleq \sum_{n=0}^{\infty} f_{(n)} z^{-n} = \mathcal{F}(z) = f_0 + f_1 z^{-1} + f_2 z^{-2} + f_3 z^{-3} + \ldots \tag{17.10}$$

We can obtain and tabulate common functions

$f_{(n)}$	$\mathcal{F}(z)$
1	$\frac{z}{z-1}$
n	$\frac{z}{(z-1)^2}$
a^n	$\frac{z}{z-a}$
$\sin \omega_* n$	$\frac{z \sin \omega_*}{z^2 - 2z\cos\omega_* + 1}$
$\cos \omega_* n$	$\frac{z(z - \cos\omega_*)}{z^2 - 2z\cos\omega_* + 1}$
$e^{-\alpha n} \sin \omega_* n$	$\frac{z\left(z - e^{-\alpha}\cos\omega_*\right)}{z^2 - 2z\cos\omega_* + 1}$

where, due to our replacing the time coordinate by $t/\Delta t$, $\omega_* \triangleq \omega\Delta t$. More extensive tables are found in mathematical handbooks.

Time-shifted functions in our difference equations are related by

f_n	$\mathcal{F}(z)$
f_{n+1}	$z\mathcal{F}(z) - zf_o$
f_{n+2}	$z^2\mathcal{F}(z) - z^2 f_o - zf_1$

Example: Applying this to the undamped Equation 16.3 with a unit-step excitation force

$$\begin{aligned}
x_{n+2} - \left(2 - (\omega_n \Delta t)^2\right) x_{n+1} + x_n &= \frac{(\omega_n \Delta t)^2}{k} F_{(t)} \\
z^2 \mathcal{X}(z) - z^2 x_o - z x_1 - \left(2 - \omega_{*n}^2\right)\left(z\mathcal{X}(z) - z x_o\right) + \mathcal{X}(z) &= \frac{\omega_{*n}^2}{k} \mathcal{F}(z)
\end{aligned}$$

which, for zero-initial-conditions x_o and x_1 reduces to

$$\begin{aligned}
\left(z^2 - \left(2 - \omega_{*n}^2\right) z + 1\right) \mathcal{X}(z) &= \frac{\omega_{*n}^2}{k} \mathcal{F}(z) \\
\mathcal{X}(z) &= \frac{1}{\left(z^2 - \left(2 - \omega_{*n}^2\right) z + 1\right)} \left(\frac{\omega_{*n}^2}{k}\right) \mathcal{F}(z)
\end{aligned}$$

As in the case of the Laplace transform, the output is the product of the system's transfer function and the excitation function; for example, for a unit step

$$\mathcal{X}(z) = \frac{1}{z^2 - \left(2 - \omega_{*n}^2\right) z + 1} \left(\frac{\omega_{*n}^2}{k}\right) \frac{z}{z-1}$$

On the other hand, the transform for an initial displacement x_o (and no excitation input) is

$$z^2 \mathcal{X}(z) - z^2 x_o - z x_o - \left(2 - \omega_{*n}^2\right)\left(z\mathcal{X}(z) - z x_o\right) + \mathcal{X}(z) = 0$$

$$\begin{aligned}
\left(z^2 - \left(2 - \omega_{*n}^2\right) z + 1\right) \mathcal{X}(z) &= \left(z^2 - z + \omega_{*n}^2 z\right) x_o \\
\mathcal{X}(z) &= \frac{z\left(z - \left(1 - \omega_{*n}^2\right)\right)}{z^2 - \left(2 - \omega_{*n}^2\right) z + 1} \cdot x_o
\end{aligned}$$

which reduces to one of the tabulated expressions

$$\begin{aligned}
\frac{z\left(z - e^{-\alpha} \cos \omega_*\right)}{z^2 - 2z \cos \omega_* + 1} &= \frac{z\left(z - \left(1 - \omega_{*n}^2\right)\right)}{z^2 - \left(2 - \omega_{*n}^2\right) z + 1} \\
\text{if we make } 2 \cos \omega_* &= 2 - \omega_{*n}^2 \\
\omega_* &= \arccos \frac{2 - (\omega_n \Delta t)^2}{2} \\
\text{and make } e^{-\alpha} \cos \omega_* &= \left(1 - \omega_{*n}^2\right) \\
\alpha &= \ln \frac{2 - (\omega_n \Delta t)^2}{2 - 2(\omega_n \Delta t)^2} \\
\text{therefore } x_{(n)} &= e^{-\alpha n} \sin \omega_* n
\end{aligned}$$

Exercise 17.6 *Find the transform $\mathcal{X}(z)$ for an initial velocity v_o.*

The location of the poles in the complex z-plane indicates stability: poles within a unit circle drawn around the origin, indicate decaying solutions; outside the unit circle, divergent solutions. As we have seen in Section 16.3.2, a finite-difference equation may be unstable (for some scaling factors Δt), even when the underlying differential equation is stable.

Initial and final values can be obtained without inverse transformation, directly from

$$f_{(0)} = \lim_{z \to \infty} \mathcal{F}(z) \tag{17.11}$$

$$f_{(\infty)} = \lim_{z \to 1} \left(\frac{z-1}{z} \right) \mathcal{F}(z) \tag{17.12}$$

17.3 Comparison

For linear systems, Laplace transforms offer a systematic method for studying stability and obtaining solutions. The meaning of s-domain expressions is not immediately intuitive, but acquires meaning with practice.

For periodically sampled systems, Zeta transforms offer a parallel analysis. It will permit us to study the effect of digitizing random data.

Problem 17.7 *Using Laplace transforms, obtain the zero-input response for an undamped spring–mass system with arbitrary initial conditions.*

Problem 17.8 *Using Laplace transforms, obtain the zero-input response for a damped spring–mass system with arbitrary initial conditions.*

Problem 17.9 *Using Laplace transforms, obtain the impulse response for an undamped spring–mass system.*

Problem 17.10 *Using Laplace transforms, obtain the impulse response for a damped spring–mass system.*

Problem 17.11 *Using Laplace transforms, obtain the steady-state sinusoidal-excitation response for an undamped spring–mass system.*

Problem 17.12 *Using Laplace transforms, obtain the steady-state sinusoidal-excitation response for a damped spring–mass system.*

Problem 17.13 *Using Laplace transforms, obtain the zero–initial-state sine-excitation response for a damped spring–mass system.*

Chapter 18

RANDOM VIBRATIONS and statistical concepts

Many vibrations are neither regular enough to be described as periodic phenomena, nor brief enough to be treated as deterministic transients. Examples range from the rumble of ball bearings to the hiss between stations on a radio. Such vibrations may be treated by statistical means.

18.1 Random Variables

There are at least two ways to define "randomness:"

1. Functions and processes which are so complex that we are unable to find deterministic patterns by means of the analytical tools at our disposal; or

2. Experiments which lead to varying outcomes even though we replicate all inputs which we are able to control.

The word "we" in both definitions alerts us to the fact that apparent randomness may the consequence either of our failure to comprehend the outcomes, or else of our inability to regulate all the inputs.

Randomness need not be total; there may be a mix of random and deterministic components. A random function could be

1. a sine-wave of fixed frequency, but unpredictably varying amplitude;

2. a sine-wave of fixed amplitude, but unpredictably varying period;

3. a sine-wave with random perturbations overlaid on it;

4. an essentially entirely random function, devoid of deterministic components;

or a mix of several of these. A process may have predictable and unpredictable components; an experimental measurement may be the sum of regular plus random effects.

Although a function may be random, it can often be described by statistical measures.

18.1.1 Probability

When we measure a force or displacement that fluctuates with time, we obtain a function which may be difficult to describe. The most compact assessment we can make of that function is the percentage of time that it is below a given magnitude, called the "percentile" of that magnitude. The **Probability Distribution** is the plot of these percentages against the magnitude of that force or displacement. A typical plot is an S-shaped curve starting at the left with 0% for very small magnitudes of the function, passing through 50% at the "median" of the function, and rising to 100% (or unity) for very large values of the function. The word "probability" is used because, if we sample the function at a random instant in time, the curve's ordinate shows the probability that the function's magnitude is below the value plotted on the abscissa.

Most engineers are more comfortable with plotting the derivative of this curve, called the **Probability Density** $p(x)$. A typical plot is a bell-shaped curve, starting and ending at zero for extremely small or large magnitudes distribution, and having a unit area under it. The abscissa is still the magnitude of the measured force or displacement (e.g., lbf or ft), but the ordinate now has the inverse dimension of the abscissa (e.g., 1/lbf or 1/ft).

A probability-density curve can be interpreted as a histogram, with the area of each bar indicating the fraction of time that the function is within the range indicated by the width of the bar, or the probability that a sampled function is within that range. In digital data acquisition, the possible values of the measured functions are divided into "bins" and a counter for each bin is incremented whenever a data point falls into the range of that bin. At the end of the run, the counts for each bin are normalized by the total number of data points, divided by the width of that bin's range, and plotted as probability density.

Exercise 18.1 *Sketch the probability distribution and the probability density of a function described by a simple sine-wave.*

A probability-density curve can have any shape, with any number of maxima; but the more complex a process, or the more random a function, the more likely it is that the resulting probability density is bell-shaped, with a single peak. (By analogy, the larger and more homogeneous a group of students is, and the more complex an examination, the more likely it is that the histogram of grades is shaped like a bell rather than like rabbit-ears.) We can describe such distributions with a few simple measures.

The "mean" or "D.C." value of the magnitude x is its average magnitude; it can be obtained from the first-moment integral of the probability density $p(x)$. This is equivalent to the arithmetic mean of the data points x_i from a sampled function

$$\begin{aligned}\overline{x} &= \int_{-\infty}^{\infty} xp(x)\,dx \\ &= \lim_{T\to\infty}\frac{1}{T}\int_0^T x(t)\,dt \\ &\cong \frac{1}{n}\sum_{i=1}^{n} x_i \end{aligned} \tag{18.1}$$

The mean-square value $\overline{x^2}$ and the root-mean-square value $x_{\text{rms}} \triangleq \sqrt{\overline{x^2}}$ can be obtained from a second-moment integral of $p(x)$; or by obtaining the arithmetic mean of the square of the sampled function

$$\begin{aligned}\overline{x^2} &= \int_{-\infty}^{\infty} x^2 p(x)\,dx \\ &= \lim_{T\to\infty}\frac{1}{T}\int_0^T x_{(t)}^2\,dt \\ &\cong \frac{1}{n}\sum_{i=1}^{n} x_i^2 \end{aligned} \tag{18.2}$$

We can subtract out the "D.C." component by replacing our coordinate with $\widetilde{x} = (x - \overline{x})$, shifting the origin so that the bell-shaped curve is centered on $\widetilde{x} = 0$. This allows us to focus on the random or "A.C." aspects of the function. The mean-square value relative to the mean value is called the "variance" and its square root is named "Standard Deviation" σ

$$\begin{aligned}\sigma^2 &\triangleq \overline{(x-\overline{x})^2} \\ &= \overline{x^2} - (\overline{x})^2 \end{aligned} \tag{18.3}$$

Exercise 18.2 *Derive the second equation from the first. Hint: this can be done either from the first- and second-moment definitions and Steiner's transfer theorem (Chapter 3), or by expanding the summation formulas.*

We often try to fit our experimental distributions to a Gaussian normal distribution

$$p(x) = \frac{1}{\sigma\sqrt{2\pi}} e^{-\widetilde{x}^2/2\sigma^2} \tag{18.4}$$

which is symmetrical about $\widetilde{x} = 0$. To fit this to experimental data, the width of the "bell" can be scaled by adjusting the standard deviation σ; and the location of the center of the "bell" can be shifted by adjusting $\overline{x}$ in the definition of the coordinate $\widetilde{x} \triangleq (x - \overline{x})$.

For absolute values of maximum amplitude A, the Rayleigh distribution is more suitable

$$\begin{aligned} p(A) &= \frac{A}{\rho^2} e^{-A^2/2\rho^2} \\ \overline{A} &= \rho\sqrt{\frac{\pi}{2}} \\ \sigma^2 &= \rho^2 \left(\frac{4-\pi}{2}\right) \end{aligned} \tag{18.5}$$

which is unsymmetrical and allows for only positive values of A.

In obtaining the probability density, we discard all information about smoothness or jitter in the function; but for vibration analysis, we need to keep some information about frequency or period. This can be done by means of the autocorrelation.

18.1.2 Autocorrelation

If we multiply the value of a measured function $x(t)$ with the value of the function at a different time $x(t \pm \tau)$, the product tells us something about the smoothness of the function with respect to the time-difference τ: a smooth function will tend to produce positive products, a jagged one will have negative products more often. The autocorrelation is the average of this product, as defined by

$$R(\tau) = \lim_{T\to\infty} \frac{1}{T} \int_0^T x(t) \cdot x(t+\tau)\, dt \tag{18.6}$$

For sampled data at regular intervals Δt, this becomes

$$R(n\Delta t) \cong \frac{1}{m} \sum_{i=1}^{m} x_i x_{i+n} \tag{18.7}$$

When reducing a fixed set of, say, 100 data points, $R(0)$ is the average of 100 products, $R(\Delta t)$ the average of 99 products, $R(2\Delta t)$ the average of 98 products, and so on. On the other hand, in real-time correlators the last n data-points are in storage at any one moment, and the products are continuously added into averaging registers for each $\tau = n\Delta t$.

It appears that a fixed set of 100 data points will permit us to obtain 100 points on the autocorrelation curve, which makes it seem that we have as much information in the autocorrelation curve as in the raw data. This is misleading: in the section on PSD we will learn that the autocorrelation contains only half as much information as the raw date. With increasing $\tau = n\Delta t$, we average-out fewer and fewer products, and the result is less and less meaningful. We might say that only 50 of the 100 points on our autocorrelation curve are reliable.

$R(\tau)$ is a symmetrical curve peaking at $\tau = 0$, where $R(\tau) = \overline{x^2}$. For a smoothly continuous function x, the autocorrelation falls off gradually with increasing τ; for a jagged function, it declines steeply. If there is non-random content, it can show up as waviness and even negative values in the autocorrelation: a sine-wave would show maxima in the autocorrelation at each multiple of its period, and minima at odd multiples of the half-period.

Exercise 18.3 *Sketch the autocorrelation of (a) a simple sine-wave; (b) a sine-wave of varying amplitude; (c) a sine-wave of varying frequency or phase; (d) a random process; and (e) a random process overlaid on a sine-wave.*

The autocorrelation contains information about smoothness and characteristic periods of the function, but not enough information to reconstruct the function of time.

18.1.3 Fourier Transform

A function of time can be converted into the frequency domain by the integral transformation

$$\mathcal{F}\left(x_{(t)}\right) \triangleq \int_{-\infty}^{\infty} x_{(t)} e^{-i\omega t} dt = \mathcal{X}(\omega) \tag{18.8}$$

$$= \int_{-\infty}^{\infty} x_{(t)} \cos(\omega t)\, dt - i \int_{-\infty}^{\infty} x_{(t)} \sin(\omega t)\, dt \tag{18.9}$$

$$\mathcal{F}^{-1}\left(\mathcal{X}(\omega)\right) \triangleq \int_{-\infty}^{\infty} \mathcal{X}(\omega)\, e^{it\omega} d\left(\frac{\omega}{2\pi}\right) = x_{(t)} \tag{18.10}$$

which is similar to the Laplace transformation of Equation 17.1, but we have changed the lower limit to $-\infty$ because we are now interested in steady-state processes rather than in step functions and transients. We have also substituted $i\omega$ for s because we find it easier to think in terms of frequency—this is not a substantive change, but merely a switch between the real and imaginary parts of the argument of the transform.

We can see a relationship to the Fourier series for periodic functions, which is identical to the Fourier transform with the infinite integration limits replaced by limits which are one period apart.

The Fourier transform of a function $x_{(t)}$ is a complex number which, for any angular frequency $\omega = 2\pi f$ tells cosine and sine components of that frequency within the function. Whether a wave shows up as a sine or a cosine is dependent on the arbitrary choice of the reference point $t = 0$, so it is sensible to convert the results to magnitude and phase, and plot each of those against the frequency. On a magnitude plot, a sine-wave shows up as a spike at ω_{ex}, and a more complex periodic function as a series of spikes at integral multiples of the base value of ω_{ex}; the phase-plot values corresponding to the spikes show the phase relationship between the harmonics; between the spikes the phase value is meaningless. Random noise shows up as a wide spectrum covering a whole range of frequencies.

The computational effort for obtaining the Fourier transform is much greater than for probability density and autocorrelation. However, if the data points are sampled at uniform intervals Δt, and if the number of data points is a power of 2 (for example 256 or 512 or 1024 or 2048 or 4096), the Cooley-Tukey Fast Fourier Transform (FFT) algorithm can be programmed in software or into computer logic. It is hard-wired into many laboratory instruments. As might be expected, 2^n values of the (real-number) function $x(t)$ result in half as many, 2^{n-1}, values of the (complex-number) transform $\mathcal{X}(\omega)$.

18.1.4 Power Spectral Density (PSD)

If we are only interested in the magnitude, and not the phase, of the frequency components of the frequency-domain representation, we can square the magnitude of the function—making the ordinate proportional to power—and plot it against frequency $f = \omega/2\pi$. This plot is particularly easy to interpret:

- Periodic functions show up as sharp spikes at the frequencies of the fundamental and the overtones.
- Unsteady periodic functions show up as wider peaks.
- Fully random functions show up as broad spectra:
  - a spectrum which is flat-topped (over a range of frequencies) is called "white noise" (within that frequency band) by analogy to visible-light spectra;
  - a broad spectrum which is higher at the low-frequency end is called "pink noise."

The "PSD" plot can be obtained in several different ways:

1. It is the square of the vector-magnitude of the complex-value function $\mathcal{X}(\omega)$.
2. It is the cosine Fourier transform of the autocorrelation $R(\tau)$.
3. It used to be obtained directly from $x(t)$ by analog instruments.

We see that the original function $x(t)$ and its transform $\mathcal{X}(\omega)$ both contain all the available information; we can convert back and forth using the integral transforms $\mathcal{F}$ and $\mathcal{F}^{-1}$. On the other hand, converting to autocorrelation $R(\tau)$ or to PSD, loses phase information; we can go back and forth between $R(\tau)$

and PSD, but we cannot ever get back to the original function $x(t)$ from them. The probability density contains even less (but somewhat different kinds of) information.

18.2 Ergodicity

Statistical measures like $p(x)$ or frequency decompositions like the PSD are useful only if they give consistent results. We have assumed that the functions and processes are stationary and ergodic; i.e., that we get substantially the same result from applying these measures to any adequately large sample.

Sampling creates some additional problems, which will be treated below.

18.3 A/D Conversion

When we convert a signal from analog to digital form, we discard some information. Whether we lose any important information depends on the sampling rate $1/\Delta t$.

18.3.1 Shannon Theorem

If we look at a simple sine wave, we can see that we need to sample it more than twice per cycle in order to capture it. Exactly twice-per-cycle sampling is a borderline case: we might just catch the peaks, or we might just happen onto the zero-crossings. Therefore *we must always sample at more than twice the rate as the highest frequency of interest.* Sampling at less than twice per cycle gives deceptive results, as we can see from the following example.

18.3.2 Nyquist Aliasing

Suppose we sample one-thousand times per second. If we are observing a sine-wave of 499 Hertz, the samples will have alternating amplitudes, slowly varying from nearly the full amplitude of the sine-wave to near-zero values and back, as the sampling slowly gains on the alternations of the sine-wave. It is almost as if the half-frequency of the sampling beats against the original sine-wave. This is the characteristic pattern in the samples by which we recognize a 499-Hertz wave.

Suppose we repeat the experiment with a 501-Hertz sine-wave. This time, the sampling slowly lags relative to the alternations of the sine-wave. The end result is a series of alternating amplitudes, waxing and waning in the same pattern as the 499-Hertz case. Looking at the data points, we cannot tell whether we had 499 or a 501 Hertz in our signal. Since we process the data as if all signals had a frequency of less that 500 Hertz, the 501-Hertz signal will show up under the alias of 499 Hertz.

Therefore *we must low-pass filter the analog signal to remove all frequencies higher than one-half the sampling rate* before we do the A/D conversion. Provision to do this is built-in with some, but not all, A/D devices! Once the data are digitized, it is no longer possible to eliminate aliases, and test information is lost forever.

Planning an experiment therefore requires three decisions, which are influenced by how "sharp" the cut-off in the low-pass filter is:

1. what the highest frequency of interest is;
2. how much higher to set the analog low-pass filter, to make sure that the frequencies of interest are not unduly attenuated or phase-shifted;
3. how much higher than twice the filter's cut-off frequency to set the sampling rate, so that the amplitude of any aliasing signals is not significant where they "fold over" signals which are of interest.

There still remains the problem of small amplitudes of very-high-frequency noise, which is present in all systems.

18.3.3 Sampling Characteristics

How sensitive sampling is to higher frequencies, depends to a great extent on the nature of the sampling. In our discussion, we have accepted the theoretical assumption that each sample is an instantaneous realization of the value of the function at the moment of sampling. In real devices, it tends to be an average over a short time span. An extreme example of alternative ways of sampling fluctuating data is meteorological data for wind strength. One traditional method is to count the revolutions of a cup anemometer for an hour, and record the total "run," which is also the average wind for that hour. A newer method is to measure the revolutions for six seconds every hour, and compute and record the "instantaneous" velocity once per hour. For the given one-hour interval between records, the older method is preferable, because it averages out short gusts.

Similarly, sampling with a duration approaching Δt is less sensitive to high-frequency noise than "instant" sampling. In practice, A/D-boards tend to have a weighted-average sensitivity during the sampling interval, biased towards the latter part of the interval.

18.3.4 Total Sample Size

In setting the parameters on an FFT-analyzer programmed for a given number of samples, there are trade-offs between high- and low-frequency, or between sampling rate and total sample size.
Example: We might limit ourselves to 1024 data points, resulting in output for 512 frequencies. If we then select an overall sample T of slightly more than

a second, we can have an sampling interval $\Delta t = T/1024$ or one-thousandth of a second long. The highest of the 512 frequencies handled is Shannon's borderline value of $f_{\text{max}} = 1/(2\Delta t)$. The frequency-points are evenly spaced, and the lowest frequency handled is $f_{\text{min}} = f_{\text{max}}/512 = 1/T$. Evidently, and not surprisingly, the lowest frequency we can recognize is the one having a period equal to the sample size. Even that is a borderline case, as we are about to see.

18.3.5 Windowing

When we analyze a run of data, we like to assume that the same kind of pattern repeats again and again. Although we are trying to obtain a reasonable approximation of the Fourier transform of the underlying function, our limited series of data points forces us to obtain a Fourier series with a fundamental frequency of $f_{\text{min}} = 1/T$. But when we take 2^n samples, we are generally not so lucky that the first and last point "close" in amplitude and slope; there will generally be a "jump" as if the were an underlying saw-tooth function (to account for the jump in magnitude) plus an underlying parabolic function (to explain the discontinuity in slope). Therefore the FFT analysis will produce some spurious low frequencies—if it affected only the first point on the plot, we would just throw that away, but obviously there will also be harmonics at multiples of that lowest frequency.

A partial solution to this problem is to multiply all the points in a run of data by a weighting function such as the half-sine-shaped or Hamming window, which gradually attenuates the points near each end of the run, so that the values at the ends approach zero. This process, which is built into many FFT analyzers, reduces spurious frequencies, but also loses some information.

18.3.6 Ensemble Averaging

The demand for high sampling rate on the one hand, and for long total sample size on the other hand, can be solved by processing a larger number of samples. On the other hand, this leads to steep increase in the computing time and/or hardware cost required.

A partial solution is to average the PSDs or autocorrelations from many sampling runs, hoping to average out some (but not all) of the consequences of short samples, and obtaining results which more nearly represent the expectation from a large sampling run.

The data points in a typical PSD show a lot of scatter, so that it is difficult to distinguish between "flyers" due to the limited total sample we are using, and spikes indicating sinusoidal content. The eye tries to pick out the real spikes by "smearing" (replacing each data point by a weighted average of itself and adjacent points), a fundamentally unreliable process. A better process is to average the PSDs of several runs of samples, so that the spectral-plot features which represent real characteristics are maintained, but the scatter which is the result of the limited data in each PSD averages out.

This approach to the scatter in data processing, has a parallel in the repetition of experiments and the averaging the results in order to wash out the random effects from the deterministic results. Except for cases where we can average the output after a synchronizing event, this approach does not work in the time domain, because the very random vibrations we are interested in would be averaged away. However, we use it in the frequency domain to reduce the random effects of sampling runs that are not long enough to obtain the stationary state.

18.4 Transfer Functions

The Fourier transform serves the same function in steady-state random excitation as the Laplace transform does for transients. We can write the response as the product of the transfer function times the (random) excitation

$$\mathcal{X}(\omega) = \mathcal{H}(\omega)\,\mathcal{F}(\omega)$$

where the transfer function of a system is the same as the periodic-excitation response—that is, the Laplace transfer functions with $i\omega$ substituted for s.

Transfer functions for a one-degree-of-freedom system are, for force input,

$$\begin{aligned}
\mathcal{F}(x) &= \frac{\omega_n^2}{-\omega^2 + i\omega 2\zeta\omega_n + \omega_n^2} \cdot \frac{\mathcal{F}(F_{ex})}{k} \\
\mathcal{F}(F_{base}) &= \frac{i\omega 2\zeta\omega_n + \omega_n^2}{-\omega^2 + i\omega 2\zeta\omega_n + \omega_n^2} \cdot \mathcal{F}(F_{ex})
\end{aligned}$$

Transfer functions for base motion input

$$\begin{aligned}
\mathcal{F}(x) &= \frac{i\omega 2\zeta\omega_n + \omega_n^2}{-\omega^2 + i\omega 2\zeta\omega_n + \omega_n^2} \cdot \mathcal{F}(y_{base}) \\
\mathcal{F}(z_{rel}) &= \frac{-\omega^2}{-\omega^2 + i\omega 2\zeta\omega_n + \omega_n^2} \cdot \mathcal{F}(y_{base}) \\
\mathcal{F}(z_{rel}) &= \frac{-1}{-\omega^2 + i\omega 2\zeta\omega_n + \omega_n^2} \cdot \mathcal{F}\left(\ddot{y}_{base}\right) \\
\mathcal{F}(F_{base}) &= \frac{-\omega^2\left(i\omega 2\zeta\omega_n + \omega_n^2\right)}{\left(-\omega^2 + i\omega 2\zeta\omega_n + \omega_n^2\right)\omega_n^2} \cdot k\mathcal{F}(y_{base})
\end{aligned}$$

Transfer functions for mass excitation by an inner mass m moving a distance

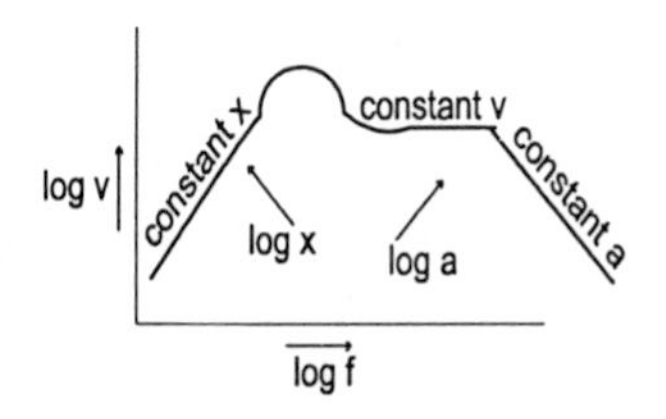

Figure 18.1: Vibration Specification

e_{rel} within the total mass M

$$
\begin{aligned}
\mathcal{F}(x) &= \frac{\omega^2}{-\omega^2 + i\omega 2\zeta\omega_n + \omega_n^2} \cdot \frac{m\mathcal{F}(e_{rel})}{M} \\
\mathcal{F}(z_{rel}) &= \frac{-\omega_n^2}{-\omega^2 + i\omega 2\zeta\omega_n + \omega_n^2} \cdot \frac{m\mathcal{F}(e_{rel})}{M} \\
\mathcal{F}(F_{\text{base}}) &= \frac{\omega^2\left(i\omega 2\zeta\omega_n + \omega_n^2\right)}{\left(-\omega^2 + i\omega 2\zeta\omega_n + \omega_n^2\right)\omega_n^2} \cdot \frac{km}{M}\mathcal{F}(e_{rel}) \\
\mathcal{F}(F_{\text{base}}) &= \frac{i\omega 2\zeta\omega_n + \omega_n^2}{-\omega^2 + i\omega 2\zeta\omega_n + \omega_n^2} \cdot \mathcal{F}(F_{\text{equiv}})
\end{aligned}
$$

If the damping ζ is small, the response of the system tends to "pick out" the resonant frequency content of the random input. The cantilever elements in reed tachometers vibrate when their tuned frequency is present. Analog frequency analyzers operate by introducing the signal to a tuned circuit which has a response-width of about one or even only one-third octave. The resonant circuit filters-out all but near-resonant frequency content. A detector circuit measures magnitude of the response, which is used as an indicator of the PSD of the source signal. When using such an instrument, one observes that the power at any frequency fluctuates constantly, and must be visually averaged to get results.

18.5 Vibration Spectra

The input and output functions can be represented by a plot similar to the PSD, but based on the log of the amplitude, of the velocity, or of the acceleration. (Figure 18.1) If the ordinate is the log of velocity, and the abscissa the log of frequency, then the log of displacement can be plotted in on a coordinate pointing to the upper left, and the log of acceleration to the upper right. Input functions like earthquake motion can be shown and compared on such a plot; output functions and specified limits on motion can be shown on similar plots.

Problem 18.4 *For the modulated oscillation of Figure 7.6 on page 77, sketch* *the Probability Density, the Probability Distribution, the Autocorrelation, and the Power Spectral Density.*

Problem 18.5 *You wish to analyze a voltage signal; you are confident that the frequencies of interest are in the range of* 5 *to* 5000 *Hz. Your plan is to record the experimental data digitally, for later processing with various types of statistical and engineering software. Specify the analog filters, sampling rate, and length of record which are required, and explain your reasoning.*

Part III

Multi-Degree-of-Freedom Systems

Chapter 19

TWO-DIRECTIONAL MOTION and principal coordinates

We can formulate multi-degree-of-freedom systems using Newton's law. In linear systems, we can obtain the mass and spring matrices more quickly using the concept of influence coefficients.

19.1 Newton's Law

Let us look at our first example of a concentrated mass, free to move in a vertical plane, restrained by two vertical springs k_1 and k_2, a horizontal spring k_3, and a spring k_4 angled at α radians from the horizontal (Figure 19.1). We can enclose the mass in a closed control volume, and write $\overrightarrow{F} = m\overrightarrow{a}/g_c$. To obtain the forces in the springs, let us express the displacement vector in terms of a horizontal component x and a vertical component y. For small deflections—small enough to avoid substantially altering the orientation of the springs—the

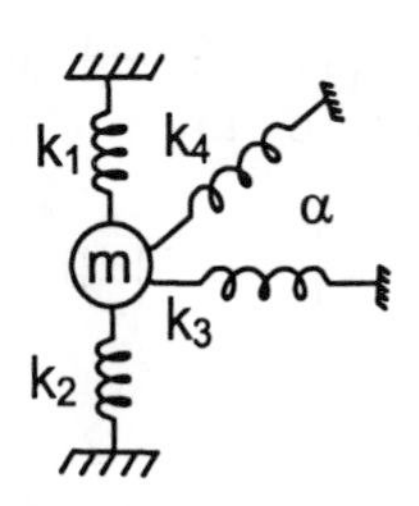

Figure 19.1: Two-Directional Example

force in each spring is proportional to the deflection component which acts along the axis of the spring:

$$\begin{aligned} F_1 &= -k_1 y \\ F_2 &= -k_2 y \\ F_3 &= -k_3 x \\ F_4 &= -k_4 (x\cos\alpha + y\sin\alpha) \end{aligned}$$

The vectorial sum of these forces, plus gravity, broken up into x- and y-components, is

$$\begin{aligned} F_x &= F_3 + F_4\cos\alpha \\ &= -k_3 x - k_4(x\cos\alpha + y\sin\alpha)\cos\alpha \\ F_y &= F_1 + F_2 + F_4\sin\alpha - mg/g_c \\ &= -k_1 y - k_2 y - k_4(x\cos\alpha + y\sin\alpha)\sin\alpha - mg/g_c \end{aligned}$$

Inserting these into Newton's law $F_x = m\ \ddot{x}\ /g_c$ and $F_y = m\ \ddot{y}\ /g_c$, and arranging the terms in the usual sequence, we obtain

$$\begin{aligned} \frac{m}{g_c}\ddot{x} + \left(k_3 + k_4\cos^2\alpha\right)x + \left(k_4\sin\alpha\cos\alpha\right)y &= 0 \\ \frac{m}{g_c}\ddot{y} + \left(k_4\sin\alpha\cos\alpha\right)x + \left(k_1 + k_2 + k_4\sin^2\alpha\right)y &= -\frac{mg}{g_c} \end{aligned}$$

An orderly way to write this uses matrix notation

$$\frac{1}{g_c}\begin{bmatrix} m & 0 \\ 0 & m \end{bmatrix}\begin{Bmatrix} \ddot{x} \\ \ddot{y} \end{Bmatrix} + \begin{bmatrix} k_3{+}k_4\cos^2\alpha & k_4\sin\alpha\cos\alpha \\ k_4\sin\alpha\cos\alpha & k_1{+}k_2{+}k_4\sin^2\alpha \end{bmatrix}\begin{Bmatrix} x \\ y \end{Bmatrix} = \begin{Bmatrix} 0 \\ \frac{-mg}{g_c} \end{Bmatrix}$$

which we see is identical to our written-out equations if we know the matrix-multiplication rule

$$\begin{bmatrix} k_{11} & k_{12} \\ k_{21} & k_{22} \end{bmatrix} \times \begin{Bmatrix} x \\ y \end{Bmatrix} \equiv \begin{Bmatrix} (k_{11}x + k_{12}y) \\ (k_{21}x + k_{22}y) \end{Bmatrix} \tag{19.1}$$

Note that the matrix notation does not add any information; it just arranges the terms systematically so that we can write compactly

$$\frac{1}{g_c}[M]\left\{\ddot{x}\right\} + [K]\{x\} = \{F(t)\} \tag{19.2}$$

where the "mass matrix" is

$$[M] \triangleq \begin{bmatrix} m_{11} & m_{12} \\ m_{21} & m_{22} \end{bmatrix} = \begin{bmatrix} m & 0 \\ 0 & m \end{bmatrix}$$

and the "stiffness matrix" is

$$[K] \triangleq \begin{bmatrix} k_{11} & k_{12} \\ k_{21} & k_{22} \end{bmatrix} = \begin{bmatrix} (k_3 + k_4 \cos^2 \alpha) & (k_4 \sin \alpha \cos \alpha) \\ (k_4 \sin \alpha \cos \alpha) & (k_1 + k_2 + k_4 \sin^2 \alpha) \end{bmatrix}$$

and the unknown or dependent variable is the displacement vector

$$\{x\} \triangleq \left\{ \begin{array}{c} x \\ y \end{array} \right\}$$

19.2 Equilibrium Solution

The excitation vector is

$$\{F(t)\} \triangleq \left\{ \begin{array}{c} F_x(t) \\ F_y(t) \end{array} \right\} = \left\{ \begin{array}{c} 0 \\ \frac{-mg}{g_c} \end{array} \right\}$$

which in our example is a constant, so that this linear system of equations is only trivially non-homogeneous. We can solve for the equilibrium position as we did in Section 4.7 by breaking the displacement up into steady and variable components $x = \overline{x} + \widetilde{x}$ and $y = \overline{y} + \widetilde{y}$. The resulting algebraic equilibrium-position equations look like the governing equations with the derivative terms left out

$$\begin{aligned} (k_3 + k_4 \cos^2 \alpha)\, \overline{x} + (k_4 \sin \alpha \cos \alpha)\, \overline{y} &= 0 \\ (k_4 \sin \alpha \cos \alpha)\, \overline{x} + (k_1 + k_2 + k_4 \sin^2 \alpha)\, \overline{y} &= -\frac{mg}{g_c} \end{aligned}$$

which in matrix notation involves only the stiffness matrix $[K]$ and gravity forces $\{\overline{F}\}$

$$\begin{bmatrix} (k_3 + k_4 \cos^2 \alpha) & (k_4 \sin \alpha \cos \alpha) \\ (k_4 \sin \alpha \cos \alpha) & (k_1 + k_2 + k_4 \sin^2 \alpha) \end{bmatrix} \left\{ \begin{array}{c} \overline{x} \\ \overline{y} \end{array} \right\} = \left\{ \begin{array}{c} 0 \\ \frac{-mg}{g_c} \end{array} \right\}$$

or in compact notation

$$\begin{aligned} [K]\{\overline{x}\} &= \{\overline{F}\} \\ &= \left\{ \frac{mg_x}{g_c} \right\} \end{aligned} \tag{19.3}$$

This is the Standard Form for expressing simultaneous algebraic equations. The tedious solution process is programmed into calculators and computers: we enter the matrix $[K]$ and obtain its "inverse" matrix. For our algebraic example, *Maple* (Section 1.5) quickly finds

$$[K]^{-1} = \frac{1}{\det [K]} \begin{bmatrix} (k_1 + k_2 + k_4 \sin^2 \alpha) & -(k_4 \sin \alpha \cos \alpha) \\ -(k_4 \sin \alpha \cos \alpha) & (k_3 + k_4 \cos^2 \alpha) \end{bmatrix}$$

where

$$\begin{aligned}\det[K] &= k_{11}k_{22} - k_{21}k_{12} \\ &= k_3\left(k_1 + k_2 + k_4\sin^2\alpha\right) + (k_1 + k_2)\,k_4\cos^2\alpha\end{aligned}$$

We can verify that this is indeed the inverse, by multiplying it with the stiffness matrix, following the rule of matrix multiplication

$$\left[\begin{array}{cc} k_{11} & k_{12} \\ k_{21} & k_{22} \end{array}\right] \times \left[\begin{array}{cc} a_{11} & a_{12} \\ a_{21} & a_{22} \end{array}\right] \equiv \left[\begin{array}{cc} (k_{11}a_{11} + k_{12}a_{21}) & (k_{11}a_{12} + k_{12}a_{22}) \\ (k_{21}a_{11} + a_{22}a_{21}) & (k_{21}a_{12} + k_{22}a_{22}) \end{array}\right] \tag{19.4}$$

The definition of the inverse is that this multiplication results in the "identity matrix"

$$[K]\,[K]^{-1} = [K]^{-1}\,[K] = [I] \triangleq \left[\begin{array}{cc} 1 & 0 \\ 0 & 1 \end{array}\right] \tag{19.5}$$

Exercise 19.1 *Verify that the inverse obtained in our example is indeed correct.*

The equilibrium solution is now obtained by pre-multiplying each term in the equilibrium equation with the inverse of the stiffness matrix

$$[K]^{-1}\,[K]\,\{\overline{x}\} = [K]^{-1}\,\{\overline{F}\}$$

which reduces to

$$\{\overline{x}\} = [K]^{-1}\,\{\overline{F}\}$$

Written out for our example

$$\left\{\begin{array}{c} \overline{x} \\ \overline{y} \end{array}\right\} = [K]^{-1}\left\{\begin{array}{c} 0 \\ \frac{-mg}{g_c} \end{array}\right\}$$

which we can expand, using the rules of multiplication,

$$\begin{aligned}\overline{x} &= \frac{(k_4\sin\alpha\cos\alpha)}{k_1k_3 + k_2k_3 + k_3k_4\sin^2\alpha + (k_1 + k_2)\,k_4\cos^2\alpha}\cdot\frac{mg}{g_c} \\ \overline{y} &= \frac{-\left(k_3 + k_4\cos^2\alpha\right)}{k_1k_3 + k_2k_3 + k_3k_4\sin^2\alpha + (k_1 + k_2)\,k_4\cos^2\alpha}\cdot\frac{mg}{g_c}\end{aligned}$$

The complexity of these expressions suggests that manual solution of the simultaneous equations would be very tedious and prone to error; we have demonstrated the benefits of using matrix notation and computer algebra.

Once we have the equilibrium solution, we can replace x and y with $\widetilde{x} \triangleq x - \overline{x}$ and $\widetilde{y} \triangleq y - \overline{y}$, which in our example will make the governing equation homogeneous.

Exercise 19.2 *Demonstrate that this substitution leads to the equations*

$$\frac{1}{g_c}\left[\begin{array}{cc} m & 0 \\ 0 & m \end{array}\right]\left\{\begin{array}{c} \ddot{\widetilde{x}} \\ \ddot{\widetilde{y}} \end{array}\right\} + \left[\begin{array}{cc} k_3 + k_4\cos^2\alpha & k_4\sin\alpha\cos\alpha \\ k_4\sin\alpha\cos\alpha & k_1 + k_2 + k_4\sin^2\alpha \end{array}\right]\left\{\begin{array}{c} \widetilde{x} \\ \widetilde{y} \end{array}\right\} = \left\{\begin{array}{c} 0 \\ 0 \end{array}\right\}$$

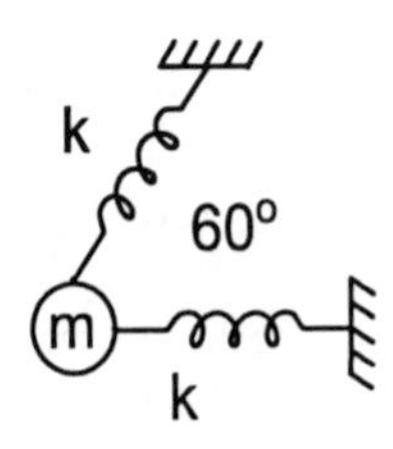

Figure 19.2: Second Example

19.3 Definition of Influence Coefficients

The general equation for a multi-degree-of-freedom system is

$$\frac{1}{g_c}[M]\left\{\ddot{x}\right\}+[C]\left\{\dot{x}\right\}+[K]\{x\}=\left\{F_{(t)}\right\} \tag{19.6}$$

Writing the elements of the $[K]$ matrix as k_{ij}, where i stands for the row and j for the column, we can obtain each element from the following rules:

1a: k_{ij} is the force required at i to maintain a unit deflection at j while maintaining zero deflection everywhere else.

If there are dampers in addition to the springs, we can add a second rule:

2: c_{ij} is the force required at i to maintain unit velocity at j and zero velocity elsewhere.

Indeed, we could define the mass matrix the same way:

3: m_{ij}/g_c is the force required at i to obtain unit acceleration at j and zero velocity elsewhere.

If you are wondering why we marked the first rule "1a.", there is a complementary Rule 1b: a_{ij} is the displacement resulting at i from a unit force applied at j and zero force everywhere else. This results in the flexibility matrix $[A]$ which is the inverse of the stiffness matrix $[K]$.

Example: Consider a second example (Figure 19.2), a concentrated mass m moving in a horizontal plane and restrained by a spring k in the x-direction, and a second, identical spring at a sixty-degree angle to it. The mass matrix is

$$[M]=\left[\begin{array}{cc} m & 0 \\ 0 & m \end{array}\right]$$

and, following *Rule 1a*, the first column of the stiffness matrix is obtained by assuming a unit deflection in the x-direction and zero deflection in the y-direction.

The deflections of the springs are 1 and 0.5, respectively. The forces in the springs are k and $0.5k$, respectively. The x and y components of the force required to oppose the springs and maintain the prescribed displacement, are $k_{11} = (k + 0.25k)$ and $k_{21} = 0.433k$.

The second column of the stiffness matrix follows from assuming a unit deflection in the y-direction instead. Now, the spring deflections are 0 and 0.866, respectively. The forces in the springs are 0 and $0.866k$, respectively. The x and y components of the force required to oppose the springs and maintain the prescribed displacement are $k_{12} = 0.433k$ and $k_{22} = 0.75k$. Thus, the stiffness matrix is

$$[K] = \begin{bmatrix} 1.25k & 0.433k \\ 0.433k & 0.75k \end{bmatrix}$$

19.4 Transformation to Simple Systems

The analysis of multi-directional problems revolves around identifying characteristic values representing the natural frequencies of the system, and characteristic vectors representing the natural directions of motion of the mass.

19.4.1 Eigenvalues

This simple two-spring example leads to the governing equation

$$\frac{1}{g_c}\begin{bmatrix} m & 0 \\ 0 & m \end{bmatrix}\begin{Bmatrix} \ddot{x} \\ \ddot{y} \end{Bmatrix} + \begin{bmatrix} 1.25k & 0.433k \\ 0.433k & 0.75k \end{bmatrix}\begin{Bmatrix} x \\ y \end{Bmatrix} = \begin{Bmatrix} 0 \\ 0 \end{Bmatrix} \tag{19.7}$$

We can write purely numerical matrices, by factoring out a reference mass and a reference spring. We could use the largest mass and spring term in each matrix

$$\frac{m}{g_c}\begin{bmatrix} 1 & 0 \\ 0 & 1 \end{bmatrix}\begin{Bmatrix} \ddot{x} \\ \ddot{y} \end{Bmatrix} + 1.25k\begin{bmatrix} 1.00 & 0.346 \\ 0.346 & 0.60 \end{bmatrix}\begin{Bmatrix} x \\ y \end{Bmatrix} = \begin{Bmatrix} 0 \\ 0 \end{Bmatrix}$$

or any other reference value

$$\frac{m}{g_c}\begin{bmatrix} 1 & 0 \\ 0 & 1 \end{bmatrix}\begin{Bmatrix} \ddot{x} \\ \ddot{y} \end{Bmatrix} + k\begin{bmatrix} 1.25 & 0.433 \\ 0.433 & 0.75 \end{bmatrix}\begin{Bmatrix} x \\ y \end{Bmatrix} = \begin{Bmatrix} 0 \\ 0 \end{Bmatrix}$$

In a practical problem, the reference mass could be one kg or one lbm or one slug; and the reference spring either one N/m or one lbf/ft.

In an undamped problem, we expect sinusoidal solutions, for which $\ddot{x} = -\omega_n^2 x$, so we can write

$$-\omega_n^2\frac{m}{g_c k}k\begin{bmatrix} 1 & 0 \\ 0 & 1 \end{bmatrix}\begin{Bmatrix} x \\ y \end{Bmatrix} + k\begin{bmatrix} 1.25 & 0.433 \\ 0.433 & 0.75 \end{bmatrix}\begin{Bmatrix} x \\ y \end{Bmatrix} = \begin{Bmatrix} 0 \\ 0 \end{Bmatrix}$$

Defining the dimensionless "eigenvalue"

$$\lambda \triangleq \omega_n^2 \frac{m}{kg_c} \tag{19.8}$$

we write

$$k \begin{bmatrix} -\lambda & 0 \\ 0 & -\lambda \end{bmatrix} \begin{Bmatrix} x \\ y \end{Bmatrix} + k \begin{bmatrix} 1.25 & 0.433 \\ 0.433 & 0.75 \end{bmatrix} \begin{Bmatrix} x \\ y \end{Bmatrix} = \begin{Bmatrix} 0 \\ 0 \end{Bmatrix}$$

Adding the matrices, we get the Standard Form eigenvalue problem

$$k \begin{bmatrix} 1.25 - \lambda & 0.433 \\ 0.433 & 0.75 - \lambda \end{bmatrix} \begin{Bmatrix} x \\ y \end{Bmatrix} = \begin{Bmatrix} 0 \\ 0 \end{Bmatrix} \tag{19.9}$$

which stands for two simultaneous equations

$$(1.25 - \lambda)\, x + (0.433)\, y = 0 \tag{19.10}$$
$$(0.433)\, x + (0.75 - \lambda)\, y = 0 \tag{19.11}$$

This pair of simultaneous equations can be solved by Cramer's rule

$$x = \frac{\det \begin{vmatrix} 0 & 0.433 \\ 0 & 0.75 - \lambda \end{vmatrix}}{\det \begin{vmatrix} 1.25 - \lambda & 0.433 \\ 0.433 & 0.75 - \lambda \end{vmatrix}}; \; y = \frac{\det \begin{vmatrix} 1.25 - \lambda & 0 \\ 0.433 & 0 \end{vmatrix}}{\det \begin{vmatrix} 1.25 - \lambda & 0.433 \\ 0.433 & 0.75 - \lambda \end{vmatrix}}$$

The numerator in both cases is zero, so that we can have (indeterminate) non-zero values for x and y only if the denominator is also zero. Therefore the characteristic values are obtained by setting the determinant of the matrix equal to zero, leading to the characteristic equation

$$\begin{aligned} (1.25 - \lambda)(0.75 - \lambda) - (0.433)^2 &= 0 \\ \lambda^2 - 2\lambda + 0.75 &= 0 \\ \lambda &= \frac{2 \mp \sqrt{4 - 3}}{2} \\ \lambda_{I,II} &= 0.5,\ 1.5, \text{ respectively} \qquad (19.12) \\ \omega_{I,II} &= \sqrt{0.5\frac{kg_c}{m}},\ \sqrt{1.5\frac{kg_c}{m}}, \text{ resp.} \end{aligned}$$

Note that we obtain two natural frequencies ω_n, and call the *smaller* one ω_I.

19.4.2 Eigenvectors

The magnitudes of x and y are indeterminate (awaiting information about initial conditions), but the relative magnitudes are fixed for each eigenvalue. We can solve for the amplitude ratio x/y by plugging the eigenvalues back into Equation 19.10

$$\begin{aligned}(1.25-\lambda)\,x+(0.433)\,y &= 0 \qquad (19.13)\\ \left(\frac{y}{x}\right)_I &= \frac{\lambda_I-1.25}{0.433}=-1.732\\ \left(\frac{y}{x}\right)_{II} &= \frac{\lambda_{II}-1.25}{0.433}=0.577\end{aligned}$$

Note that we did not need the other Equation, 19.11, because extracting the eigenvalue information made it redundant; indeed, we could have skipped the first Equation, 19.10, and used the Equation 19.11 in its stead and obtained the same results

$$\begin{aligned}(0.433)\,x+(0.75-\lambda)\,y &= 0\\ \left(\frac{y}{x}\right)_I &= \frac{0.433}{\lambda_I-0.75}=-1.732\\ \left(\frac{y}{x}\right)_{II} &= \frac{0.433}{\lambda_{II}-0.75}=0.577\end{aligned}$$

We see that an eigenvalue is a value of λ which makes the two Equations, 19.10 and 19.11, agree with each other.

The two ratios $(y/x)_{I,II} = -1.732,\ 0.577$, respectively, each describe a straight line; and each is at right angles (orthogonal) to the other, because one slope is the negative of the inverse of the other slope.

Therefore, our example has two modes: a low-frequency motion with a negative sixty-degree slope across the tips of the two springs, and the high-frequency motion bisecting the springs with a thirty-degree slope. Because they are at right angles to each other, any initial displacement (and any initial velocity) of x and y can be resolved into initial conditions for each of the two modes.

19.4.3 Modes as Coordinate Systems

In our simple example, we see that the "mode shapes" represented by the characteristic ratios $(y/x)_{I,II}$ can be regarded as an orthogonal coordinate system. If we had anticipated this and rotated the original coordinate system by sixty degrees, so that the new y-axis bisected the space between the springs, we would have obtained the mass and stiffness matrices

$$\frac{1}{g_c}\begin{bmatrix} m & 0\\ 0 & m\end{bmatrix}\begin{Bmatrix}\ddot{x}_{new}\\ \ddot{y}_{new}\end{Bmatrix}+\begin{bmatrix}0.5k & 0\\ 0 & 1.5k\end{bmatrix}\begin{Bmatrix}x_{new}\\ y_{new}\end{Bmatrix}=\begin{Bmatrix}0\\ 0\end{Bmatrix} \qquad (19.14)$$

Both mass and spring matrices would have been diagonal, and the two equations would have been uncoupled

$$\ddot{x}_{new}+\frac{0.5kg_c}{m}x_{new} = 0 \qquad (19.15)$$

$$\ddot{y}_{new}+\frac{1.5kg_c}{m}y_{new} = 0 \qquad (19.16)$$

with a natural frequency ω_n of $\sqrt{0.5kg_c/m}$ in the new x-direction and $\sqrt{1.5kg_c/m}$ in the new y-direction.

Exercise 19.3 *Using the rotated coordinate system, obtain the K-matrix for our example, by means of Newton's law and/or influence coefficients.*

This demonstrates that the right choice of coordinate system—the "principal" coordinates—uncouples the equations. In general, we have to solve the eigenvector problem in order to find the principal coordinates. Any other orientation of the coordinates x and y gives us off-diagonal terms in the K-matrix.

Exercise 19.4 *Show that the stiffness of any two-directional linear-spring support for a compact mass can be sketched as an ellipse similar to a stress ellipse; that the intercepts of the ellipse in an arbitrary coordinate systems show the diagonal terms of the stiffness matrix, and that non-orthogonal intercepts of the stiffness ellipse at the coordinates indicate coupling terms.*

19.4.4 Modal Matrix

A good way to display the mode shapes is to form a matrix of the vectors of relative magnitudes:

$$[u] \triangleq \left[\left\{ \begin{array}{c} 1 \\ y/x \end{array} \right\}_I \; \left\{ \begin{array}{c} 1 \\ y/x \end{array} \right\}_{II} \right] = \left[\begin{array}{cc} 1 & 1 \\ -1.732 & 0.577 \end{array} \right]$$

Note that the scalar dot-product of the two column vectors is zero: this is how we detect orthogonality. We have chosen to make the first element of each column vector unity. Since eigenvectors only represent relative magnitudes, they could have been normalized in other ways, for example, we could just as well have made the second component of each vector unity

$$[u] \triangleq \left[\left\{ \begin{array}{c} x/y \\ 1 \end{array} \right\}_I \; \left\{ \begin{array}{c} x/y \\ 1 \end{array} \right\}_{II} \right] = \left[\begin{array}{cc} -0.577 & 1.732 \\ 1 & 1 \end{array} \right]$$

and this version of the modal matrix serves just as well, as long as it is *used consistently*. A third approach is to make the biggest component of each vector unity

$$[u] = \left[\begin{array}{cc} -0.577 & 1 \\ 1 & 0.577 \end{array} \right]$$

but the *standard* way to normalize vectors is to make the sum of the squares of the elements of each column vector equal to unity, by dividing each term by the square-root of the sum of the squares of the column vector it is in; and to make the terms on the diagonal positive, by changing all the signs in a column vector if necessary:

$$[u] = \left[\begin{array}{cc} 0.5 & 0.866 \\ -0.866 & 0.5 \end{array} \right]$$

The resulting orthogonal and normalized modal matrix is "orthonormal" and has the convenient property that its inverse $[u]^{-1}$ is easy to obtain because it is simply the transpose

$$[u]^T = \begin{bmatrix} 0.5 & -0.866 \\ 0.866 & 0.5 \end{bmatrix}$$

that is, the matrix terms are "flipped" across the diagonal axis.

Exercise 19.5 *Check that, when using orthogonal eigenvectors and appropriate normalization,* $[u]\,[u]^T = [u]^T\,[u] = [I]$

The modal matrix $[u]$ is the "transformation matrix" which converts our principal coordinates to the original coordinates; its inverse transforms the original coordinates to principal coordinates.

19.4.5 Coordinate Transformation

We note that we can transform any position from our original x and y coordinates to our principal coordinates x_{new} and y_{new} by the transformations

$$\left\{ \begin{array}{c} x \\ y \end{array} \right\} = [u] \left\{ \begin{array}{c} x_{new} \\ y_{new} \end{array} \right\} = \begin{bmatrix} 0.5 & 0.866 \\ -0.866 & 0.5 \end{bmatrix} \left\{ \begin{array}{c} x_{new} \\ y_{new} \end{array} \right\} \tag{19.17}$$

$$\left\{ \begin{array}{c} x_{new} \\ y_{new} \end{array} \right\} = [u]^{-1} \left\{ \begin{array}{c} x \\ y \end{array} \right\} = \begin{bmatrix} 0.5 & -0.866 \\ 0.866 & 0.5 \end{bmatrix} \left\{ \begin{array}{c} x \\ y \end{array} \right\} \tag{19.18}$$

where we can obtain the inverted matrix $[u]^{-1}$ from an applications program like *Matlab* (Section 1.5) and verify it by showing that the product of $[u]$ and $[u]^{-1}$ is the identity matrix. We observe that the inverse matrix $[u]^{-1}$ turns out to be simply the transpose $[u]^T$; this is a consequence of using a properly normalized matrix of orthogonal eigenvectors. The procedure would also work with the *consistent* use of a differently normalized u-matrix (for which the inverse is *not* the transpose); only the distance scales along the new coordinates would be different.

Therefore we can modify the original equations by substituting $[u]$ times the principal-coordinate vectors for the original-coordinate vectors in Equation 19.7

$$\frac{1}{g_c} \begin{bmatrix} m & 0 \\ 0 & m \end{bmatrix} \times \begin{bmatrix} 0.5 & 0.866 \\ -0.866 & 0.5 \end{bmatrix} \left\{ \begin{array}{c} \ddot{x}_{new} \\ \ddot{y}_{new} \end{array} \right\}$$
$$+ \begin{bmatrix} 1.25k & 0.433k \\ 0.433k & 0.75k \end{bmatrix} \times \begin{bmatrix} 0.5 & 0.866 \\ -0.866 & 0.5 \end{bmatrix} \left\{ \begin{array}{c} x_{new} \\ y_{new} \end{array} \right\} = \left\{ \begin{array}{c} 0 \\ 0 \end{array} \right\}$$

which multiplies out to

$$\frac{1}{g_c} \begin{bmatrix} 0.5m & 0.866m \\ -0.866m & 0.5m \end{bmatrix} \left\{ \begin{array}{c} \ddot{x}_{new} \\ \ddot{y}_{new} \end{array} \right\} + \begin{bmatrix} 0.25k & 1.299k \\ -0.433k & 0.75k \end{bmatrix} \left\{ \begin{array}{c} x_{new} \\ y_{new} \end{array} \right\} \ldots$$

which is not in an attractive form. However, if we now pre-multiply each term by the inverse (or, in this case, also the transpose) of the u-matrix, we get

$$\frac{1}{g_c}\begin{bmatrix} 0.5 & -0.866 \\ 0.866 & 0.5 \end{bmatrix}\begin{bmatrix} 0.5m & 0.866m \\ -0.866m & 0.5m \end{bmatrix}\begin{Bmatrix} \ddot{x}_{new} \\ \ddot{y}_{new} \end{Bmatrix}$$
$$+\begin{bmatrix} 0.5 & -0.866 \\ 0.866 & 0.5 \end{bmatrix}\begin{bmatrix} 0.25k & 1.299k \\ -0.433k & 0.75k \end{bmatrix}\begin{Bmatrix} x_{new} \\ y_{new} \end{Bmatrix} = \begin{Bmatrix} 0 \\ 0 \end{Bmatrix}$$

which multiplies out to

$$\frac{1}{g_c}\begin{bmatrix} m & 0 \\ 0 & m \end{bmatrix}\begin{Bmatrix} \ddot{x}_{new} \\ \ddot{y}_{new} \end{Bmatrix} + \begin{bmatrix} 0.5k & 0 \\ 0 & 1.5k \end{bmatrix}\begin{Bmatrix} x_{new} \\ y_{new} \end{Bmatrix} = \begin{Bmatrix} 0 \\ 0 \end{Bmatrix}$$

identical to the result obtained for this coordinate system in Section 19.4.3, showing that this mathematical transformation procedure gives us the governing equation in terms of the characteristic modes. Since this is the principal coordinate system, K- and M-matrices are both diagonal: there are no coupling terms. Therefore, **each direction of motion behaves like a one-degree-of-freedom system**, independent of the other coordinate.

19.5 Initial Conditions

We can use the matrix $[u]^{-1}$ to transform the initial conditions to principal coordinates, apply them to the separate one-degree-of-freedom equations in those coordinates, and then use $[u]$ to transform the solutions to the original coordinates. In our simple example, given the displacement Initial Conditions $x_{(0)}$ and $y_{(0)}$, the principal-coordinate Initial Conditions obtained by means of $[u]^{-1}$ are

$$\begin{aligned} x_{new(0)} &= 0.5x_{(0)} - 0.866y_{(0)} \\ y_{new(0)} &= 0.866x_{(0)} + 0.5y_{(0)} \end{aligned}$$

Applying these initial displacements to Equations 19.15 and 19.16, we write one-degree-of-freedom solutions from Equations 4.19 and 4.20 for each mode, in terms of the natural frequency of each mode

$$\begin{aligned} x_{new}(t) &= x_{new}(0) \cdot \cos\omega_I t = \left(0.5x_{(0)} - 0.866y_{(0)}\right)\cos\omega_I t \\ y_{new}(t) &= y_{new}(0) \cdot \cos\omega_{II} t = \left(0.866x_{(0)} + 0.5y_{(0)}\right)\cos\omega_{II} t \end{aligned}$$

This solution converted back to the original coordinates by means of $[u]$ is

$$\begin{aligned}
x &= 0.5x_{new}(t) + 0.866y_{new}(t) \\
&= 0.5\left(0.5x_{(0)} - 0.866y_{(0)}\right)\cos\omega_I t + 0.866\left(0.866x_{(0)} + 0.5y_{(0)}\right)\cos\omega_{II} t \\
&= \left(0.25x_{(0)} - 0.433y_{(0)}\right)\cos\omega_I t + \left(0.75x_{(0)} + 0.433y_{(0)}\right)\cos\omega_{II} t \\
y &= -0.866x_{new}(t) + 0.5y_{new}(t) \\
&= -0.866\left(0.5x_{(0)} - 0.866y_{(0)}\right)\cos\omega_I t + 0.5\left(0.866x_{(0)} + 0.5y_{(0)}\right)\cos\omega_{II} t \\
&= \left(-0.433x_{(0)} + 0.75y_{(0)}\right)\cos\omega_I t + \left(0.433x_{(0)} + 0.25y_{(0)}\right)\cos\omega_{II} t
\end{aligned}$$

so that each of the original coordinates experiences motion which is a combination of two different frequencies.

Exercise 19.6 *Describe the envelope of the path taken by the mass—what is its maximum excursion in any particular direction? What if the higher frequency had turned out to be an even multiple of the lower one?*

19.6 A Special Case

What if the two identical springs had been at right angles to each other? Then every coordinate system would have been a principal coordinate system. This occurs whenever the two natural frequencies are identical; it is called "physical decoupling" (as distinguished from the mathematical decoupling which depends on a judicious choice of coordinate system).

19.7 Summary

We have introduced the concept of multi-degree-of-freedom systems with two examples of two-directional motion of a compact mass, so that $[M] = m[I]$. We have shown how to set up the undamped free-vibration governing equation

$$\frac{m}{g_c}[I]\left\{\ddot{x}\right\} + [K]\{x\} = \{0\}$$

using either Newton's law or else influence coefficients to obtain $[K]$. Simple manipulations allowed us to obtain the characteristic matrix $\frac{g_c}{m}[K]$, the eigenvectors (for frequencies), and the eigenvectors (for modes).

We have used a simple example to illustrate the orthogonality of modes, and to show the use of the modal matrix $[u]$ to transform to principal coordinates

$$\frac{m}{g_c}[I][u]\left\{\ddot{x}_{new}\right\} + [K][u]\{x_{new}\} = \{0\}$$

and (for a uniform diagonal mass matrix) pre-multiplication by the inverse $[u]^{-1}$ to rediagonalize the mass matrix

$$\frac{m}{g_c}\left[[u]^{-1}[I][u]\right]\left\{\ddot{x}_{new}\right\} + \left[[u]^{-1}[K][u]\right]\{x_{new}\} = \{0\}$$

Since we have used matrix notation, our methods can be extended to the three-dimensional motion of a compact mass. Multiple masses, leading to non-uniform terms in the mass matrix $[M]$, require additional steps described in Chapter 20.

Problem 19.7 *For the system shown below, find the characteristic matrix, the natural frequencies, the orientation of the principal coordinates, and the response to a unit initial deflection in the x-direction.*

Problem 19.8 *For the system shown below, find the characteristic matrix, the natural frequencies, the orientation of the principal coordinates, and the response to a unit initial deflection in the y-direction.*

Problem 19.9 *For the system shown below, find the characteristic matrix, the natural frequencies, the orientation of the principal coordinates, and the response to a unit initial velocity in the x-direction.*

Problem 19.10 *For the system shown below, find the characteristic matrix, the natural frequencies, the orientation of the principal coordinates, and the response to a unit initial velocity in the y-direction.*

Problem 19.11 *For the system shown below, find the characteristic matrix, the natural frequencies, the orientation of the principal coordinates, and the response to a unit initial deflection in the x-direction.*

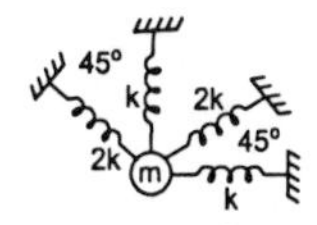

Chapter 20

MULTI-MASS SYSTEMS from Newton's law

Many systems cannot be modeled as one-mass systems, but can be represented by two or more lumped masses (Figure 20.1).

20.1 Problem Formulation

Starting with a "boxcar" system with two masses, two springs to ground, and a coupling spring, we can define two coordinates x_1 and x_2, and write Newton's equation for two control volumes, one around each mass:

$$\begin{aligned} \frac{m_1}{g_c}\ddot{x}_1 + k_1 x_1 + k_c(x_1 - x_2) &= 0 \\ \frac{m_2}{g_c}\ddot{x}_2 + k_2 x_2 + k_c(x_2 - x_1) &= 0 \end{aligned}$$

which we can sort out to

$$\begin{aligned} \frac{m_1}{g_c}\ddot{x}_1 + (k_1 + k_c)x_1 + (-k_c)x_2 &= 0 \\ \frac{m_2}{g_c}\ddot{x}_2 + (-k_c)x_1 + (k_2 + k_c)x_2 &= 0 \end{aligned}$$

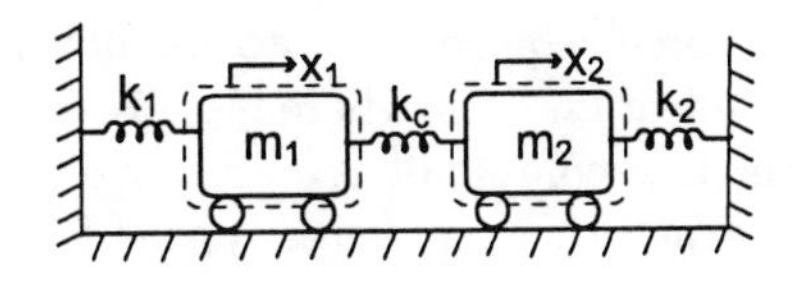

Figure 20.1: Boxcar System

Writing these two equations in matrix notation:

$$\frac{1}{g_c}\begin{bmatrix} m_1 & 0 \\ 0 & m_2 \end{bmatrix}\begin{Bmatrix} \ddot{x}_1 \\ \ddot{x}_2 \end{Bmatrix}+\begin{bmatrix} (k_1+k_c) & (-k_c) \\ (-k_c) & (k_2+k_c) \end{bmatrix}\begin{Bmatrix} x_1 \\ x_2 \end{Bmatrix}=\begin{Bmatrix} 0 \\ 0 \end{Bmatrix} \tag{20.1}$$

Note that we have not done anything mathematically sophisticated; we have merely expressed the two governing equations in matrix and vector notation. A shorthand version of the same equation is

$$\frac{1}{g_c}[M]\left\{\ddot{x}\right\}+[K]\{x\}=0 \tag{20.2}$$

We can now look at each matrix and observe the following:

- In a stable vibrational problem, the matrices can be written so that all terms on the matrix diagonals are positive. If they are not, we should change the signs on one of the equations, i.e., of one of the rows in all the matrices.
- We try to choose a coordinate system such that the mass matrix $[M]$ is diagonal; the off-diagonal terms are zero. In our example this happened because we associated each coordinate with a point mass.
- In an absolute coordinate system, the matrices are symmetrical about the diagonal; $k_{ij} = k_{ji}$. We can explain this from reciprocity.
- With our particular choice of coordinate directions, the off-diagonal "coupling" terms in the spring matrix $[K]$ happen to be negative, but by reversing one of the coordinate directions we could just as easily have made them positive.

If our matrices do not follow these rules, we should take the opportunity to reexamine the problem formulation: either we made a mistake, or this is a divergent (rather than vibrational) problem.

20.2 Formulation from Influence Coefficients

We recall from Section 19.3 that we can develop the stiffness matrix $[K]$ from influence-coefficient rules.

Rule 1a: k_{ij} is the force required at i to maintain a unit deflection at j while maintaining zero deflection everywhere else. In this boxcar-problem, that means that we obtain the first column of k-values by displacing the first mass by a unit distance and hold the other mass steady; the force required to maintain the first mass in position is $k_{11} = (k_1 + k_c)$, the force required to hold the second mass in place against the compression of the coupling spring is $k_{21} = (-k_c)$.

The second column is obtained the same way, but displacing the second mass by a unit distance and holding the first mass steady

$$[K] = \begin{bmatrix} (k_1 + k_c) & (-k_c) \\ (-k_c) & (k_2 + k_c) \end{bmatrix}$$

Rule 2: c_{ij} is the force required at i to maintain unit velocity at j and zero velocity elsewhere. If we had specified dampers c_1, c_2, and c_c next to each spring, we could have obtained each column of the damping matrix $[C]$ by making the corresponding mass have unit velocity and finding the forces required to maintain that condition

$$[C] = \begin{bmatrix} (c_1 + c_c) & (-c_c) \\ (-c_c) & (c_2 + c_c) \end{bmatrix}$$

Rule 3: m_{ij}/g_c is the force required at i to maintain a unit acceleration at j and zero acceleration elsewhere. The mass matrix $[M]$ is diagonal if we choose absolute coordinates for each concentrated mass

$$[M] = \begin{bmatrix} m_1 & 0 \\ 0 & m_2 \end{bmatrix}$$

20.3 Transformation to Simple Systems

We can adapt the approach of Chapter 19 to multi-mass problems. Let us develop the ideas again, in a slightly different way in order to deepen our understanding.

20.3.1 Natural Frequencies

Because this is an undamped linear problem, we anticipate sinusoidal-type solutions. Therefore we can set $\left\{\ddot{x}\right\} = -\omega^2 \{x\}$. Substituting these into the set of differential equations, we obtain the set of algebraic characteristic equations

$$\begin{aligned} -\omega_n^2 \frac{m_1}{g_c} x_1 + (k_1 + k_c)\, x_1 + (-k_c)\, x_2 &= 0 \\ -\omega_n^2 \frac{m_2}{g_c} x_2 + (-k_c)\, x_1 + (k_2 + k_c)\, x_2 &= 0 \end{aligned}$$

or in matrix notation

$$-\omega_n^2 \frac{1}{g_c} \begin{bmatrix} m_1 & 0 \\ 0 & m_2 \end{bmatrix} \begin{Bmatrix} x_1 \\ x_2 \end{Bmatrix} + \begin{bmatrix} (k_1 + k_c) & (-k_c) \\ (-k_c) & (k_2 + k_c) \end{bmatrix} \begin{Bmatrix} x_1 \\ x_2 \end{Bmatrix} = \begin{Bmatrix} 0 \\ 0 \end{Bmatrix}$$

which we abbreviate to

$$\left[\frac{-\omega_n^2}{g_c} [M] + [K] \right] \{x\} = 0$$

This gives us two equations in two unknowns, x_1 and x_2. They cannot be solved for actual values of the unknowns, because the equations are homogeneous; they can be solved only for the ratio of magnitudes x_2/x_1. The two equations must, of course, give the same result for this ratio; this will only occur for certain characteristic values or "eigenvalues" of ω_n^2.

The best way to find these characteristic values is to put the problem into Standard Form, by normalizing the equations to get unit values for the first coefficients. We divide the first equation by m_1/g_c and the second equation by m_2/g_c to get

$$\begin{bmatrix} 1 & 0 \\ 0 & 1 \end{bmatrix} \begin{Bmatrix} \ddot{x}_1 \\ \ddot{x}_2 \end{Bmatrix} + \begin{bmatrix} \frac{(k_1+k_c)g_c}{m_1} & \frac{(-k_c)g_c}{m_1} \\ \frac{(-k_c)g_c}{m_2} & \frac{(k_2+k_c)g_c}{m_2} \end{bmatrix} \begin{Bmatrix} x_1 \\ x_2 \end{Bmatrix} = \begin{Bmatrix} 0 \\ 0 \end{Bmatrix} \tag{20.3}$$

In matrix-notation terms the first matrix is now the identity matrix, and the second matrix contains all the information needed to describe the system. That second matrix is the "characteristic matrix" of the system.

The step of making the coefficients of the second derivatives equal to unity by dividing each equation by its mass term, is equivalent to obtaining the inverse of the mass matrix, which is easily done for a diagonal matrix

$$\begin{bmatrix} m_1 & 0 \\ 0 & m_2 \end{bmatrix}^{-1} = \begin{bmatrix} \frac{1}{m_1} & 0 \\ 0 & \frac{1}{m_2} \end{bmatrix} \tag{20.4}$$

and pre-multiplying each term in the governing equation with it

$$\begin{bmatrix} \frac{1}{m_1} & 0 \\ 0 & \frac{1}{m_2} \end{bmatrix} \begin{bmatrix} m_1 & 0 \\ 0 & m_2 \end{bmatrix} \begin{Bmatrix} \ddot{x}_1 \\ \ddot{x}_2 \end{Bmatrix}$$
$$+ g_c \begin{bmatrix} \frac{1}{m_1} & 0 \\ 0 & \frac{1}{m_2} \end{bmatrix} \begin{bmatrix} (k_1+k_c) & (-k_c) \\ (-k_c) & (k_2+k_c) \end{bmatrix} \begin{Bmatrix} x_1 \\ x_2 \end{Bmatrix} = \begin{Bmatrix} 0 \\ 0 \end{Bmatrix}$$

which leads to the same result, Equation 20.3.

Writing $\left\{\ddot{x}\right\} = -\omega_n^2 \{x\}$ again, we can combine the two matrices and write

$$\begin{bmatrix} \frac{(k_1+k_c)g_c}{m_1} - \omega_n^2 & \frac{(-k_c)g_c}{m_1} \\ \frac{(-k_c)g_c}{m_2} & \frac{(k_2+k_c)g_c}{m_2} - \omega_n^2 \end{bmatrix} \begin{Bmatrix} x_1 \\ x_2 \end{Bmatrix} = \begin{Bmatrix} 0 \\ 0 \end{Bmatrix}$$

This matrix equation can only be valid if the determinant of the matrix is equal to zero. Therefore we must satisfy the "characteristic equation"

$$\left(\frac{(k_1+k_c)\,g_c}{m_1} - \omega_n^2\right)\left(\frac{(k_2+k_c)\,g_c}{m_2} - \omega_n^2\right) - \left(\frac{(-k_c)\,g_c}{m_2}\right)^2 = 0$$

which can be satisfied by two different values of ω_n^2. (If we had chosen the coordinates differently, and had obtained positive coupling terms in the K-matrix, this characteristic equation would still be the same, because the positive or negative term is squared.)

Example: Let both masses have the same magnitude and all three springs be identical. Then the governing equation is

$$\frac{m}{g_c}\begin{bmatrix} 1 & 0 \\ 0 & 1 \end{bmatrix}\begin{Bmatrix} \ddot{x}_1 \\ \ddot{x}_2 \end{Bmatrix} + k\begin{bmatrix} 2 & -1 \\ -1 & 2 \end{bmatrix}\begin{Bmatrix} x_1 \\ x_2 \end{Bmatrix} = \begin{Bmatrix} 0 \\ 0 \end{Bmatrix}$$

and the characteristic matrix is the second matrix in

$$\begin{bmatrix} 1 & 0 \\ 0 & 1 \end{bmatrix}\begin{Bmatrix} \ddot{x}_1 \\ \ddot{x}_2 \end{Bmatrix} + \frac{kg_c}{m}\begin{bmatrix} 2 & -1 \\ -1 & 2 \end{bmatrix}\begin{Bmatrix} x_1 \\ x_2 \end{Bmatrix} = \frac{g_c}{m}\begin{Bmatrix} 0 \\ 0 \end{Bmatrix}$$

and the eigenvalue problem is

$$\frac{kg_c}{m}\begin{bmatrix} 2-\lambda & -1 \\ -1 & 2-\lambda \end{bmatrix}\begin{Bmatrix} x_1 \\ x_2 \end{Bmatrix} = \frac{g_c}{m}\begin{Bmatrix} 0 \\ 0 \end{Bmatrix}$$

where $\lambda \triangleq \frac{m}{kg_c}\omega_n^2$, so that the characteristic equation is

$$\lambda^2 - 4\lambda + (4-1) = 0$$

There are two solutions:

$$\lambda_{I,II} = \frac{4 \mp \sqrt{16-12}}{2} = 1,\ 3, \text{ respectively}$$

$$\omega_{I,II} = \sqrt{\frac{kg_c}{m}},\ \sqrt{3\frac{kg_c}{m}}, \text{ respectively.}$$

It is customary to list the *lower* frequency value first.

20.3.2 Eigenvectors

We can now solve for the relative magnitudes of the x-values by inserting an eigenvalue into the equations and solving one of them for x_1/x_2; either equation should give us the same result if we solved for the eigenvalue correctly. However, each of the two eigenvalues will give us a different result.

Example continued: For our symmetrical case of identical masses and springs,

$$\left(\frac{x_2}{x_1}\right)_{I,II} = 2 - \lambda_{I,II} = \frac{1}{2-\lambda_{I,II}} = 1,\ -1, \text{ respectively}$$

The first of these represents simultaneous motion of the two masses; the second, opposite motion of the masses. (If we defined the coordinates differently, and the coupling terms in the K-matrix were positive rather than negative, the ratios would have different signs, but would represent the same physical relationship in the respective coordinate system.)

The eigenvectors can be displayed graphically as string figures, or numerically as vectors:

$$\left\{ \begin{array}{c} x_1/x_1 \\ x_2/x_1 \end{array} \right\}_{I,II} = \left\{ \begin{array}{c} 1 \\ 1 \end{array} \right\}, \left\{ \begin{array}{c} 1 \\ -1 \end{array} \right\}, \text{ resp.}$$

One interesting property of these pairs of vectors is that they are "orthogonal" because the scalar dot-product of the two vectors is equal to zero.

Note that these vectors show only *relative* magnitude; we found it convenient to normalize them so that the first element of each column vector is unity—that is the simplest, but not necessarily the best way to normalize these vectors. A computer program might normalize them so that the sum of the squares of the elements is unity, and might also choose different signs to make the diagonal terms positive:

$$\left\{ \begin{array}{c} x_1/\sqrt{x_1^2+x_2^2} \\ x_2/\sqrt{x_1^2+x_2^2} \end{array} \right\}_{I,II} = \left\{ \begin{array}{c} 0.707 \\ 0.707 \end{array} \right\}, \left\{ \begin{array}{c} -0.707 \\ 0.707 \end{array} \right\}, \text{ resp.}$$

This is still the same set of eigenvectors, because the *ratios* of amplitudes within each column vector are the same.

20.3.3 Initial Conditions

The actual amplitudes of each mode depends on the initial conditions.

Example continued: In our symmetrical case of identical masses, we can see three possibilities for displacement initial conditions:

- If we start out with both masses displaced a distance A to the right, the initial $x_2/x_1 = 1$, matching the first natural frequency's mode shape, and the subsequent motion is entirely at the lower natural frequency, maintaining $x_2/x_1 = 1$

$$\begin{aligned} x_1 &= A\cos\omega_I t \\ x_2 &= A\cos\omega_I t \end{aligned}$$

- If we start out with displacements of magnitude A but in opposite directions, then the initial $x_2/x_1 = -1$, matching the second natural frequency's mode shape, and the subsequent motion is entirely at the higher natural frequency, maintaining $x_2/x_1 = -1$

$$\begin{aligned} x_1 &= A\cos\omega_{II} t \\ x_2 &= -A\cos\omega_{II} t \end{aligned}$$

- If we start out with some other combination, both modes are activated, and the subsequent motion involves both frequencies. For example. if $x_1(0) = A$ and $x_2(0) = 0$,

$$\begin{aligned} x_1 &= \frac{A}{2}\cos\omega_I t + \frac{A}{2}\cos\omega_{II} t \\ x_2 &= \frac{A}{2}\cos\omega_I t - \frac{A}{2}\cos\omega_{II} t \end{aligned}$$

  which satisfies the initial conditions and maintains $x_2/x_1 = 1$ for the lower-frequency terms, and $x_2/x_1 = -1$ for the higher-frequency terms.

To find the amount of each mode that is activated, we follow the procedure of Section 19.5, transforming the initial conditions to the principal coordinates.

20.3.4 Modes as Coordinates

We should note that any linearly independent set of weighted combinations of x_1 and x_2 is also a coordinate system. For example, we could define

$$\begin{aligned} y &\triangleq x_1 + x_2 \\ z &\triangleq x_2 - x_1 \end{aligned}$$

This y and z together unambiguously describe any state of the system, and the ordered set of numbers (y, z) and be converted back to (x_1, x_2) by the inverse transformation

$$\left\{ \begin{array}{c} y \\ z \end{array} \right\} = \left[\begin{array}{cc} 1 & 1 \\ -1 & 1 \end{array} \right] \left\{ \begin{array}{c} x_1 \\ x_2 \end{array} \right\}; \left\{ \begin{array}{c} x_1 \\ x_2 \end{array} \right\} = \left[\begin{array}{cc} 0.5 & -0.5 \\ 0.5 & 0.5 \end{array} \right] \left\{ \begin{array}{c} y \\ z \end{array} \right\}$$

Example continued: If we substitute $(y - z)/2$ for x_1 and $(y + z)/2$ for x_2 in our original Equation 20.1 for the equal-mass and equal-spring problem, we get

$$\frac{m}{g_c} \left[\begin{array}{cc} 1/2 & -1/2 \\ 1/2 & 1/2 \end{array} \right] \left\{ \begin{array}{c} \ddot{y} \\ \ddot{z} \end{array} \right\} + k \left[\begin{array}{cc} 1/2 & -3/2 \\ 1/2 & 3/2 \end{array} \right] \left\{ \begin{array}{c} y \\ z \end{array} \right\} = \left\{ \begin{array}{c} 0 \\ 0 \end{array} \right\}$$

which looks very odd until we combine the two equations this matrix expression represents into two new equations: one is the sum of the two original equations, and the other is the second equation minus the first equation, so that the mass matrix is diagonal

$$\frac{m}{g_c} \left[\begin{array}{cc} 1 & 0 \\ 0 & 1 \end{array} \right] \left\{ \begin{array}{c} \ddot{y} \\ \ddot{z} \end{array} \right\} + k \left[\begin{array}{cc} 1 & 0 \\ 0 & 3 \end{array} \right] \left\{ \begin{array}{c} y \\ z \end{array} \right\} = \left\{ \begin{array}{c} 0 \\ 0 \end{array} \right\}$$

The stiffness matrix is now also diagonal: there are no coupling terms. We have obtained two independent one-degree-of-freedom problems. This occurred because for our symmetrical example, the coordinates y and z are non-normalized versions of the principal coordinates. This demonstrates again that the right choice of coordinates, obtained from the modes, decouples our system.

In our example, we can stumble onto principal coordinates because of the obvious symmetry; but in general, we must find them by solving the eigenvector problem and defining, as we did in Section 19.4.5

$$\begin{aligned} \{x\} &= [u]\{p\} \\ \{p\} &\triangleq [u]^{-1}\{x\} \end{aligned}$$

Therefore we can find the initial conditions in the principal coordinates

$$\begin{aligned} \{p\}_{t=0} &\triangleq [u]^{-1}\{x\}_{t=0} \\ \left\{\dot{p}\right\}_{t=0} &\triangleq [u]^{-1}\left\{\dot{x}\right\}_{t=0} \end{aligned}$$

and solve the two equations

$$\begin{aligned} \ddot{p}_I + \omega_I^2 p_I &= 0 \\ \ddot{p}_{II} + \omega_{II}^2 p_{II} &= 0 \end{aligned}$$

as one-degree-of-freedom problems

$$\begin{aligned} p_I &= \frac{\dot{p}_I(0)}{\omega_I}\sin\omega_I t + p_I(0)\cos\omega_I t \\ p_{II} &= \frac{\dot{p}_{II}(0)}{\omega_{II}}\sin\omega_{II} t + p_{II}(0)\cos\omega_{II} t \end{aligned}$$

and convert back to the original coordinates

$$\{x\} = [u]\{p\}$$

Exercise 20.1 *Repeat the solution for our example with displacement initial conditions* $x_1(0) = A$ *and* $x_2(0) = 0$*, using the transformations with* $[u]$ *and* $[u]^{-1}$.

20.4 Weird Coordinates

What if we had chosen arbitrary non-absolute coordinates; what if we expressed our boxcar-system in terms of the coordinates x_1 and $z \triangleq (x_2 - x_1)$, for example? Going back to Equation 20.1 and replacing x_2 with $(x_1 + z)$, we get

$$\begin{aligned} \frac{m_1}{g_c}\ddot{x}_1 + k_1 x_1 + k_c(-z) &= 0 \\ \frac{m_2}{g_c}\left(\ddot{x}_1 + \ddot{z}\right) + k_2(x_1 + z) + k_c(z) &= 0 \end{aligned}$$

$$\frac{1}{g_c}\begin{bmatrix} m_1 & 0 \\ m_2 & m_2 \end{bmatrix}\begin{Bmatrix} \ddot{x}_1 \\ \ddot{z} \end{Bmatrix}+\begin{bmatrix} (k_1) & (-k_c) \\ (k_2) & (k_2+k_c) \end{bmatrix}\begin{Bmatrix} x_1 \\ z \end{Bmatrix}=\begin{Bmatrix} 0 \\ 0 \end{Bmatrix}$$

Obviously, in these coordinates the matrices are not symmetrical. Could they still be solved? Looking at our equal-masses and equal-springs example, we have

$$\frac{m}{g_c}\begin{bmatrix} 1 & 0 \\ 1 & 1 \end{bmatrix}\begin{Bmatrix} \ddot{x}_1 \\ \ddot{z} \end{Bmatrix}+k\begin{bmatrix} 1 & -1 \\ 1 & 2 \end{bmatrix}\begin{Bmatrix} x_1 \\ z \end{Bmatrix}=\begin{Bmatrix} 0 \\ 0 \end{Bmatrix}$$

To bring this into standard form, we pre-multiply each term by the inverse of the mass matrix, obtained from *Matlab* or *Maple* (Section 1.5)

$$\begin{bmatrix} 1 & 0 \\ -1 & 1 \end{bmatrix}\begin{bmatrix} 1 & 0 \\ 1 & 1 \end{bmatrix}\begin{Bmatrix} \ddot{x}_1 \\ \ddot{z} \end{Bmatrix}+\tfrac{kg_c}{m}\begin{bmatrix} 1 & 0 \\ -1 & 1 \end{bmatrix}\begin{bmatrix} 1 & -1 \\ 1 & 2 \end{bmatrix}\begin{Bmatrix} x_1 \\ z \end{Bmatrix}$$
$$=\tfrac{g_c}{m}\begin{bmatrix} 1 & 0 \\ -1 & 1 \end{bmatrix}\begin{Bmatrix} 0 \\ 0 \end{Bmatrix}$$

obtaining an identity matrix and the characteristic matrix

$$\begin{bmatrix} 1 & 0 \\ 0 & 1 \end{bmatrix}\begin{Bmatrix} \ddot{x}_1 \\ \ddot{z} \end{Bmatrix}+\frac{kg_c}{m}\begin{bmatrix} 1 & -1 \\ 0 & 3 \end{bmatrix}\begin{Bmatrix} x_1 \\ z \end{Bmatrix}=\begin{Bmatrix} 0 \\ 0 \end{Bmatrix}$$

which gives us the same eigenvalues as before. However, the eigenvectors in these coordinates are

$$\begin{Bmatrix} x_1/x_1 \\ z/x_1 \end{Bmatrix}_{I,II}=\begin{Bmatrix} 1 \\ 0 \end{Bmatrix},\ \begin{Bmatrix} 1 \\ -2 \end{Bmatrix},\text{ respectively}$$

Going back to the definition of z, we find that these eigenvectors represent the same relative motion as the solution in x_1 and x_2.

20.5 Simple Procedure for Unsymmetrical Cases

In order to keep it simple, we have used a symmetrical system for our numerical example. Systems with different-sized masses work the same way, after we take steps to convert the mass matrix to an identity matrix.

We can test this on an example from the automotive industry: the quarter-car model (Figure 20.2) with an unsprung weight $m_1 = m_{ref}$, and a sprung weight $m_2 = 10m_{ref}$; the first spring is the tire $k_1 = 10k_{ref}$, and the coupling spring is the suspension $k_c = k_{ref}$. If we neglect damping (a bad assumption for a well-maintained car!) we obtain the equation

$$\frac{m_{ref}}{g_c}\begin{bmatrix} 1 & 0 \\ 0 & 10 \end{bmatrix}\begin{Bmatrix} \ddot{x}_1 \\ \ddot{x}_2 \end{Bmatrix}+k_{ref}\begin{bmatrix} 11 & -1 \\ -1 & 1 \end{bmatrix}\begin{Bmatrix} x_1 \\ x_2 \end{Bmatrix}=\begin{Bmatrix} 0 \\ 0 \end{Bmatrix} \tag{20.5}$$

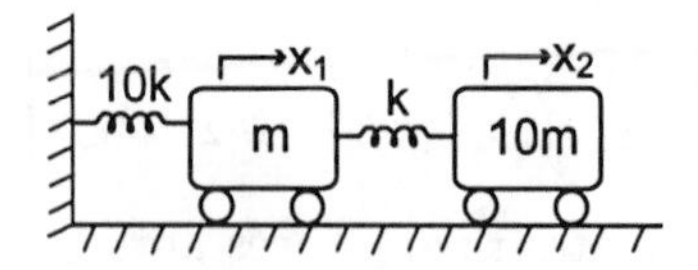

Figure 20.2: Quarter-Car Model

Pre-multiplying each term with the inverse of the mass matrix (see Equation 20.4), we obtain

$$\begin{bmatrix} 1 & 0 \\ 0 & 1 \end{bmatrix} \begin{Bmatrix} \ddot{x}_1 \\ \ddot{x}_2 \end{Bmatrix} + \frac{k_{ref} g_c}{m_{ref}} \begin{bmatrix} 11 & -1 \\ -0.1 & 0.1 \end{bmatrix} \begin{Bmatrix} x_1 \\ x_2 \end{Bmatrix} = \frac{g_c}{m_{ref}} \begin{Bmatrix} 0 \\ 0 \end{Bmatrix} \tag{20.6}$$

which differs from our symmetrical example by having a non-symmetrical characteristic matrix. Nevertheless, defining $\lambda \triangleq \left(\frac{m_{ref}}{k_{ref} g_c}\right) \omega^2$ reduces it to the standard eigenvalue problem:

$$\det \begin{bmatrix} 11-\lambda & -1 \\ -0.1 & 0.1-\lambda \end{bmatrix} = 0$$

leading to $\lambda_{I,II} = 0.090,833,\ 11.009$, respectively. Solving for

$$\frac{x_2}{x_1} = 11 - \lambda = \frac{0.1}{0.1 - \lambda} = 10.909,\ 0.009\,166, \text{ resp.}$$

and the normalized eigenvectors are

$$[u] = \left[\begin{Bmatrix} 0.091 \\ 0.996 \end{Bmatrix}_I, \begin{Bmatrix} -1.00 \\ 0.009 \end{Bmatrix}_{II} \right]$$

and, unlike those in our symmetrical example, are not orthogonal. Nevertheless, we can use this matrix of eigenvectors to transform Equation 20.6 to principal coordinates, and obtain

$$\begin{bmatrix} 1 & 0 \\ 0 & 1 \end{bmatrix} \begin{bmatrix} 0.091 & -1.00 \\ 0.996 & 0.009 \end{bmatrix} \begin{Bmatrix} \ddot{q}_I \\ \ddot{q}_{II} \end{Bmatrix}$$

$$+ \frac{k_{ref} g_c}{m_{ref}} \begin{bmatrix} 11 & -1 \\ -0.1 & 0.1 \end{bmatrix} \begin{bmatrix} 0.091 & -1.00 \\ 0.996 & 0.009 \end{bmatrix} \begin{Bmatrix} q_I \\ q_{II} \end{Bmatrix} = \frac{g_c}{m_{ref}} \begin{Bmatrix} 0 \\ 0 \end{Bmatrix}$$

which multiplies out to

$$\begin{bmatrix} 0.091 & -1.00 \\ 0.996 & 0.009 \end{bmatrix} \begin{Bmatrix} \ddot{q}_I \\ \ddot{q}_{II} \end{Bmatrix} + \frac{k_{ref} g_c}{m_{ref}} \begin{bmatrix} 0.008 & -11.01 \\ 0.090 & 0.101 \end{bmatrix} \begin{Bmatrix} q_I \\ q_{II} \end{Bmatrix} = \begin{Bmatrix} 0 \\ 0 \end{Bmatrix}$$

If we pre-multiply by the inverse of the first matrix, we obtain

$$\begin{bmatrix} 0.009 & 1.003 \\ -.999 & 0.092 \end{bmatrix} \begin{bmatrix} 0.091 & -1.00 \\ 0.996 & 0.009 \end{bmatrix} \left\{ \begin{array}{c} \ddot{q}_I \\ \ddot{q}_{II} \end{array} \right\}$$
$$+ \frac{k_{ref} g_c}{m_{ref}} \begin{bmatrix} 0.009 & 1.003 \\ -.999 & 0.092 \end{bmatrix} \begin{bmatrix} 0.008 & -11.01 \\ 0.090 & 0.101 \end{bmatrix} \left\{ \begin{array}{c} q_I \\ q_{II} \end{array} \right\} =$$
$$= \frac{g_c}{m_{ref}} \left\{ \begin{array}{c} 0 \\ 0 \end{array} \right\}$$

which multiplies out to

$$\begin{bmatrix} 1.000 & 10^{-5} \\ 10^{-5} & 1.000 \end{bmatrix} \left\{ \begin{array}{c} \ddot{q}_1 \\ \ddot{q}_2 \end{array} \right\} + \frac{k_{ref} g_c}{m_{ref}} \begin{bmatrix} 0.0908 & 10^{-5} \\ 10^{-5} & 11.009 \end{bmatrix} \left\{ \begin{array}{c} q_1 \\ q_2 \end{array} \right\} = \frac{g_c}{m_{ref}} \left\{ \begin{array}{c} 0 \\ 0 \end{array} \right\}$$

The modal matrix is *not* orthonormal, so we had to obtain the inverse for the reverse transformation (and couldn't make do with the transpose). If we had multiplied each element by the square root of the mass associated with it, we would have obtained orthogonal vectors

$$\left[\left\{ \begin{array}{c} 0.091\sqrt{1} \\ 0.996\sqrt{10} \end{array} \right\}_I , \left\{ \begin{array}{c} -1.00\sqrt{1} \\ 0.009\sqrt{10} \end{array} \right\}_{II} \right] = \left[\left\{ \begin{array}{c} 0.091 \\ 3.15 \end{array} \right\}_I , \left\{ \begin{array}{c} -1.00 \\ 0.029 \end{array} \right\}_{II} \right]$$
$$= \left[\left\{ \begin{array}{c} 0.029 \\ 1.00 \end{array} \right\}_I , \left\{ \begin{array}{c} -1.00 \\ 0.029 \end{array} \right\}_{II} \right]$$

We will use this idea for a more elegant procedure in Section 20.6.

An interesting feature of this particular system is that the two natural frequencies are widely separated (by a factor of about 11), and that each mode's motion is associated mainly with only one of the coordinates. In such an extreme case, motion of the car itself at a low frequency can be analyzed approximately as if the small unsprung mass was forced to follow the sprung mass in quasi-static proportional motion; in other words, as if we had a single mass $m_1 = 10m_{ref}$ on a pure spring $k_{eff} = 0.909k_{ref}$. Similarly, motion of the unsprung mass at high frequency can be treated approximately as if the car itself were stationary; in other words, as if we had single mass $m_2 = m_{eff}$ on a combined spring $k_{eff} = 11k_{ref}$. When we add damping, this situation will change dramatically (Chapter 25).

Exercise 20.2 *Solve a boxcar-system with $m_2 = 2m_1$, $k_2 = 2k_1$, and $k_c = 4k_1$.*

If you have followed our procedure using a computer, you may have found that the unsymmetrical characteristic matrix led to inaccuracies in the eigenvectors, and that round-off errors caused non-zero off-diagonal terms in the supposedly decoupled equations. One solution is to compute with double accuracy.

A more elegant procedure keeps the characteristic equation symmetrical and the modal matrix orthonormal.

20.6 Elegant Procedure

Let us repeat the quarter-car example of the previous Section 20.5, rescaling the original coordinate system with the square-root of the mass associated with each coordinate. We replace the displacement vector by

$$\begin{bmatrix} \sqrt{\frac{1}{m_1}} & 0 \\ 0 & \sqrt{\frac{1}{m_2}} \end{bmatrix} \begin{Bmatrix} \sqrt{m_1} x_1 \\ \sqrt{m_2} x_2 \end{Bmatrix} = \begin{Bmatrix} x_1 \\ x_2 \end{Bmatrix}$$

where the first matrix is $[M]^{\frac{-1}{2}}$, the inverse square-root of the mass matrix, defined as that matrix which, multiplied by itself, yields the inverse of the mass matrix. For a diagonal mass matrix, inverse and square-root and inverse square-root can be written down term-by-term without effort; but when we encounter off-diagonal terms in the mass matrix (in Chapter 22), substantial computing effort will be required to obtain the square-root.

Inserting this weighted coordinate system into Equation 20.5

$$\begin{bmatrix} 1 & 0 \\ 0 & 10 \end{bmatrix} \begin{bmatrix} 1 & 0 \\ 0 & \sqrt{\frac{1}{10}} \end{bmatrix} \begin{Bmatrix} \ddot{x}_1 \\ \sqrt{10}\, \ddot{x}_2 \end{Bmatrix}$$
$$+ \frac{k_{ref} g_c}{m_{ref}} \begin{bmatrix} 11 & -1 \\ -1 & 1 \end{bmatrix} \begin{bmatrix} 1 & 0 \\ 0 & \sqrt{\frac{1}{10}} \end{bmatrix} \begin{Bmatrix} x_1 \\ \sqrt{10} x_2 \end{Bmatrix} = \frac{g_c}{m_{ref}} \begin{Bmatrix} 0 \\ 0 \end{Bmatrix}$$

and pre-multiplying each term with the same inverse of square-root of the mass matrix, we obtain

$$\begin{bmatrix} 1 & 0 \\ 0 & \sqrt{\frac{1}{10}} \end{bmatrix} \begin{bmatrix} 1 & 0 \\ 0 & 10 \end{bmatrix} \begin{bmatrix} 1 & 0 \\ 0 & \sqrt{\frac{1}{10}} \end{bmatrix} \begin{Bmatrix} \ddot{x}_1 \\ \sqrt{10}\, \ddot{x}_2 \end{Bmatrix}$$
$$+ \frac{k_{ref} g_c}{m_{ref}} \begin{bmatrix} 1 & 0 \\ 0 & \sqrt{\frac{1}{10}} \end{bmatrix} \begin{bmatrix} 11 & -1 \\ -1 & 1 \end{bmatrix} \begin{bmatrix} 1 & 0 \\ 0 & \sqrt{\frac{1}{10}} \end{bmatrix} \begin{Bmatrix} x_1 \\ \sqrt{10} x_2 \end{Bmatrix} = \frac{g_c}{m_{ref}} \begin{Bmatrix} 0 \\ 0 \end{Bmatrix}$$

which multiplies out to

$$\begin{bmatrix} 1 & 0 \\ 0 & 1 \end{bmatrix} \begin{Bmatrix} \ddot{x}_1 \\ \sqrt{10}\, \ddot{x}_2 \end{Bmatrix} + \frac{k_{ref} g_c}{m_{ref}} \begin{bmatrix} 11 & \frac{-\sqrt{10}}{10} \\ \frac{-\sqrt{10}}{10} & \frac{1}{10} \end{bmatrix} \begin{Bmatrix} x_1 \\ \sqrt{10} x_2 \end{Bmatrix} = \frac{g_c}{m_{ref}} \begin{Bmatrix} 0 \\ 0 \end{Bmatrix} \tag{20.7}$$

which has a symmetrical characteristic matrix. Apparently we have found a better way of converting the mass matrix to an identity matrix. In our simple procedure of Section 20.5, we had divided equations (rows) by mass terms; but we see now that we can also rescale coordinates and divide columns by mass terms. By splitting the adjustment equally between altering rows and altering columns, we keep the off-diagonal terms symmetrical.

Defining $\lambda \triangleq \left(\frac{m_{ref}}{k_{ref}g_c}\right)\omega^2$ gives us a standard eigenvalue problem again (but this time with a symmetrical characteristic matrix)

$$\det\begin{bmatrix} 11-\lambda & \frac{-\sqrt{10}}{10} \\ \frac{-\sqrt{10}}{10} & \frac{1}{10}-\lambda \end{bmatrix} = 0$$

leading to the same characteristic equation as before, and the same $\lambda_{I,II} = 0.090,833,\ 11.009$, respectively. The normalized eigenvectors are now orthogonal and, when properly normalized, orthonormal.

$$[u] = \left[\left\{\begin{array}{c} 0.028,975 \\ 0.999,58 \end{array}\right\}_I \left\{\begin{array}{c} -0.999,58 \\ 0.028,975 \end{array}\right\}_{II}\right]$$

We can use this matrix of eigenvectors to transform the Equation 20.7 to principal coordinates by replacing $\{\sqrt{m}x\}$ with $[u]\{p\}$, and pre-multiplying each term by $[u]^{-1}$ to renormalize the first matrix; conveniently, the inverse of the modal matrix is now its transpose $[u]^T$

$$[u]^T\begin{bmatrix} 1 & 0 \\ 0 & 1 \end{bmatrix}[u]\left\{\begin{array}{c} \ddot{p}_1 \\ \ddot{p}_2 \end{array}\right\} + \frac{k_{ref}g_c}{m_{ref}}\left[[u]^T\begin{bmatrix} 11 & \frac{-\sqrt{10}}{10} \\ \frac{-\sqrt{10}}{10} & \frac{1}{10} \end{bmatrix}[u]\right]\left\{\begin{array}{c} p_1 \\ p_2 \end{array}\right\} = \left\{\begin{array}{c} 0 \\ 0 \end{array}\right\}$$

$$\begin{bmatrix} 1 & 0 \\ 0 & 1 \end{bmatrix}\left\{\begin{array}{c} \ddot{p}_1 \\ \ddot{p}_2 \end{array}\right\} + \frac{k_{ref}g_c}{m_{ref}}\begin{bmatrix} 0.090\,833 & 10^{-6} \\ 10^{-6} & 11.009 \end{bmatrix}\left\{\begin{array}{c} p_1 \\ p_2 \end{array}\right\} = \left\{\begin{array}{c} 0 \\ 0 \end{array}\right\}$$

The first matrix is the identity matrix again; the second matrix is called the "spectral matrix," because its elements contain the characteristic frequencies $\omega_{I,II} = \sqrt{0.090\,833\frac{k_{ref}g_c}{m_{ref}}};\ \sqrt{11.009\frac{k_{ref}g_c}{m_{ref}}}$, respectively. The off-diagonal terms are zero, except for small round-off errors.

20.7 Multi-Mass Problems

We have used the matrix notation because it allows us to extend all our procedures to any number of connected masses. For example, if we have a system of three identical flywheels connected to each other and to ground by five identical torsion bars, our governing equation is

$$\frac{J}{g_c}\begin{bmatrix} 1 & 0 & 0 \\ 0 & 1 & 0 \\ 0 & 0 & 1 \end{bmatrix}\left\{\begin{array}{c} \ddot{\theta}_1 \\ \ddot{\theta}_2 \\ \ddot{\theta}_3 \end{array}\right\} + k_T\begin{bmatrix} 2 & -1 & 0 \\ -1 & 2 & -1 \\ 0 & -1 & 2 \end{bmatrix}\left\{\begin{array}{c} \theta_1 \\ \theta_2 \\ \theta_3 \end{array}\right\} = \left\{\begin{array}{c} 0 \\ 0 \\ 0 \end{array}\right\}$$

Assuming sinusoidal solutions, $\ddot{\theta} = \omega^2\theta$, and defining $\lambda \triangleq \left(\frac{J}{k_T g_c}\right)\omega^2$

$$\frac{k_T g_c}{J}\begin{bmatrix} 2-\lambda & -1 & 0 \\ -1 & 2-\lambda & -1 \\ 0 & -1 & 2-\lambda \end{bmatrix}\begin{Bmatrix} \theta_1 \\ \theta_2 \\ \theta_3 \end{Bmatrix} = \begin{Bmatrix} 0 \\ 0 \\ 0 \end{Bmatrix}$$

This is a Standard Form eigenvalue problem; entering the characteristic matrix into *Matlab* or *Maple* will show that the eigenvalues are $\lambda_{I,II,III} = 0.5858$, 2.0, 3.4142, respectively.

We can now discard any one of the three equations, and solve the remaining two for the eigenvectors θ_2/θ_1 and θ_3/θ_1. *Matlab* or *Maple* will obtain the (orthonormal) eigenvectors

$$[u] = \left[\begin{Bmatrix} 0.5 \\ 0.707 \\ 0.5 \end{Bmatrix}_I \begin{Bmatrix} -0.707 \\ 0 \\ 0.707 \end{Bmatrix}_{II} \begin{Bmatrix} 0.5 \\ -0.707 \\ 0.5 \end{Bmatrix}_{III}\right]$$

If we draw these as string figures, we note that the first mode has all the flywheels moving in the same direction; the second mode has a node at the center, and the third mode has two nodes (between the flywheels).

Converting the characteristic matrix to principal coordinates

$$\begin{bmatrix} 0.5 & 0.707 & 0.5 \\ -0.707 & 0 & 0.707 \\ 0.5 & -0.707 & 0.5 \end{bmatrix}\begin{bmatrix} 2.0 & -1 & 0 \\ -1 & 2 & -1 \\ 0 & -1 & 2 \end{bmatrix}\begin{bmatrix} 0.5 & -0.707 & 0.5 \\ 0.707 & 0 & -0.707 \\ 0.5 & 0.707 & 0.5 \end{bmatrix}$$

we obtain the spectral matrix and the governing equation becomes

$$\begin{bmatrix} 1 & 0 & 0 \\ 0 & 1 & 0 \\ 0 & 0 & 1 \end{bmatrix}\begin{Bmatrix} \ddot{p}_1 \\ \ddot{p}_2 \\ \ddot{p}_3 \end{Bmatrix} + \frac{k_T g_c}{J}\begin{bmatrix} 0.5858 & 0 & 0 \\ 0 & 2.0 & 0 \\ 0 & 0 & 3.4142 \end{bmatrix}\begin{Bmatrix} p_1 \\ p_2 \\ p_3 \end{Bmatrix} = \begin{Bmatrix} 0 \\ 0 \\ 0 \end{Bmatrix}$$

Exercise 20.3 *Solve a system with four equal masses and springs.*

20.8 Semi-Definite Problems

What happens if there are no springs to ground? If the coupling spring in Equation 20.1 is the only spring, we get

$$\frac{1}{g_c}\begin{bmatrix} m_1 & 0 \\ 0 & m_2 \end{bmatrix}\begin{Bmatrix} \ddot{x}_1 \\ \ddot{x}_2 \end{Bmatrix} + \begin{bmatrix} k_c & -k_c \\ -k_c & k_c \end{bmatrix}\begin{Bmatrix} x_1 \\ x_2 \end{Bmatrix} = \begin{Bmatrix} 0 \\ 0 \end{Bmatrix}$$

which normalizes to

$$\frac{1}{g_c}\begin{bmatrix} 1 & 0 \\ 0 & 1 \end{bmatrix}\begin{Bmatrix} \ddot{x}_1 \\ \ddot{x}_2 \end{Bmatrix} + \begin{bmatrix} \frac{k_c}{m_1} & \frac{-k_c}{m_1} \\ \frac{-k_c}{m_2} & \frac{k_c}{m_2} \end{bmatrix}\begin{Bmatrix} x_1 \\ x_2 \end{Bmatrix} = \begin{Bmatrix} 0 \\ 0 \end{Bmatrix}$$

for a characteristic equation

$$\left(\omega_n^2\right)^2 - \left(\frac{k_c}{m_1} + \frac{k_c}{m_2}\right)\omega_n^2 + 0 = 0$$

with the roots $\omega_n^2 = 0, \left(\frac{k_c}{m_1} + \frac{k_c}{m_2}\right)$, and the eigenvectors

$$[u]_{I,II} = \left[\left\{\begin{array}{c} 1/\sqrt{2} \\ 1/\sqrt{2} \end{array}\right\}\left\{\begin{array}{c} m_2/\sqrt{m_1^2+m_2^2} \\ m_1/\sqrt{m_1^2+m_2^2} \end{array}\right\}\right]$$

The zero frequency represents the center-of-gravity of the combined boxcars either standing still, or else rolling away at constant speed.

20.9 Summary

We have demonstrated some properties of undamped multi-degree-of-freedom systems on a simple boxcar-style model. To take care of non-uniform terms in the mass matrix, we can use the simple procedure of pre-multiplying all terms of the governing equation by the inverse of the mass matrix

$$\begin{aligned} [M]^{-1}[M]\left\{\ddot{x}\right\} + g_c [M]^{-1}[K]\{x\} &= \{0\} \\ [I]\left\{\ddot{x}\right\} + g_c \left[M^{-1}K\right]\{x\} &= \{0\} \end{aligned}$$

generating the (unsymmetrical) characteristic matrix $\left[M^{-1}K\right]$; from which we can obtain the eigenvalues and the (non-orthogonal) eigenvectors We use of the modal matrix $[u]$ to transform to principal coordinates

$$[I][u]\left\{\ddot{q}\right\} + g_c\left[M^{-1}K\right][u]\{q\} = \{0\}$$

and its inverse $[u]^{-1}$ to reestablish the identity matrix

$$\left[[u]^{-1}[I][u]\right]\left\{\ddot{q}\right\} + g_c\left[[u]^{-1}\left[M^{-1}K\right][u]\right]\{q\} = \{0\}$$

and find the spectral matrix $\left[[u]^{-1}\left[M^{-1}K\right][u]\right]$.

If the mass matrix is diagonal, we may prefer use the more elegant procedure of using the square-root of the inverse of the mass matrix

$$\begin{aligned} [M]^{\frac{-1}{2}}[M][M]^{\frac{-1}{2}}\left\{\sqrt{m}\,\ddot{x}\right\} + g\,[M]^{\frac{-1}{2}}[K][M]^{\frac{-1}{2}}\left\{\sqrt{m}x\right\} &= \{0\} \\ [I]\left\{\sqrt{m}\,\ddot{x}\right\} + g_c\left[M^{\frac{-1}{2}}KM^{\frac{-1}{2}}\right]\left\{\sqrt{m}x\right\} &= \{0\} \end{aligned}$$

generating the symmetrical characteristic matrix $\left[M^{\frac{-1}{2}} K M^{\frac{-1}{2}}\right]$, from which we can obtain the eigenvalues and orthonormal eigenvectors. We use the modal matrix $[u]$ to transform to principal coordinates

$$[I]\,[u]\left\{\ddot{q}\right\} + g_c\left[\left[M^{\frac{-1}{2}} K M^{\frac{-1}{2}}\right]\right][u]\,\{q\} = \{0\}$$

and its inverse $[u]^{-1}$, which is now equal to its transpose $[u]^{T}$, to reestablish the identity matrix

$$\left[[u]^{T}\,[I]\,[u]\right]\left\{\ddot{q}\right\} + g_c\left[[u]^{T}\left[M^{\frac{-1}{2}} K M^{\frac{-1}{2}}\right][u]\right]\{q\} = \{0\}$$

and find the spectral matrix $\left[[u]^{T}\left[M^{\frac{-1}{2}} K M^{\frac{-1}{2}}\right][u]\right]$.

There are as many natural frequencies as degrees-of-freedom, and a mode shape associated with each frequency. Whether all these modes occur depends on the initial conditions.

Torsional systems are often boxcar-type systems, with moments-of-inertia substituted for the masses, torsional spring constants for the springs, angles for the displacements, and moments for the forces.

Problem 20.4 *For the system shown below, find the characteristic matrix, the natural frequencies, the modes, and the response to a unit initial deflection of the x_1-coordinate.*

Problem 20.5 *For the system shown below, find the characteristic matrix, the natural frequencies, the modes, and the response to a unit initial deflection of the x_2-coordinate.*

Problem 20.6 *For the system shown below, find the characteristic matrix, the natural frequencies, the modes, and the response to a unit initial velocity of the x_1-coordinate.*

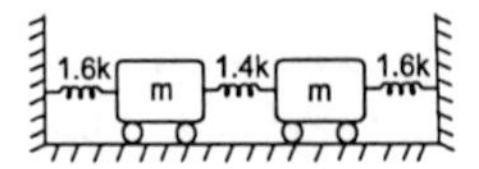

Problem 20.7 *For the system shown below, find the characteristic matrix, the natural frequencies, the modes, and the response to a unit initial velocity of the x_2-coordinate.*

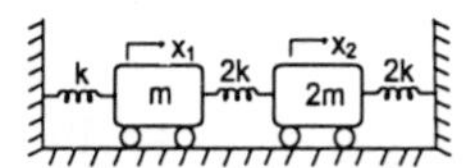

Problem 20.8 *For the system shown below, find the characteristic matrix, the natural frequencies, the modes, and the response to a unit initial deflection of the x_1-coordinate.*

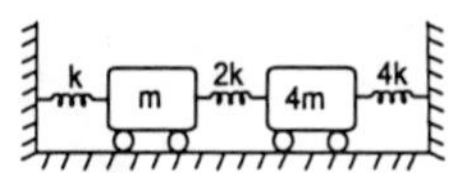

Problem 20.9 *For the system shown below, find the characteristic matrix, the natural frequencies, the modes, and the response to a unit initial deflection of the x_1-coordinate.*

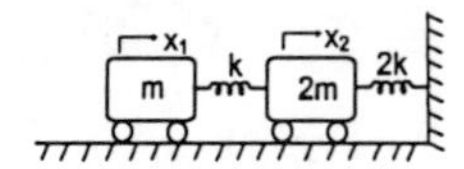

Chapter 21

COMBINED TRANSLATION AND ROTATION and mass coupling

In the preceding chapters, we showed that the principal coordinates are those having no coupling coefficients. We will now examine the coupling terms in a variety of other coordinate systems, using the example of an extensive but rigid mass which is free to move in one translational direction as well as in rotation (Figure 21.1).

21.1 Mass Coupling

Let us look at the problem of a yardstick of mass m, suspended by identical springs k at each end. If we associate a coordinate x with each spring, the stiffness matrix (obtained from Rule 1a in Section 19.3) is diagonal, but the mass matrix (obtained from Rule 3) has off-diagonal terms which couple these

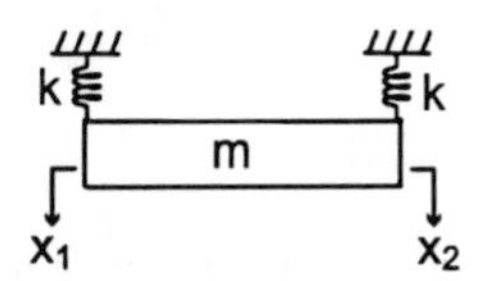

Figure 21.1: Mass Coupling

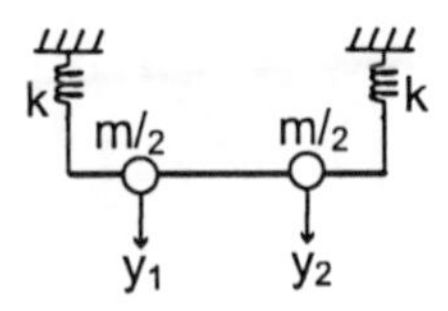

Figure 21.2: Spring Coupling

coordinates:

$$\frac{m}{g_c}\begin{bmatrix} 1/3 & 1/6 \\ 1/6 & 1/3 \end{bmatrix}\begin{Bmatrix} \ddot{x}_1 \\ \ddot{x}_2 \end{Bmatrix} + k\begin{bmatrix} 1 & 0 \\ 0 & 1 \end{bmatrix}\begin{Bmatrix} x_1 \\ x_2 \end{Bmatrix} = \begin{Bmatrix} 0 \\ 0 \end{Bmatrix}$$

To put this into the standard form for solution, we need to make the first matrix the identity matrix. We can do this by multiplying each term by the inverse of the mass matrix

$$\begin{bmatrix} 4 & -2 \\ -2 & 4 \end{bmatrix}\begin{bmatrix} 1/3 & 1/6 \\ 1/6 & 1/3 \end{bmatrix}\begin{Bmatrix} \ddot{x}_1 \\ \ddot{x}_2 \end{Bmatrix} + \frac{kg_c}{m}\begin{bmatrix} 4 & -2 \\ -2 & 4 \end{bmatrix}\begin{bmatrix} 1 & 0 \\ 0 & 1 \end{bmatrix}\begin{Bmatrix} x_1 \\ x_2 \end{Bmatrix}$$

to obtain

$$\begin{bmatrix} 1 & 0 \\ 0 & 1 \end{bmatrix}\begin{Bmatrix} \ddot{x}_1 \\ \ddot{x}_2 \end{Bmatrix} + \frac{kg_c}{m}\begin{bmatrix} 4 & -2 \\ -2 & 4 \end{bmatrix}\begin{Bmatrix} x_1 \\ x_2 \end{Bmatrix} = \begin{Bmatrix} 0 \\ 0 \end{Bmatrix}$$

We can now solve for the eigenvalues $\lambda_{I,II} \triangleq \omega_{I,II}^2 m/kg_c = 2, 6$, respectively, and the eigenvectors

$$[u] = \left[\begin{Bmatrix} 0.707 \\ 0.707 \end{Bmatrix} \begin{Bmatrix} -0.707 \\ 0.707 \end{Bmatrix} \right]$$

For this symmetrical problem, the first mode is parallel motion, and the second mode is rotation about the center-of-gravity.

Exercise 21.1 *Can you find any other pairs of coordinates which result in a diagonal stiffness matrix? Hint: specify one coordinate, say the displacement of the twelve-inch-mark on the yardstick, and then find the point on the stick which stays stationary when a steady force is applied to this point.*

21.2 Spring Coupling

We could have modeled the yardstick in this problem as a dumbbell consisting of two masses $m/2$ on a weightless rigid bar. (Figure 21.2) In order to give us the correct moment of inertia, these masses would have to be a distance $r_g = L/\sqrt{12}$

from the center of the yardstick, at the 7.6 and 28.4-inch marks. If we place our coordinates y at these locations, the mass matrix will be diagonal, because these locations are a pair of centers-of-percussion, but the stiffness matrix will have coupling terms which we obtain from **Rule 1a** in Section 19.3.

For the first column of terms in $[K]$ we apply a unit deflection at y_1 and a zero deflection at y_2. A sketch of the deflections shows that the left end of the yardstick moves a distance $(1+\sqrt{3})/2 = 1.366$, and the right end $(1-\sqrt{3})/2 = -0.366$. A sketch of the forces shows a bar with a force of $-1.366k$ at the left end, and $+0.366k$ at the right end. A torque balance around the 28.4-inch-mark requires a force at the 7.6-inch-mark of $k_{11} = (1.336 \times 1.366k + 0.366 \times 0.366k) = 2.0k$. A torque balance around the 7.6-inch-mark requires a force at the 28.4-inch-mark of $k_{21} = (0.336 \times 1.366k + 1.366 \times 0.366k) = -1.0k$. A force balance can be used to double-check that the sum of all four forces is zero.

For the second column of terms in $[K]$ we apply a unit deflection at y_2 and a zero deflection at y_1. Because of the symmetry of this example, the deflection sketch and the force sketch are mirror images of the first-column analysis; $k_{12} = -1.0k$ and $k_{22} = 2.0k$.

$$\frac{m}{g_c}\begin{bmatrix} \frac{1}{2} & 0 \\ 0 & \frac{1}{2} \end{bmatrix}\begin{Bmatrix} \ddot{y}_1 \\ \ddot{y}_2 \end{Bmatrix} + k\begin{bmatrix} 2 & -1 \\ -1 & 2 \end{bmatrix}\begin{Bmatrix} y_1 \\ y_2 \end{Bmatrix} = \begin{Bmatrix} 0 \\ 0 \end{Bmatrix}$$

Pre-multiplying each term by the inverse of the mass matrix

$$\begin{bmatrix} 2 & 0 \\ 0 & 2 \end{bmatrix}\begin{bmatrix} \frac{1}{2} & 0 \\ 0 & \frac{1}{2} \end{bmatrix}\begin{Bmatrix} \ddot{y}_1 \\ \ddot{y}_2 \end{Bmatrix} + \frac{kg_c}{m}\begin{bmatrix} 2 & 0 \\ 0 & 2 \end{bmatrix}\begin{bmatrix} 2 & -1 \\ -1 & 2 \end{bmatrix}\begin{Bmatrix} y_1 \\ y_2 \end{Bmatrix} = \begin{Bmatrix} 0 \\ 0 \end{Bmatrix}$$

$$\begin{bmatrix} 1 & 0 \\ 0 & 1 \end{bmatrix}\begin{Bmatrix} \ddot{y}_1 \\ \ddot{y}_2 \end{Bmatrix} + \frac{kg_c}{m}\begin{bmatrix} 4 & -2 \\ -2 & 4 \end{bmatrix}\begin{Bmatrix} y_1 \\ y_2 \end{Bmatrix} = \begin{Bmatrix} 0 \\ 0 \end{Bmatrix}$$

We get the same eigenvalues $\lambda_{I,II} \triangleq \omega_{I,II}^2 m/kg_c = 2,\ 6$, respectively. The eigenvectors should look different in the new coordinate system, but because of the symmetry have the same appearance. We have demonstrated that the same physical problem can appear to be mass-coupled or spring-coupled, depending on the choice of coordinate system.

Exercise 21.2 *Coordinates located at centers of percussion automatically give decoupled mass matrices. Can you find other pairs of centers-of-percussion for a yardstick? Hint: look for asymmetrical dumbbells consisting of two concentrated masses which meet the following criteria: the masses must add up to the correct total mass m; their locations must be such that the center of gravity is in the right place; and spacing must also be such that their moment of inertia J_{cg} equals $mL^2/12$.*

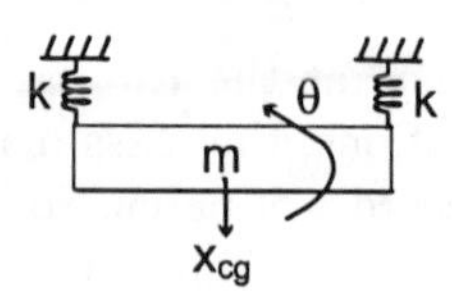

Figure 21.3: Preferred Coordinates

21.3 Preferred Raw Coordinates

If you verified the mass matrix in Section 21.1, you will have observed that using Rule 3 from Section 19.3 to obtain off-diagonal terms can be tedious. In general, the preferred coordinate system for two-dimensional rigid bodies uses the displacement of the center-of-gravity as one coordinate (or two, if y-direction motion is also possible), and the rotation angle as an additional coordinate. (Figure 21.3) Applying Rule 3 with appropriate consideration for the different dimensionality (meters and radians) of the two coordinates, we easily find that associating the displacement with the center-of-gravity yields a diagonal mass matrix

$$\frac{1}{g_c}\begin{bmatrix} m & 0 \\ 0 & J_{cg} \end{bmatrix}\begin{Bmatrix} \ddot{x}_{cg} \\ \ddot{\theta} \end{Bmatrix}$$

but we must still apply Rule 1 to obtain the stiffness matrix.

The preferred coordinate system can be extended to six-dimensional motion of an extended rigid body, resulting in

$$\frac{1}{g_c}\begin{bmatrix} m & 0 & 0 & 0 & 0 & 0 \\ 0 & m & 0 & 0 & 0 & 0 \\ 0 & 0 & m & 0 & 0 & 0 \\ 0 & 0 & 0 & J_{xx} & J_{xy} & J_{xz} \\ 0 & 0 & 0 & J_{xy} & J_{yy} & J_{yz} \\ 0 & 0 & 0 & J_{xz} & J_{yz} & J_{zz} \end{bmatrix}\begin{Bmatrix} \ddot{x}_{cg} \\ \ddot{y}_{cg} \\ \ddot{z}_{cg} \\ \ddot{\theta} \\ \ddot{\phi} \\ \ddot{\varphi} \end{Bmatrix}$$

which, if the coordinates line up with the principal axes of the second moments of the body, is a diagonal matrix.

21.4 Principal Coordinates

For our example of a yardstick supported by springs at the ends, the preferred coordinate system yields

$$\frac{1}{g_c}\begin{bmatrix} m & 0 \\ 0 & mL^2/12 \end{bmatrix}\begin{Bmatrix} \ddot{x}_{cg} \\ \ddot{\theta} \end{Bmatrix} + \begin{bmatrix} 2k & 0 \\ 0 & kL^2/2 \end{bmatrix}\begin{Bmatrix} x_{cg} \\ \theta \end{Bmatrix} = \begin{Bmatrix} 0 \\ 0 \end{Bmatrix}$$

Because this coordinate system happens to be identical to our modes for this symmetrical system, we have stumbled onto the principal coordinates: both coefficient matrices turn out to be diagonal, and the equations are decoupled. In general, there is a whole set of coordinate systems which decouple the mass matrix, and another set which decouple the stiffness matrix, but only one coordinate system which does both at the same time, and which is found by solving for the eigenvectors. A special exception is demonstrated in the next section.

21.5 Physical Decoupling

Let us modify the previous system by placing an additional support spring of stiffness $4k$ at the center of the yardstick. In our various coordinate systems, the governing equations now become

$$\frac{m}{g_c}\begin{bmatrix} 1/3 & 1/6 \\ 1/6 & 1/3 \end{bmatrix}\begin{Bmatrix} \ddot{x}_1 \\ \ddot{x}_2 \end{Bmatrix} + k\begin{bmatrix} 2 & 1 \\ 1 & 1 \end{bmatrix}\begin{Bmatrix} x_1 \\ x_2 \end{Bmatrix} = \begin{Bmatrix} 0 \\ 0 \end{Bmatrix}$$

$$\frac{m}{g_c}\begin{bmatrix} 1/2 & 0 \\ 0 & 1/2 \end{bmatrix}\begin{Bmatrix} \ddot{y}_1 \\ \ddot{y}_2 \end{Bmatrix} + k\begin{bmatrix} 3 & 0 \\ 0 & 3 \end{bmatrix}\begin{Bmatrix} y_1 \\ y_2 \end{Bmatrix} = \begin{Bmatrix} 0 \\ 0 \end{Bmatrix}$$

$$\frac{1}{g_c}\begin{bmatrix} m & 0 \\ 0 & mL^2/12 \end{bmatrix}\begin{Bmatrix} \ddot{x}_{cg} \\ \ddot{\theta} \end{Bmatrix} + \begin{bmatrix} 6k & 0 \\ 0 & kL^2/2 \end{bmatrix}\begin{Bmatrix} x_{cg} \\ \theta \end{Bmatrix} = \begin{Bmatrix} 0 \\ 0 \end{Bmatrix}$$

Evidently the latter two formulations are decoupled. It is not immediately apparent that the first one is, but by multiplying every term by the inverse of one of the matrices they do become diagonal. The characteristic matrices for the three cases are:

$$\begin{bmatrix} 1 & 0 \\ 0 & 1 \end{bmatrix}\begin{Bmatrix} \ddot{x}_1 \\ \ddot{x}_2 \end{Bmatrix} + \frac{kg_c}{m}\begin{bmatrix} 6 & 0 \\ 0 & 6 \end{bmatrix}\begin{Bmatrix} x_1 \\ x_2 \end{Bmatrix} = \begin{Bmatrix} 0 \\ 0 \end{Bmatrix}$$

$$\begin{bmatrix} 1 & 0 \\ 0 & 1 \end{bmatrix}\begin{Bmatrix} \ddot{y}_1 \\ \ddot{y}_2 \end{Bmatrix} + \frac{kg_c}{m}\begin{bmatrix} 6 & 0 \\ 0 & 6 \end{bmatrix}\begin{Bmatrix} y_1 \\ y_2 \end{Bmatrix} = \begin{Bmatrix} 0 \\ 0 \end{Bmatrix}$$

$$\frac{1}{g_c}\begin{bmatrix} 1 & 0 \\ 0 & 1 \end{bmatrix}\begin{Bmatrix} \ddot{x}_{cg} \\ \ddot{\theta}\,L \end{Bmatrix} + \frac{kg_c}{m}\begin{bmatrix} 6 & 0 \\ 0 & 6 \end{bmatrix}\begin{Bmatrix} x_{cg} \\ \theta L \end{Bmatrix} = \begin{Bmatrix} 0 \\ 0 \end{Bmatrix}$$

where the mixed dimensions in the last case have been corrected by associating one factor of the reference dimension L with the angle (so that the second column of each matrix is divided by L), and then dividing the second equation by the remaining factor of the reference dimension L (so that the second row of each matrix is divided by L). Evidently all coordinate systems are principal coordinate systems now. We achieved this by making the two natural frequencies identical; we noted that our vertical "jounce" frequency originally was lower than our angular "pitch" frequency, and placed an appropriate spring at the center-of-gravity to stiffen up the jounce motion.

In an automobile, we find it desirable to decouple the jounce and pitch oscillations, which are excited when the vehicle passes over a bump in the road. The idea is that an initial disturbance should play itself out only as a motion corresponding to the initial displacement, and not as the complex interaction of two modes. Like the yardstick in our example above, most cars have a higher natural frequency in pitch than in bounce. Obviously this could be corrected by shortening the wheelbase to twice the radius-of-gyration; however, a longer wheelbase is generally specified either for directional stability or else to maximize interior space.

In the mid-twentieth-century, a few ingenious designers of light cars accomplished decoupling by means of spring systems which stiffen jounce, either by means of spring-loaded rods connecting front and rear suspensions (in the Citroën 2CV), or else by hoses connecting the liquid filling of front and rear rubber springs (in the Austin/Morris/MG 1100). Decoupling could also be achieved by adding longitudinal Z-shaped torsion elements (the exact opposite of U-shaped anti-roll bars used in the transverse direction) to restrain bounce, while reducing the stiffness of the coil springs at each wheel, which are still needed to control pitch. The force transferred by the Z-bar when a step in the road deflects the front wheels, helps to maintain the riding height at the rear wheels, and opposes the tendency of the rear suspension to squat down and then be hit extra hard when the step in the road reaches the rear wheels.

More primitively, designers of some luxury cars achieved comparable ride behavior by placing extra mass into long overhangs front and rear. This places a pair of centers-of-percussion near the front and rear wheels, so the suspension of the rear wheel is unaffected when the front wheel is impacted.

Exercise 21.3 *Analyze a yardstick supported by a spring k at one end, and a spring $3k$ at the two-foot mark, one foot from the other end. Why is this system decoupled? How can you find other two-spring systems which are decoupled?*

21.6 Review

We now have all the conceptual elements to formulate and solve multi-degree-of-freedom problems. Let us demonstrate each step on the example of a yardstick supported asymmetrically by a stiff spring $2k$ at the left end, and a softer spring k at the right end.

By choosing the preferred coordinate system $\{x_{cg}, \theta\}$ of Section 21.3, we immediately obtain the mass matrix

$$[M] = \begin{bmatrix} m & 0 \\ 0 & J_{cg} \end{bmatrix} = \begin{bmatrix} m & 0 \\ 0 & mr_g^2 \end{bmatrix} = \begin{bmatrix} m & 0 \\ 0 & mL^2/12 \end{bmatrix}$$

To obtain the first column of the stiffness matrix from **Rule 1a**, we apply a unit translation to the entire bar, without rotation. This leads to a force $-2k$ at the left end and a restoring force $-k$ at the right end. To balance these, we need a force of $k_{11} = (2k + k)$ at the center, and a torque $k_{21} = (-2kL/2 + kL/2)$. (We have drawn an upward displacement and a counter-clockwise rotation as positive; other combinations could reverse the sign of k_{12}.)

To obtain the second column of the stiffness matrix, we apply a unit rotation for the entire bar, without displacement at the center. This leads to a deflection $-L/2$ at the left end, and $L/2$ at the right end, resulting in a force $2kL/2$ at the left end and a force $-kL/2$ at the right end. (We have written these displacements for a unit deflection of one radian; at the same time we have assumed that the deflections are small and do not distort the geometry. This may seem contradictory, but is the correct way to apply Rule 1a.) To balance these forces, we need a force of $k_{12} = (-2kL/2 + kL/2)$ at the center, and a torque $k_{22} = \left(2kL^2/4 + kL^2/4\right)$. The resulting matrix is

$$[k] = \begin{bmatrix} 3k & \frac{-kL}{2} \\ \frac{-kL}{2} & \frac{3kL^2}{4} \end{bmatrix}$$

We should get the units out of the matrices, as we did in Section 21.5. The elegant procedure (Section 20.6) of using the square-root inverse matrix $[M]^{\frac{-1}{2}}$, which is easily obtained for a diagonal mass matrix, accomplishes this automatically

$$\begin{bmatrix} \sqrt{\frac{1}{m}} & 0 \\ 0 & \sqrt{\frac{12}{mL^2}} \end{bmatrix} \begin{bmatrix} m & 0 \\ 0 & \frac{mL^2}{12} \end{bmatrix} \begin{bmatrix} \sqrt{\frac{1}{m}} & 0 \\ 0 & \sqrt{\frac{12}{mL^2}} \end{bmatrix} \begin{Bmatrix} \ddot{x}_{cg} \\ \frac{L}{\sqrt{12}}\ddot{\theta} \end{Bmatrix}$$
$$+g_c \begin{bmatrix} \sqrt{\frac{1}{m}} & 0 \\ 0 & \sqrt{\frac{12}{mL^2}} \end{bmatrix} \begin{bmatrix} 3k & \frac{-kL}{2} \\ \frac{-kL}{2} & \frac{3kL^2}{4} \end{bmatrix} \begin{bmatrix} \sqrt{\frac{1}{m}} & 0 \\ 0 & \sqrt{\frac{12}{mL^2}} \end{bmatrix} \begin{Bmatrix} x_{cg} \\ \frac{L}{\sqrt{12}}\theta \end{Bmatrix} = \begin{Bmatrix} 0 \\ 0 \end{Bmatrix}$$

which multiplies out to

$$\begin{bmatrix} 1 & 0 \\ 0 & 1 \end{bmatrix} \begin{Bmatrix} \ddot{x}_{cg} \\ \frac{L}{\sqrt{12}}L\ddot{\theta} \end{Bmatrix} + \frac{kg_c}{m} \begin{bmatrix} 3 & -\sqrt{3} \\ -\sqrt{3} & 9 \end{bmatrix} \begin{Bmatrix} x_{cg} \\ \frac{L}{\sqrt{12}}\theta \end{Bmatrix} = \begin{Bmatrix} 0 \\ 0 \end{Bmatrix}$$

Maple or *Matlab* yield the eigenvalues $\lambda_{I,II} = 6 \mp 2\sqrt{3} = 2.536,\ 9.464$, respectively, and the orthonormal eigenvectors

$$[u] = \left[\begin{Bmatrix} 0.965,93 \\ 0.258,81 \end{Bmatrix}_I \begin{Bmatrix} -0.258,81 \\ 0.965,93 \end{Bmatrix}_{II} \right]$$

Each of these represents a combination of displacement and rotation, as described in Section 21.7 below.

Defining the principal coordinates using this orthonormal modal matrix and its inverse

$$\begin{aligned}\left\{\begin{array}{c}\sqrt{m}x_{cg}\\ \sqrt{\frac{m}{12}}L\theta\end{array}\right\} &= \left[\begin{array}{cc}0.96593 & -0.25881\\ 0.25881 & 0.96593\end{array}\right]\left\{\begin{array}{c}p_I\\ p_{II}\end{array}\right\}\\ \left\{\begin{array}{c}p_I\\ p_{II}\end{array}\right\} &= \left[\begin{array}{cc}0.96593 & 0.25881\\ -0.25881 & 0.96593\end{array}\right]\left\{\begin{array}{c}x_{cg}\\ \frac{L}{\sqrt{12}}\theta\end{array}\right\}\end{aligned}$$

we substitute into the governing equation, and renormalize to obtain

$$\begin{gathered}\left[\begin{array}{cc}0.96593 & 0.25881\\ -0.25881 & 0.96593\end{array}\right]\left[\begin{array}{cc}1 & 0\\ 0 & 1\end{array}\right]\left[\begin{array}{cc}0.96593 & -0.25881\\ 0.25881 & 0.96593\end{array}\right]\left\{\begin{array}{c}\ddot{p}_I\\ \ddot{p}_{II}\end{array}\right\}\\ +\frac{kg_c}{m}\left[\begin{array}{cc}0.96593 & 0.25881\\ -0.25881 & 0.96593\end{array}\right]\left[\begin{array}{cc}3 & -\sqrt{3}\\ -\sqrt{3} & 9\end{array}\right]\left[\begin{array}{cc}0.96593 & -0.25881\\ 0.25881 & 0.96593\end{array}\right]\left\{\begin{array}{c}p_I\\ p_{II}\end{array}\right\}\end{gathered}$$

which multiplies out to

$$\left[\begin{array}{cc}1 & 0\\ 0 & 1\end{array}\right]\left\{\begin{array}{c}\ddot{p}_I\\ \ddot{p}_{II}\end{array}\right\}+\frac{kg_c}{m}\left[\begin{array}{cc}2.536 & 0\\ 0 & 9.464\end{array}\right]\left\{\begin{array}{c}p_I\\ p_{II}\end{array}\right\}=\left\{\begin{array}{c}0\\ 0\end{array}\right\}$$

21.7 Nodes

A combined rotation and linear translation can be described in terms of the "node," that is, the place where the yardstick continues to intersect its rest position. For small angles (i.e., when $\tan\theta \cong \theta \cong \sin\theta$), the distance Y of a node from the center-of-gravity can be obtained from $\theta = x_{cg}/Y$. In this example, the first eigenvector describes a translation, relative to the rotation, of

$$\begin{aligned}\left(\frac{\sqrt{12}x_{cg}}{L\theta}\right)_I &= \frac{\sqrt{3}}{3-\lambda_I} = \frac{9-\lambda_I}{\sqrt{3}} = \sqrt{3}+2 = 3.732 = \frac{0.965,93}{0.258,81}\\ Y_I &= \left(\frac{x_{cg}}{\theta}\right)_I = \frac{3.732}{\sqrt{12}}L = 1.077L = 38.78 \text{ inches}\end{aligned}$$

so that the lower frequency corresponds to a rotation around a point 20.78 inches beyond the right end of the yardstick. The second eigenvector describes a translation

$$\begin{aligned}\left(\frac{\sqrt{12}x_{cg}}{L\theta}\right)_{II} &= \frac{\sqrt{3}}{3-\lambda_{II}} = \frac{9-\lambda_{II}}{\sqrt{3}} = \sqrt{3}-2 = -0.268 = \frac{-0.258,81}{0.965,93}\\ Y_{II} &= \left(\frac{x_{cg}}{\theta}\right)_{II} = \frac{-0.268}{\sqrt{12}}L = -0.077L = -2.78 \text{ inches}\end{aligned}$$

so that the higher frequency corresponds to a rotation around the 15.22-inch-mark of the yardstick.

In the early days of automobile design, engineers observed that the ride of a car was perceived to be better if the lower-frequency node was some distance ahead of the car, putting the higher-frequency node slightly behind the center of gravity. This could be accomplished by adjusting the relative stiffness of front and rear springs. When it was properly done, the passengers in a classic luxury-car ended up sitting near the neutral point of the second-mode, avoiding the unpleasant higher-frequency motions. A more modern solution would be to reduce the pitching frequency, using techniques discussed in Section 21.5.

Exercise 21.4 *Given a yardstick supported by a spring $2k$ at the left end and a second spring k at the right end, find the response of a velocity Initial Condition consisting of a rotation $\dot{\theta}(0)$ around the left end of the yardstick. Procedure: find the corresponding $\dot{x}_{cg}(0)$ for the given initial motion; then convert the Initial Conditions to principal coordinates. The solution will consist of a sine term in each principal coordinate. Finally, convert back to convenient coordinates. Be sure to be consistent in the transformations you use.*

21.8 Summary

We have demonstrated different kinds of coupling, and both mathematical and actual physical decoupling of the governing equations. The example of the yardstick resembles the problem of an automobile driving over a bump, and the management of jounce and pitch modes has been illustrated.

Problem 21.5 *For the system shown below, find the characteristic matrix, the natural frequencies, the nodes, and the response to a unit initial deflection of the x_{cg}-coordinate.*

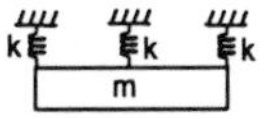

k k k
m

Problem 21.6 *For the system shown below, find the characteristic matrix, the natural frequencies, the nodes, and the response to a "unit" initial deflection of the θ-coordinate.*

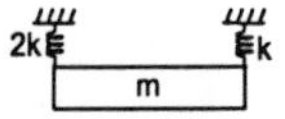

2k k
m

Problem 21.7 *For the system shown below, find the characteristic matrix, the natural frequencies, the nodes, and the response to a unit initial velocity of the x_{cg}-coordinate.*

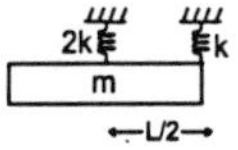

2k k
m
L/2

Problem 21.8 *For the system shown below, find the characteristic matrix, the natural frequencies, the nodes, and the response to a unit initial velocity of the θ-coordinate.*

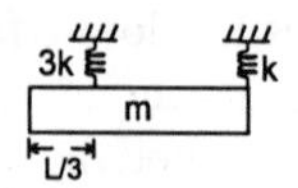

Chapter 22

LAGRANGIAN METHODS and equivalent coupling

In the preceding chapters, we obtained the governing equations of multi-degree-of-freedom systems from Newton's law. In complex physical situations, methods based on Hamilton's principle, first introduced in Section 5.6, have advantages.

22.1 Lagrange's Equation

If we can obtain the potential energy U of a system in terms of a set of generalized displacements x_i, and the kinetic energy in a system as a function of the displacements and velocities x_i and v_i (where we write v_i for $\dot{x}_i$ in order to emphasize that we are talking about velocity as a state variable), we can write a set of Lagrange's equations

$$\frac{d}{dt}\left(\frac{\partial T}{\partial v_i}\right) - \frac{\partial T}{dx_i} + \frac{\partial U}{dx_i} = 0 \tag{22.1}$$

There are the following restrictions to the use of these equations:

- The coordinate system must be holonomic. In other words, the state of the system must be entirely known from the state variables. For example, giving the *location* variables of a billiard ball rolling around in a bowl is not enough; the *orientation* is part of its state description, and depending on its path it could arrive at a given location with different orientations.
- We have left out dissipative terms due to damping. We will need to introduce these later.

On the other hand, excitation from moving boundaries does not create the problems that the use of Conservation of Energy did in Section 5.5.

22.2 Flywheel Governor

Sometimes it is not possible to find quadratic terms in the expressions for T and U, and it is necessary to use Lagrange's equations to obtain the (nonlinear) governing equations. Let us return to the simplified flyball governor of Chapter 5, where a vertical shaft supports the hinge of a lever which holds a concentrated mass at its tip (see Figure 5.3), which is pulled out and upward by the centrifugal force, and down and inward by gravity. To add some realism, the shaft does not rotate at perfectly constant speed, but is connected to a flywheel J rotating at angular velocity $\dot{\Phi}$. We can easily write the kinetic and potential energies

$$\begin{aligned} T &= \frac{1}{2}\frac{m}{g_c}l^2\left(\dot{\theta}^2 + \dot{\Phi}^2\sin^2\theta\right) + \frac{1}{2}\frac{J}{g_c}\dot{\Phi}^2 \\ U &= \frac{mg}{g_c}l\left(1-\cos\theta\right) \end{aligned}$$

The first equation we write is

$$\begin{aligned} \frac{d}{dt}\left(\frac{\partial T}{\partial \dot{\theta}}\right) - \frac{\partial T}{d\theta} + \frac{\partial U}{d\theta} &= 0 \\ \frac{m}{g_c}l^2\ddot{\theta} - \frac{m}{g_c}l^2\dot{\Phi}^2\sin\theta\cos\theta + \frac{mg}{g_c}l\sin\theta &= 0 \end{aligned}$$

The second equation is

$$\begin{aligned} \frac{d}{dt}\left(\frac{\partial T}{\partial \dot{\Phi}}\right) - \frac{\partial T}{d\Phi} + \frac{\partial U}{d\Phi} &= 0 \\ \left(\frac{m}{g_c}l^2\sin^2\theta + \frac{J}{g_c}\right)\ddot{\Phi} + 2\frac{m}{g_c}l^2\dot{\theta}\dot{\Phi}\sin\theta\cos\theta &= 0 \end{aligned}$$

The two resulting equations simplify to

$$\begin{aligned} \ddot{\theta} - \dot{\Phi}^2\sin\theta\cos\theta + \frac{g}{l}\sin\theta &= 0 \\ \left(\sin^2\theta + \frac{J}{ml^2}\right)\ddot{\Phi} + 2\dot{\theta}\dot{\Phi}\sin\theta\cos\theta &= 0 \end{aligned}$$

These equations describe free rotation at constant energy and angular momentum. To solve them, for small oscillations, we would need to solve the first equation for the equilibrium first (the second equation has only derivatives in it and therefore does not figure in the equilibrium calculation)

$$\left(\dot{\Phi}^2\right)_{avg} \sin\overline{\theta}\cos\overline{\theta} - \frac{g}{l}\sin\overline{\theta} = 0$$

We can insert this value of $\overline{\theta}$ into our equations and linearize them for small values of $\widetilde{\theta} = \theta - \overline{\theta}$.

22.3 Equivalent Masses and Springs

If the system is linear and homogeneous, and the coordinates are absolute rather than relative, kinetic energy will take on quadratic forms incorporating the influence coefficients

$$T = \frac{1}{2}\left(\frac{m_{11}}{g_c}\right)v_1^2 + \frac{1}{2}\left(\frac{m_{12}}{g_c} + \frac{m_{21}}{g_c}\right)v_1 v_2 + \frac{1}{2}\left(\frac{m_{22}}{g_c}\right)v_2^2 \tag{22.2}$$

where $m_{12} = m_{21}$ because of the absolute coordinate system; and potential energy will take the form

$$\begin{aligned} U &= \frac{1}{2}(k_{11})x_1^2 + \frac{1}{2}(k_{12} + k_{21})x_1 x_2 + \frac{1}{2}(k_{22})x_2^2 \\ &\quad + W_1 x_1 + W_2 x_2 \end{aligned} \tag{22.3}$$

where $k_{12} = k_{21}$. All the terms are quadratic, except for the linear gravitational terms $W_1 x_1$ and $W_2 x_2$

Applying Lagrange's equations to quadratic expressions gives us our linear equations

$$\frac{1}{g_c}\begin{bmatrix} m_{11} & m_{12} \\ m_{21} & m_{22} \end{bmatrix}\begin{Bmatrix} \ddot{x}_1 \\ \ddot{x}_2 \end{Bmatrix} + \begin{bmatrix} k_{11} & k_{12} \\ k_{21} & k_{22} \end{bmatrix}\begin{Bmatrix} x_1 \\ x_2 \end{Bmatrix} = \begin{Bmatrix} W_1 \\ W_2 \end{Bmatrix}$$

and the gravitational terms show up as constants on the right sides of the equations.

Therefore, when we obtain quadratic expressions in the kinetic and potential energy, we can immediately recognize that whatever shows up in the parentheses of Equation 22.2 is the equivalent mass associated with the adjacent coordinate(s); and whatever shows up in the parentheses of Equation 22.3 is the equivalent stiffness term. This is the multi-degree-of-freedom version of the equivalent masses and springs introduced in Section 5.3.

However, if non-quadratic terms (other than gravitational terms) show up in the potential energy, the system is non-linear.

22.4 Double Pendulum

We can set up the equations for a double pendulum with concentrated masses by writing the kinetic and potential energies (using the law of cosines for oblique triangles):

$$T = \frac{1}{2}\frac{m_1}{g_c}\ell_1^2\,\dot{\theta}_1^2 + \frac{1}{2}\frac{m_2}{g_c}\left(\ell_1^2\,\dot{\theta}_1^2 + \ell_2^2\,\dot{\theta}_2^2 - 2\ell_1\,\dot{\theta}_1\,\ell_2\,\dot{\theta}_2\cos(\theta_2-\theta_1)\right)$$

$$U = \frac{m_1 g}{g_c}\ell_1\,(1-\cos\theta_1) + \frac{m_2 g}{g_c}\left(\ell_1\,(1-\cos\theta_1) + \ell_2\,(1-\cos\theta_2)\right)$$

We can linearize the equations by reducing these expressions to quadratic forms. For small amplitudes, $\cos(\theta_2-\theta_1) \cong 1$, so that

$$T \cong \frac{1}{2}\left(\frac{m_1}{g_c}\ell_1^2 + \frac{m_2}{g_c}\ell_1^2\right)\dot{\theta}_1^2 + \frac{2}{2}\left(\frac{-m_2}{g_c}\ell_1\ell_2\right)\dot{\theta}_2\dot{\theta}_1 + \frac{1}{2}\left(\frac{m_2}{g_c}\ell_2^2\right)\dot{\theta}_2^2$$

We identify the contents of the brackets as the equivalent masses in this coordinate system, m_{11}/g_c in the first bracket, m_{12}/g_c and m_{21}/g_c in the second bracket, and m_{22}/g_c in the third:

$$[M] \cong \begin{bmatrix} (m_1\ell_1^2 + m_2\ell_1^2) & (-m_2\ell_1\ell_2) \\ (-m_2\ell_1\ell_2) & (m_2\ell_2^2) \end{bmatrix}$$

Also, for small amplitudes, $(1-\cos\theta) \cong \theta^2/2$, so that

$$U \cong \frac{1}{2}\left(\frac{m_1 g}{g_c}\ell_1 + \frac{m_2 g}{g_c}\ell_1\right)\theta_1^2 + \frac{2}{2}(0)\,\theta_1\theta_2 + \frac{1}{2}\left(\frac{m_2 g}{g_c}\ell_2\right)\theta_2^2$$

We identify the contents of the brackets as equivalent spring constants in this coordinate system, k_{11} in the first bracket, k_{12} and k_{21} equal to zero according to the second bracket, and k_{22} in the third bracket:

$$[K] \cong \frac{1}{g_c}\begin{bmatrix} (m_1 g\ell_1 + m_2 g\ell_1) & (0) \\ (0) & (m_2 g\ell_2) \end{bmatrix}$$

Therefore, if we look at a problem in which the two lengths and the two mass are equal,

$$\frac{m\ell^2}{g_c}\begin{bmatrix} 2 & -1 \\ -1 & 1 \end{bmatrix}\begin{Bmatrix} \ddot{\theta}_1 \\ \ddot{\theta}_2 \end{Bmatrix} + \frac{mg\ell}{g_c}\begin{bmatrix} 2 & 0 \\ 0 & 1 \end{bmatrix}\begin{Bmatrix} \theta_1 \\ \theta_2 \end{Bmatrix} \cong \begin{Bmatrix} 0 \\ 0 \end{Bmatrix}$$

Exercise 22.1 *Develop the equations if the masses are not concentrated, but have non-infinitesimal values of J_{cg}.*

Exercise 22.2 *Develop the linearized equations in other absolute or relative coordinate systems. Compare the results obtained by Newton's law, influence coefficients, and Lagrange's equations; transform coordinates to show the equivalence of the different expressions.*

22.5 Equivalent Mass of Springs

What if our springs in a boxcar system, like our two-mass problem of Chapter 20, are not perfectly weightless? In an earlier discussion we found that about one-third of the mass of each spring connected to ground should be assigned to the mass it restrains. By this reasoning, one-third of the mass of the coupling spring in a boxcar system should be assigned to each of the masses to which it connects. But what about the remaining third of the mass?

If we assume knowledge of the deflections within the spring, we can include it in our computation of kinetic energy. For linear deflection (i.e., a deflection distribution like the static deflection), the kinetic energy of the spring k_1 attached at the left end to ground is

$$T = \frac{1}{2}\int_0^L \frac{m_{k1}}{Lg_c}\left(\frac{\xi}{L}\dot{x}_1\right)^2 d\xi = \frac{1}{2}\left(\frac{m_{k1}}{3g_c}\right)\dot{x}_1^2$$

so that we should add one-third of the mass of the spring k_1 to m_{11}; the added-mass term for that one spring is

$$\begin{bmatrix} \frac{1}{3}m_{k1} & 0 \\ 0 & 0 \end{bmatrix}$$

The added mass for the spring k_2 on the right can be obtained similarly

$$\begin{bmatrix} 0 & 0 \\ 0 & \frac{1}{3}m_{k2} \end{bmatrix}$$

On the other hand, the kinetic energy of the coupling spring k_c is

$$\begin{aligned}
T &= \frac{1}{2}\int_0^L \frac{m_{kc}}{Lg_c}\left(\dot{x}_1 + \frac{\xi}{L}\left(\dot{x}_2 - \dot{x}_1\right)\right)^2 d\xi \\
&= \frac{1}{2}\frac{m_{kc}}{g_c}\int_0^1 \left(\frac{\xi^2}{L^2}\left(\dot{x}_2 - \dot{x}_1\right)^2 + 2\frac{\xi}{L}\left(\dot{x}_2 - \dot{x}_1\right)\dot{x}_1 + \dot{x}_1^2\right) d\frac{\xi}{L} \\
&= \frac{1}{2}\left[\frac{m_{kc}}{g_c}\left(\frac{1}{3}\left(\dot{x}_2 - \dot{x}_1\right)^2 + \left(\dot{x}_2 - \dot{x}_1\right)\dot{x}_1 + \dot{x}_1^2\right)\right] \\
&= \frac{1}{2}\left(\frac{m_{kc}}{3g_c}\right)\dot{x}_1^2 + \frac{2}{2}\left(\frac{m_{kc}}{6g_c}\right)\dot{x}_1\dot{x}_2 + \frac{1}{2}\left(\frac{m_{kc}}{3g_c}\right)\dot{x}_1^2
\end{aligned}$$

The added mass terms from the coupling spring k_c are

$$\begin{bmatrix} \frac{1}{3}m_{kc} & \frac{1}{6}m_{kc} \\ \frac{1}{6}m_{kc} & \frac{1}{3}m_{kc} \end{bmatrix}$$

For our two-mass problem of Chapter 20 the final mass matrix now is

$$[M] = \begin{bmatrix} \left(m_1 + \frac{1}{3}m_{k1} + \frac{1}{3}m_{kc}\right) & \left(\frac{1}{6}m_{kc}\right) \\ \left(\frac{1}{6}m_{kc}\right) & \left(m_2 + \frac{1}{3}m_{k2} + \frac{1}{3}m_{kc}\right) \end{bmatrix}$$

If we use our symmetrical example with $m_1 = m_2$, $k_1 = k_2$, and $m_{k1} = m_{k2}$, the eigenvectors are

$$[u] = \left[\left\{ \begin{array}{c} 0.707 \\ 0.707 \end{array} \right\} \left\{ \begin{array}{c} -0.707 \\ 0.707 \end{array} \right\} \right]$$

and the system equations in principal coordinates are

$$\frac{1}{g_c} \left[\begin{array}{cc} \left(m + \frac{m_k}{3} + \frac{m_{kc}}{2}\right) & 0 \\ 0 & \left(m + \frac{m_k}{3} + \frac{m_{kc}}{6}\right) \end{array} \right] \left\{ \begin{array}{c} \ddot{y} \\ \ddot{z} \end{array} \right\} + k \left[\begin{array}{cc} (k) & 0 \\ 0 & (k + 2k_c) \end{array} \right] \left\{ \begin{array}{c} y \\ z \end{array} \right\} = \left\{ \begin{array}{c} 0 \\ 0 \end{array} \right\}$$

Transformed to principal coordinates, we find that the entire mass m_{kc} of the coupling spring in our simplest system participates in the lower-frequency mode, and about one-sixth of it in the can be assigned to each adjacent mass in the higher-frequency mode.

22.6 Hydrodynamic Inertia Coupling

Similarly, if we have tubes or webs immersed in a liquid, we can calculate kinetic energy from an approximate flow distribution and express the result in terms of equivalent mass and mass-coupling coefficients.

For example, in an earlier chapter, we obtained the hydrodynamic added mass for a web in a duct from a kinetic-energy integral. This would be the term we add to the mass of the web in order to form $m_{11} = m_{web} + m_{hydro}$.

To determine m_{21}, we consider that only the mass of the thin web has a displacement; everything else has no net motion; therefore the added forces must come from reaction on the outer duct, and $m_{21} = -m_{hydro}$. (If the web had a significant volume, then we would have to expand $m_{11} = m_{web}\left(1 - \rho_{fluid}/\rho_{web}\right) + m_{web}\left(\rho_{fluid}/\rho_{web}\right) + m_{hydro}$, and the coupling coefficient modified to $m_{21} = -\left(m_{hydro} + m_{web}\left(\rho_{fluid}/\rho_{web}\right)\right)$.)

The mass matrix should be symmetrical because we are using absolute coordinates in an inertial coordinate system, and $m_{12} = m_{21}$.

To determine m_{22}, we note that for uniform motion, we can argue from Newton's law or from kinetic energy that

$$m_{11} + m_{21} + m_{12} + m_{22} = m_{case} + m_{fluid} + m_{web}$$

Therefore, substituting from above,

$$m_{22} = m_{case} + \left(m_{fluid} + m_{hydro}\right)$$

In other words, the outer case has its own added hydrodynamic inertia, which is the mass of the enclosed fluid plus the hydrodynamic inertia computed for the inner web. (If the web had significant volume, this would become

$$m_{22} = m_{case} + \left(m_{fluid} + m_{web}\left(\rho_{fluid}/\rho_{web}\right)\right) + \left(m_{hydro} + m_{web}\left(\rho_{fluid}/\rho_{web}\right)\right)$$

to allow for the buoyancy of the web.)

We could have obtained m_{22} directly by doing Southwell's method for a stream function with the outside boundary moving and the central web stationary; this would give the same result as superimposing a opposing unit velocity on the previous problem. The result would be that the hydrodynamic inertia for the outer web is the same as that for the inner web plus the superposed velocity of the entire fluid mass—the same result as above.

Therefore, the mass matrix is, for the thin web,

$$\left[\begin{bmatrix} m_{web} & 0 \\ 0 & m_{case} \end{bmatrix} + \begin{bmatrix} m_{hydro} & -m_{hydro} \\ -m_{hydro} & (m_{fluid} + m_{hydro}) \end{bmatrix}\right] \begin{Bmatrix} \ddot{x}_{web} \\ \ddot{x}_{case} \end{Bmatrix}$$

or, for a thick web, the added-mass terms become

$$\begin{bmatrix} (m_{hydro}) & -\left(m_{hydro} + m_{web}\frac{\rho_{fluid}}{\rho_{web}}\right) \\ -\left(m_{hydro} + m_{web}\frac{\rho_{fluid}}{\rho_{web}}\right) & \left(m_{fluid} + m_{hydro} + 2\left(m_{hydro} + m_{web}\frac{\rho_{fluid}}{\rho_{web}}\right)\right) \end{bmatrix}$$

The same relationship can be applied to any annulus problem, once we have found m_{hydro} for the inner member for the case when the outer member is stationary.

Problem 22.3 *Consider the problem of a spring-restrained mass in a radial slot on a turntable. The turntable rotates freely with angular velocity $\dot{\Phi}$, with a moment-of-inertia J, and the spring force is zero when the mass m is at $r = r_o$. Find the governing equations and the equilibrium position.*

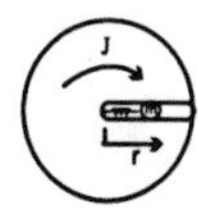

Hint:

$$\begin{aligned} T &= \frac{1}{2}\frac{m}{g_c}\left(\dot{r}^2 + \left(\dot{\Phi}\, r\right)^2\right) + \frac{1}{2}\frac{J}{g_c}\,\dot{\Phi}^2 \\ U &= \frac{1}{2}k\,(r - r_o)^2 \end{aligned}$$

Problem 22.4 *Find the mass matrix $[M]$ for a "boxcar" system with three identical masses M, and four identical springs k, each with mass m. Solve for the inverse $[M]^{-1}$, the square root $[M]^{1/2}$, and the inverse of the square root $[M]^{-1/2}$.*

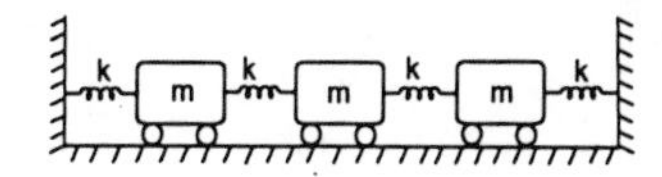

Chapter 23

FLEXIBILITY FORMULATION and estimation methods

In many problems, such as masses supported by beams, the stiffness matrix is difficult to calculate directly. In those cases, the flexibility matrix is often easier to obtain.

23.1 Flexibility Coefficients

Influence coefficients in structures are defined in a manner complementary to Rule 1a in Section 19.3:

Rule 1b: a_{ij} is the displacement resulting at i from a unit force applied at j and zero force everywhere else.

This results in the flexibility matrix $[A]$, which is the inverse of the stiffness matrix $[K]$. We can demonstrate *Rule 1a* by taking the equilibrium equation

$$[K]\left\{\begin{array}{c} x_1 \\ x_2 \end{array}\right\} = \left\{\begin{array}{c} F_1 \\ F_2 \end{array}\right\} \tag{23.1}$$

and substituting deflections $x_1 = 1$ and $x_2 = 0$ in it

$$\left[\begin{array}{cc} k_{11} & k_{12} \\ k_{21} & k_{22} \end{array}\right]\left\{\begin{array}{c} 1 \\ 0 \end{array}\right\} = \left\{\begin{array}{c} k_{11} \\ k_{21} \end{array}\right\} \tag{23.2}$$

showing that forces are the first column of the K-matrix.

We can show that the inverse $[A] = [K]^{-1}$ expresses the reverse relationship by pre-multiplying the equilibrium equation by it

$$[A]\,[K]\left\{\begin{array}{c} x_1 \\ x_2 \end{array}\right\} = [A]\left\{\begin{array}{c} F_1 \\ F_2 \end{array}\right\} \tag{23.3}$$

writing $[A]\,[K] = [I]$ and switching sides to get

$$[A]\left\{\begin{array}{c} F_1 \\ F_2 \end{array}\right\} = \left\{\begin{array}{c} x_1 \\ x_2 \end{array}\right\} \tag{23.4}$$

And, we can demonstrate **Rule 1b** by substituting forces $F_1 = 1$ and $F_2 = 0$ in that relationship, and showing that the resulting displacements are the first column of the A-matrix

$$\left[\begin{array}{cc} a_{11} & a_{12} \\ a_{21} & a_{22} \end{array}\right]\left\{\begin{array}{c} 1 \\ 0 \end{array}\right\} = \left\{\begin{array}{c} a_{11} \\ a_{21} \end{array}\right\} \tag{23.5}$$

In boxcar and torsion-shaft problems, **Rule 1a** (for obtaining the K-matrix directly) is easier to use than **Rule 1b**; but in beam problems **Rule 1b** is much easier to apply.

Exercise 23.1 *Determine the $[A]$-matrix corresponding to the stiffness matrix*

$$K = \left[\begin{array}{cc} k & -k \\ -k & k \end{array}\right]$$

Why do you run into difficulties in trying to do that?

23.2 Beam Deflections

The deflection of slender beams under specified loadings can be obtained by integrating the forces to obtain shear, integrating shear to obtain bending moments, dividing moments by EI to obtain beam curvature, integrating curvature to obtain slope, integrating slope to obtain beam deflection, and fitting the integration constants to the support conditions.

Example: What is the deflection of a simply-supported uniform beam under its own weight?

- Defining y positive upward (opposite to gravity), and calling the mass-per-unit length $\mu = m/L$, the upward load-per-unit-length is

$$f_y = \frac{F_y}{L} = \frac{-\mu g}{g_c}$$

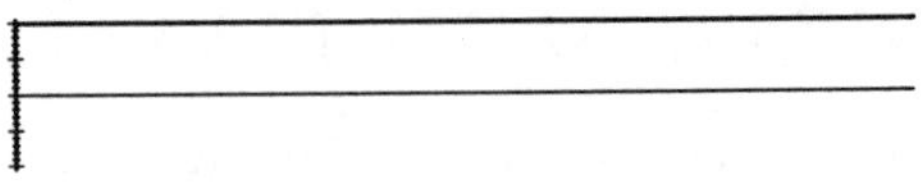

uniformly over the whole span; the reaction force at the endpoints is $-F_y/2 = mg/2g_c$.

- Integrating the constant f_y along the beam, the shear is

$$V = \int f_y dx = f_y x + C_1 = f_y \left(x - \frac{L}{2} \right) = \frac{\mu g}{g_c} \left(\frac{L}{2} - x \right)$$

where the integration constant $C_1 = -f_y L/2$ reflects the fact that value of the integral jumps from zero to $-F_y/2$ as it passes the concentrated reaction force at the left end, and jumps the same amount, back to zero, as it passes the similar singularity at the right-end support.

- Integrating the shear along the beam, the moment is

$$\text{Moment} = \int f_y \left(x - \frac{L}{2} \right) dx = \frac{f_y x (x - L)}{2} + C_2 = \frac{\mu g}{g_c} \cdot \frac{x (L - x)}{2}$$

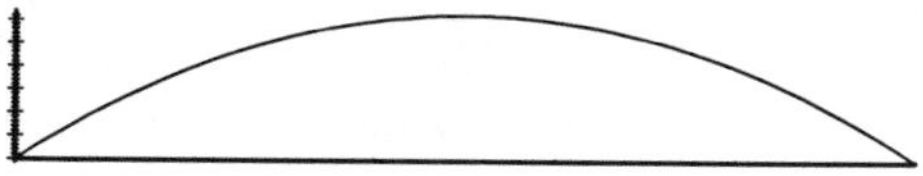

where the integration constant $C_2 = 0$ to make the moment zero at the ends, because simple supports do not sustain moments.

- Dividing the moment by the uniform stiffness, the curvature is

$$\frac{d^2 y}{dx^2} = \frac{f_y x (x - L)}{2EI} = \frac{\mu g}{g_c} \cdot \frac{x (L - x)}{2EI}$$

- Integrating the curvature, the slope is

$$\frac{dy}{dx} = \int \frac{f_y x (x - L)}{2EI} dx = \frac{f_y}{EI} \left(\frac{x^3}{6} - \frac{Lx^2}{4} \right) + C_3$$

where the integration constant cannot be evaluated yet because the simple supports do not restrain the slope.

- Integrating the slope, the displacement is

$$y = \int \left(\frac{f_y}{EI} \left(\frac{x^3}{6} - \frac{Lx^2}{4} \right) + C_3 \right) dx = \frac{f_y}{EI} \left(\frac{x^4}{24} - \frac{Lx^3}{12} \right) + C_3 x + C_4$$

where $C_4 = 0$ to fit the Boundary Condition of $y = 0$ at $x = 0$; and $C_3 = f_y L^3/24EI$ to fit the Boundary Condition of $y = 0$ at $x = L$

$$y = \frac{f_y}{EI} \left(\frac{x^4}{24} - \frac{Lx^3}{12} + \frac{L^3 x}{24} \right) = \frac{-\mu g}{EI g_c} \left(\frac{x^4}{24} - \frac{Lx^3}{12} + \frac{L^3 x}{24} \right)$$

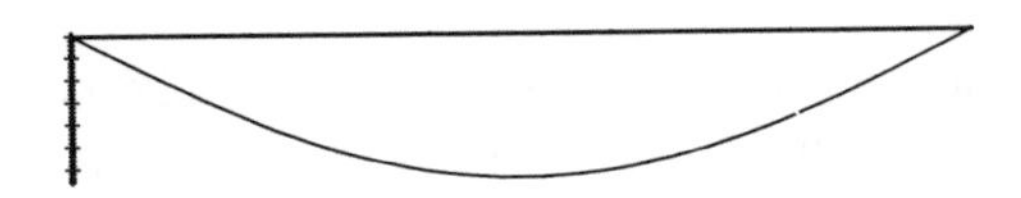

with the deflection at $x = L/2$

$$y_{ctr} = \frac{5}{384} \frac{f_y L^4}{EI} = \frac{-5}{384} \frac{\mu g L^4}{EI g_c}$$

Exercise 23.2 *Find the deflection of a uniform simply-supported beam with a concentrated weight $F_y = -W = -Mg/g_c$ at the center. Hint: The shear jumps as it passes the force at the center, instead of rising smoothly as it does with a distributed load*

$$\begin{aligned} V &= -F_y/2 = W/2 \text{ for } x < L/2 \\ &= F_y/2 = -W/2 \text{ for } x > L/2 \end{aligned}$$

Answer: Third-order polynomials in x

$$\begin{aligned} y &= \frac{F_y}{48EI}\left(3L^2 x - 4x^3\right) = \frac{-W}{48EI}\left(3L^2 x - 4x^3\right) \text{ for } x \le L/2 \\ &= \frac{F_y L^3}{48EI} = \frac{-WL^3}{48EI} \text{ at } x = L/2 \\ &= \frac{F_y}{48EI}\left(4x^3 - 12Lx^2 + 9L^2 x - L^3\right) \text{ for } x \ge L/2 \end{aligned}$$

For uniform beams, this derivation can usually be avoided, because the solutions for many standard loadings (point forces as well as distributed loads) applied to many standard support conditions (simple supports as well as cantilevers) are given in reference books.[1] Young's moduli of elasticity E for steel

[1] *e.g.,* Warren C. Young, *Roark's Formulas for Stress and Strain,* 6th Edition, McGraw-Hill, New York, 1989, ISBN 0-07-072541-1.

and other materials are listed in engineering handbooks,[2] and sectional second moments I for standard construction beams have been published by the American Institute of Steel Construction (AISC).[3]

Example: Let us consider a horizontal cantilever supporting two concentrated masses, m_1 at a distance ℓ_1 from the wall, and m_2 at a distance of $L = \ell_1 + \ell_2$ from the wall. If we associate the coordinates y_1 and y_2 with the vertical motions of the respective masses, the mass matrix is diagonal

$$[M] = \begin{bmatrix} m_1 & 0 \\ 0 & m_2 \end{bmatrix}$$

The flexibility matrix is most easily obtained from the influence coefficients **Rule 1b**: a_{ij} is the displacement resulting at i from a unit force applied at j and zero force everywhere else. For a_{11} we place a unit load at the tip of a short cantilever of length ℓ_1; tabulated formulas give us the deflection at m_1

$$a_{11} = \frac{\ell_1^3}{3EI}$$

Using the formulas tabulated in most engineering books, you will find that you have to plug in $F = -1$ for the load, because the civil engineering convention is to use a downward coordinate for (gravitational) loads, but an upward (altitude) coordinate for position; but in dynamics, our convention is to take forces and deflection coordinates in the same direction (Section 2.1). Therefore *the diagonal terms of the matrix should be positive.*

We can also differentiate the polynomial for the deflection of the cantilever as a function of the distance from the wall x. [Some books measure x from the tip of the cantilever instead; to get the formulas below, you will need to substitute $(\ell - x)$ for the x in those books.] Evaluating at $x = \ell_1$, we obtain the displacement and slope at that location

$$\begin{aligned} y(x) &= \frac{3\ell_1 x^2 - x^3}{6EI}; \; y(\ell_1) = \frac{\ell_1^3}{3EI} \\ \frac{dy}{dx} &= \frac{6\ell_1 x - 3x^2}{6EI}; \; \left(\frac{dy}{dx}\right)_{x=\ell_1} = \frac{\ell_1^2}{2EI} \end{aligned}$$

For a_{21} we keep the unit load at $x = \ell_1$. The section of beam between the two masses is not loaded, so it proceeds in a straight line form the deflection and slope given above, resulting in a deflection at $x = \ell_1 + \ell_2$

$$a_{21} = a_{11} + \ell_2 \left(\frac{dy}{dx}\right)_{x=\ell_1} = \frac{\ell_1^3}{3EI} + \ell_2 \frac{\ell_1^2}{2EI} = \frac{\ell_1^2 (2\ell_1 + 3\ell_2)}{6EI}$$

[2] *e.g.*, Ray E. Bolz and George L. Tuve, editors, *CRC Handbook of Tables for Applied Engineering Science,* Second Edition, CRC Press, Boca Raton, Florida, 1973, ISBN 0-8493-0252-8.

[3] *e.g.*, Manual of Steel Construction: *Load and Resistance Factor Design,* Volume I., "Structural Members, Specifications, and Codes," Second Edition, AISC, Chicago, Illinois, 1994, ISBN 1-56424-041-X.

which completes the first column of the flexibility matrix.

For a_{22} we place a unit load at the tip of a long cantilever of length $\ell_1 + \ell_2$; tabulated formulas give us the deflection at m_2

$$a_{22} = \frac{(\ell_1 + \ell_2)^3}{3EI}$$

The polynomial for the deflection of the cantilever as a function of the distance from the wall x for this loading is

$$\begin{aligned} y(x) &= \frac{3(\ell_1 + \ell_2)x^2 - x^3}{6EI} \\ a_{12} &= y_{(x=\ell_1)} = \frac{3(\ell_1 + \ell_2)\ell_1^2 - \ell_1^3}{6EI} = \frac{\ell_1^2(2\ell_1 + 3\ell_2)}{6EI} \\ a_{22} &= y_{(\ell_1+\ell_2)} = \frac{3(\ell_1 + \ell_2)^3 - (\ell_1 + \ell_2)^3}{6EI} = \frac{(\ell_1 + \ell_2)^3}{3EI} \end{aligned}$$

verifying the second column of the flexibility matrix.

The resulting flexibility matrix has positive terms on the diagonal, and is symmetrical because of reciprocity

$$[A] = \frac{1}{6EI}\begin{bmatrix} 2\ell_1^3 & \ell_1^2(2\ell_1 + 3\ell_2) \\ \ell_1^2(2\ell_1 + 3\ell_2) & 2(\ell_1 + \ell_2)^3 \end{bmatrix}$$

and the stiffness matrix is its inverse, calculated by *Maple* (Section 1.5) as

$$[K] = \frac{6EI}{3\ell_1^4\ell_2^2 + 4\ell_1^3\ell_2^3}\begin{bmatrix} 2(\ell_1 + \ell_2)^3 & -\ell_1^2(2\ell_1 + 3\ell_2) \\ -\ell_1^2(2\ell_1 + 3\ell_2) & 2\ell_1^3 \end{bmatrix}$$

Exercise 23.3 *Consider a horizontal beam of length $L = 3\ell$ on simple (pinned) supports at its ends, supporting two concentrated masses, m_1 at a distance ℓ from the left end, and m_2 at a distance of ℓ from the right end. Associate the coordinates y_1 and y_2 with the vertical motions of the respective masses, so that the mass matrix is diagonal*

$$[M] = \begin{bmatrix} m_1 & 0 \\ 0 & m_2 \end{bmatrix}$$

Using influence coefficients Rule 1b, and an engineering book available to you, show that the flexibility matrix is

$$[A] = \frac{\ell^3}{18EI}\begin{bmatrix} 8 & 7 \\ 7 & 8 \end{bmatrix} = \frac{L^3}{486EI}\begin{bmatrix} 8 & 7 \\ 7 & 8 \end{bmatrix}$$

and its inverse is

$$[K] = \frac{18EI}{15\ell^3}\begin{bmatrix} 8 & -7 \\ -7 & 8 \end{bmatrix} = \frac{162EI}{5L^3}\begin{bmatrix} 8 & -7 \\ -7 & 8 \end{bmatrix}$$

23.3 Flexibility-Matrix Methods

Once we have the flexibility matrix $[A]$, we can obtain the stiffness matrix $[K] = [A]^{-1}$ by matrix inversion, and solve the problem as before. However, there are several interesting methods that work directly from the $[A]$-matrix. These are particularly appropriate if we have associated the coordinate system with the centers of gravity of lumped masses (and/or rotations about appropriate axis) so that the mass matrix is diagonal, because then the characteristic matrix is simply the $[K]$-matrix with each row divided by the associated mass:

$$g_c \left[M^{-1}K\right] = g_c \begin{bmatrix} \frac{k_{11}}{m_1} & \frac{k_{12}}{m_1} & \frac{k_{13}}{m_1} \\ \frac{k_{21}}{m_2} & \frac{k_{22}}{m_2} & \frac{k_{23}}{m_2} \\ \frac{k_{31}}{m_3} & \frac{k_{32}}{m_3} & \frac{k_{33}}{m_3} \end{bmatrix}$$

In that special diagonal-mass-matrix case, the inverse of the characteristic matrix is the $[A]$-matrix with each column multiplied by the associated mass:

$$\frac{1}{g_c}\left[M^{-1}K\right]^{-1} = \frac{1}{g_c}\left[[A]\,[M]\right] = \frac{1}{g_c}\begin{bmatrix} a_{11}m_1 & a_{12}m_2 & a_{13}m_3 \\ a_{21}m_1 & a_{22}m_2 & a_{23}m_3 \\ a_{31}m_1 & a_{32}m_2 & a_{33}m_3 \end{bmatrix}$$

The inverse of our characteristic matrix is also a Standard Form eigenvalue problem, but the eigenvalues are $\lambda \triangleq 1/\omega_n^2$.

23.4 Dunkerley's Equation

If we look at the characteristic equation associated with an eigenvalue problem, we find that the elements on the diagonal of the matrix provide an upper bound to the eigenvalues:

$$\lambda \leq \frac{1}{g_c}\sum_i a_{ii}m_i$$

and therefore a lower bound to the natural frequencies:

$$\omega_n^2 \geq \frac{g_c}{\sum_i a_{ii}m_i}$$

We can easily demonstrate this on a two-by-two problem:

$$\det \begin{bmatrix} a_{11}\frac{m_1}{g_c} - \lambda & a_{12}\frac{m_2}{g_c} \\ a_{21}\frac{m_1}{g_c} & a_{22}\frac{m_2}{g_c} - \lambda \end{bmatrix} = 0$$

$$\lambda^2 g_c - (a_{11}m_1 + a_{22}m_2)\,\lambda + a_{11}m_1 \cdot a_{22}m_2 - a_{21}m_1 \cdot a_{12}m_2 = 0$$

$$\lambda = \frac{(a_{11}m_1 + a_{22}m_2) \pm \sqrt{(a_{11}m_1 + a_{22}m_2)^2 - 4\,(a_{11}m_1 \cdot a_{22}m_2 - a_{21}m_1 \cdot a_{12}m_2)}}{2g_c}$$

If the original matrix was positive-definite, then the eigenvalue must be real. This tells us that the content of the square-root is positive, and

$$(a_{11}m_1 + a_{22}m_2)^2 \leq 4\,(a_{11}m_1 \cdot a_{22}m_2 - a_{21}m_1 \cdot a_{12}m_2)$$

Therefore

$$\lambda \leq \frac{(a_{11}m_1 + a_{22}m_2) + \sqrt{(a_{11}m_1 + a_{22}m_2)^2}}{2g_c} = \frac{(a_{11}m_1 + a_{22}m_2)}{g_c}$$

$$\omega_n^2 \geq \frac{g_c}{(a_{11}m_1 + a_{22}m_2)}$$

The practical importance of Dunkerley's expression is very great, for two reasons:

1. Only the diagonal terms a_{ii} are needed. We save not only the computational effort of solving for actual eigenvalues, but also the very substantial analytical effort of obtaining off-diagonal influence coefficients. In problems with 3, 4, 5, 6, or n dimensions, this saves us the calculation of 3, 6, 10, 15, or $\left(n^2 - n\right)/2$ pairs of terms.

2. Working from the flexibility-matrix formulation, the theorem provides a *lower bound* on the lowest natural frequency. The lowest natural frequency is usually the most important one—sometimes it is the *only* important one. In many practical problems, from earthquakes to flow-induced vibrations, we specify a minimum natural frequency as a measure of dynamic rigidity.

If we are concerned that Dunkerley's lower bound might be excessively conservative as an estimate of the dynamic rigidity, we can also develop an upper bound through Rayleigh's method.

23.5 Rayleigh's Method

If the entire flexibility matrix is known and the mass-matrix is diagonal, we also have a quick means of obtaining the static deflection at each point by applying a load $m_i g/g_c$ at each coordinate. This amounts to summing the rows of the $[AM]$-matrix

$$\overline{x}_i = \frac{g}{g_c} \sum_j a_{ij} m_j$$

In our example,

$$\begin{aligned}\overline{x}_1 &= \tfrac{g}{g_c}\left(a_{11}m_1 + a_{12}m_2 + a_{13}m_3\right)\\ \overline{x}_2 &= \tfrac{g}{g_c}\left(a_{21}m_1 + a_{22}m_2 + a_{23}m_3\right)\\ \overline{x}_3 &= \tfrac{g}{g_c}\left(a_{31}m_1 + a_{32}m_2 + a_{33}m_3\right)\end{aligned}$$

These static deflections (which often have already been computed for static structural analysis, before the dynamic analysis began) can be used as a guess for the lowest-frequency mode shape in Rayleigh's method. Since our matrix model is a linear (or linearized) model, we know that the Initial-Condition response is harmonic, $x_i = -\omega_n^2 x_i$, and the maximum kinetic energy is

$$T_{\max} = \frac{1}{2}\sum_i \frac{m_i}{g_c}\omega_I^2 \overline{x}_i^2 = \frac{1}{2}\cdot\frac{\omega_I^2}{g_c}\left(m_1\overline{x}_1^2 + m_2\overline{x}_2^2 + m_3\overline{x}_3^2\right)$$

Since our matrix model is based on linear (or linearized) springs *and* our mode shape is obtained from an equilibrium deflection, we know that the maximum potential energy is one-half of the final load $m_i g/g_c$ times the displacement $\overline{x}_i$ through which that load acted:

$$U_{\max} = \frac{1}{2}\sum_i \frac{m_i g}{g_c}\overline{x}_i = \frac{1}{2}\cdot\frac{g}{g_c}\left(m_1\overline{x}_1 + m_2\overline{x}_2 + m_3\overline{x}_3\right)$$

Setting $T_{\max} = U_{\max}$, we can solve for the natural frequency

$$\omega_I^2 \leq \frac{g\sum_i m_i\overline{x}_i}{\sum_i m_i\overline{x}_i^2} = g\cdot\frac{\left(m_1\overline{x}_1 + m_2\overline{x}_2 + m_3\overline{x}_3\right)}{\left(m_1\overline{x}_1^2 + m_2\overline{x}_2^2 + m_3\overline{x}_3^2\right)}$$

The inequality reminds us that Rayleigh's method, because of the assumed mode shape, generally gives us an answer which is somewhat higher than the that obtained with the actual fundamental mode. Compared to Dunkerley's equation, we note some important differences:

- Rayleigh's method requires more information than Dunkerley's equation—all the flexibility terms, not just the diagonal ones. As a consequence, it has the potential of coming closer to the exact answer.
- Rayleigh's method provides an *upper bound* for the lowest natural frequency. Together with Dunkerley's lower bound, we have a *bounded estimate.* In many cases, the upper and lower bounds are close together, and further computation is not required.

It must be remembered that our reasoning required us to use a static deflection curve caused by gravity; our simple expression $U = mg\overline{x}/2g_c$ will not work for arbitrary mode-shapes. While the *magnitude* of gravity must be the same throughout (and ultimately cancels out), we can use our judgment to manipulate the *direction* of gravity in a mental experiment in such a way that the imagined static deflection looks as much like the fundamental mode as possible. For example, if we have a beam on three supports, we may invent a gravity field which acts downward in one span, and upward in the other span. *That is equivalent to choosing the direction of the vertical coordinates in such a way that all the elements in the A-matrix are positive.* This mental experiment will lead to a more realistic guessed mode and therefore a lower, closer value for the upper bound of ω_I. The next section will show a way of obtaining successively better mode shapes.

23.6 Matrix Iteration

If we have obtained the full $[A]$ and $[M]$ matrices, we are in a position to solve for all of the eigenvalues. However, if we are mainly interested in the fundamental mode, we can save a great deal of computation by rearranging the governing equation

$$-\omega_n^2 \{x\} + \left[M^{-1}K\right]\{x\} = 0$$

into the relationship

$$\begin{aligned} \{x\}_I &= \omega_I^2 \left[M^{-1}K\right]^{-1} \{x\}_I \\ &= \omega_I^2 \left[[A]\,[M]\right] \{x\}_I \end{aligned}$$

This is an implicit expression for $\{x\}_I$, but the right-hand side of the equation is a comparatively weak function of $\{x\}_I$. A guess at the mode-shape $\{x\}_I$ can be inserted in the right side of the equation, and then multiplied by the $[AM]$ matrix to obtain an improved guess. This successive-approximation procedure converges on the lowest frequency and mode-shape. We will proceed as follows:

1. Plug a guess at the mode shape $\{x\}_I$ into the right-hand side of the equation. The static-deflection curve from the previous section would be a good guess, but a bad guess will work too—the procedure will just take a few steps longer to converge. The column vector must be normalized in some way, for example by making the first element unity.

2. Carry out the matrix multiplication to get an improved mode shape.

3. Obtain the value of ω_I^2 needed to bring the column vector into the same normalization (e.g., make the first element unity).

4. Recognize the resulting normalized vector as an improved mode shape; return to the first step and repeat the process, using this improved guess.

When the mode shape $\{x\}_I$ and the eigenvalue ω_I^2 no longer change from one iteration to the next, the equation is satisfied and the lowest mode has been obtained.

A matrix-iteration procedure can also be developed using the expression $\omega_n^2 \{x\} = [M^{-1}K] \{x\}$; unfortunately, it converges on the highest frequency and mode, which is generally of little interest.

23.7 Lumped-Mass Models

Many systems which are actually continuous systems can be analyzed by breaking their masses up into lumped elements, while treating the structures as continuous beams. For example, if we have a beam which supports several masses, but which also has a non-negligible mass of its own, we can choose coordinates which are located at the main masses, and lump the remaining masses with these coordinates. Thus we artificially obtain a lumped-mass model with a diagonal mass matrix.

If we can then obtain the flexibility matrix for the same coordinates, we can solve for natural frequencies and mode-shapes.

Theoretically, we will obtain as many frequencies as the number of coordinates we have chosen to describe the system. In practice, only about half of these frequencies are physically meaningful: the lumping process destroys the accuracy of the higher-frequency modes. Therefore we generally use at least twice as many coordinates as the number of modes we want to obtain.

Problem 23.4 *Consider two simple (pinned) supports, a distance L apart, and an uniform beam of length $L + R$ which overhangs the right-hand support by a length R. The ratio R/L is given in the table below. The right-hand tip of the beam holds a concentrated mass. The properties of the beam are mass-per-unit-length μ and a stiffness EI. The tip-mass is $M = 0.57\mu L$. Divide the beam into a few segments, sufficient to model the fundamental vibration, and associate a y-coordinate with each segment's center-of-gravity. Write the mass matrix of the segments and the tip mass (a segment at a support may be ignored). Write the flexibility matrix, using Rule 1b and tabulated beam-deflection formulas. Obtain the fundamental frequency from Dunkerley's lower bound, from Rayleigh's upper bound, from matrix iteration, and from eigenvalue analysis. Select the parameter R/L for your calculation from the values for the second letter of your last name in the following Table:*

	R/L		*R/L*		*R/L*		*R/L*		*R/L*
a	*0.20*	*f*	*0.55*	*l*	*0.80*	*q*	*1.05*	*v*	*1.30*
b	*0.30*	*g*	*0.60*	*m*	*0.85*	*r*	*1.10*	*w*	*1.40*
c	*0.40*	*h*	*0.65*	*n*	*0.90*	*s*	*1.15*	*x*	*1.50*
d	*0.45*	*i,j*	*0.70*	*o*	*0.95*	*t*	*1.20*	*y*	*1.70*
e	*0.50*	*k*	*0.75*	*p*	*1.00*	*u*	*1.25*	*z*	*2.00*

Express the lowest natural frequency in the form

$$f_1 \cong \frac{1}{2\pi} \cdot ? \cdot \frac{1}{\ell^2} \cdot \sqrt{\frac{EIg_c}{\mu}}$$

Chapter 24

FORCED EXCITATION and modal analysis

Particular solutions for periodic excitation can be obtained in the original coordinates. However, for complete solutions or for arbitrary excitations, we transform the governing equations, including the forcing functions, to principal coordinates. This allows us to treat each mode as a one-degree-of-freedom system and solve it by the methods of Part I.

24.1 Periodic Excitation

Let us add a harmonic excitation force to the simple symmetrical two-mass example in Section 20.3.1. Both masses have the same magnitude and all three springs will be identical (Figure 24.1). Then the governing equation is

$$\frac{m}{g_c}\begin{bmatrix} 1 & 0 \\ 0 & 1 \end{bmatrix}\begin{Bmatrix} \ddot{x}_1 \\ \ddot{x}_2 \end{Bmatrix} + k\begin{bmatrix} 2 & -1 \\ -1 & 2 \end{bmatrix}\begin{Bmatrix} x_1 \\ x_2 \end{Bmatrix} = \begin{Bmatrix} F_o \sin\omega_{ex}t \\ 0 \end{Bmatrix} \qquad (24.1)$$

Because this set of equations is not homogeneous, we can obtain the particular solution without decomposing the problem into principal coordinates: to cancel

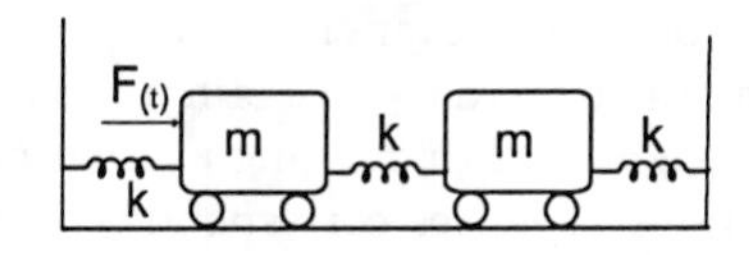

Figure 24.1: Forced Excitation on a 2-D.o.F. System

the term on the right side, the solution must have the form

$$\begin{aligned} x_{p1} &= C_1 \sin \omega_{ex} t + D_1 \cos \omega_{ex} t = X_1 \sin(\omega_{ex} t - \phi_1) \\ x_{p2} &= C_2 \sin \omega_{ex} t + D_2 \cos \omega_{ex} t = X_2 \sin(\omega_{ex} t - \phi_2) \end{aligned} \tag{24.2}$$

For the undamped case, it is even simpler: D and ϕ are zero and $C = X$

$$\begin{aligned} x_{p1} &= X_1 \sin \omega_{ex} t \\ \ddot{x}_{p1} &= -X_1 \omega_{ex}^2 \sin \omega_{ex} t \end{aligned}$$

which we can plug back into the matrix equation to obtain

$$\begin{aligned} \left(-\omega_{ex}^2 \frac{m}{g_c} \left\{ \begin{array}{c} X_1 \\ X_2 \end{array} \right\} + k \left[\begin{array}{cc} 2 & -1 \\ -1 & 2 \end{array} \right] \left\{ \begin{array}{c} X_1 \\ X_2 \end{array} \right\} \right) \sin \omega_{ex} t &= \left\{ \begin{array}{c} F_o \\ 0 \end{array} \right\} \sin \omega_{ex} t \\ \left[\begin{array}{cc} 2 - r^2 & -1 \\ -1 & 2 - r^2 \end{array} \right] \left\{ \begin{array}{c} X_1 \\ X_2 \end{array} \right\} &= \left\{ \begin{array}{c} \frac{F_o}{k} \\ 0 \end{array} \right\} \end{aligned}$$

where

$$r^2 \triangleq \omega_{ex}^2 \frac{m}{k g_c} = \frac{\omega_{ex}^2}{\omega_{ref}^2}$$

This is a pair of algebraic equations which we can solve by Cramer's rule, i.e., by pre-multiplying each term by the inverse of the matrix of coefficients

$$\left[\begin{array}{cc} 1 & 0 \\ 0 & 1 \end{array} \right] \left\{ \begin{array}{c} X_1 \\ X_2 \end{array} \right\} = \left[\begin{array}{cc} \frac{2-r^2}{3-4r^2+r^4} & \frac{1}{3-4r^2+r^4} \\ \frac{1}{3-4r^2+r^4} & \frac{2-r^2}{3-4r^2+r^4} \end{array} \right] \left\{ \begin{array}{c} \frac{F_o}{k} \\ 0 \end{array} \right\}$$

so that the first mass has an amplitude

$$\frac{X_1}{F_o/k} = \frac{2 - r^2}{3 - 4r^2 + r^4} = \frac{-(r^2 - 2)}{(r^2 - 1)(r^2 - 3)}$$

We can see that $X_1 k/F$ has two resonances: it starts from a value of 2/3 at $r = 0$, rises up towards $+\infty$ at $r = 1$, returning from $-\infty$, crossing 0 at $r = \sqrt{2}$, rising towards $+\infty$ again at $r = 3$, returning from $-\infty$ again, and sloping up towards zero with increasing r. The second mass has the amplitude

$$\frac{X_2}{F_o/k} = \frac{1}{3 - 4r^2 + r^4} = \frac{1}{(r^2 - 1)(r^2 - 3)}$$

On the other hand, $X_2 k/F$ starts from a value of 1/3 at $r = 0$, rises up towards $+\infty$ at $r = 1$, returning from $-\infty$, never crosses the axis but has a maximum of -1 at $r = \sqrt{2}$, and then falls towards $-\infty$ again at $r = 3$, returning from $+\infty$, and sloping down towards zero with increasing r. As we expect from our earlier analysis in Chapter 20, the resonances correspond to our free-vibration modes, with the two masses moving in the same direction $X_2/X_1 = 1$ for the first resonance, and in opposite directions $X_2/X_1 = -1$ for the second resonance.

Exercise 24.1 *Add damping $c_{11} = 0.1 = c_{22}$ to Equation 24.1 and obtain the particular solution (which will also be the steady-state solution). You will have to solve not only for C_1 and C_2, but also for D_1 and D_2, and $X_1 = \sqrt{C_1^2 + D_1^2}$, etc.*

To get the complete solution, we now need to add the initial-condition solution from Chapter 20, which was in terms of the modes. It might have been advantageous to switch to principal coordinates in the first place!

24.2 Simple Transformation

The general governing equation for a force-excited system is

$$\frac{1}{g_c}[M]\left\{\ddot{x}\right\} + [K]\{x\} = \{F(t)\} \tag{24.3}$$

where the elements of the column vector $\{F(t)\}$ are the excitation forces acting at each coordinate in the column vector $\{x\}$ (or if there are base excitations $y_i(t)$, we can convert them to force excitation $k_i \cdot y_i(t)$ through the linear supporting springs connecting the base and the coordinate x_i).

After computing the inverse of the mass matrix, we find the principal coordinates $\{p\}$ by pre-multiplying each term by $[M]^{-1}$ in order to put the equation into the Standard Form

$$[I]\left\{\ddot{x}\right\} + g_c\left[[M]^{-1}[K]\right]\{x\} = g_c[M]^{-1}\{F(t)\} \tag{24.4}$$

where $[I]$ stands for the identity matrix. After obtaining the eigenvalues and eigenvectors of the characteristic matrix $\left[g_c[M]^{-1}[K]\right]$, we can write the modal matrix

$$[u] \triangleq [\{x_{rel}\}_n] \tag{24.5}$$

It is desirable, but not essential, to normalize the eigenvectors (the scale of the principal coordinates will depend on that normalization). We can use this modal matrix as the transformation matrix from the principal coordinates $\{p\}$ to the primitive coordinates $\{x\}$, and substitute $[u]\{p\}$ for $\{x\}$

$$[u]\left\{\ddot{p}\right\} + g_c\left[M^{-1}K\right][u]\{p\} = g_c[M]^{-1}\{F(t)\} \tag{24.6}$$

In order to separate the equations, getting rid of the coupling terms, we pre-multiply by the inverse of the modal matrix:

$$[I]\left\{\ddot{p}\right\} + g_c[u]^{-1}\left[M^{-1}K\right][u]\{p\} = g_c[u]^{-1}[M]^{-1}\{F(t)\} \tag{24.7}$$

The characteristic matrix in p-coordinates, $g_c \left[[u]^{-1}\left[M^{-1}K\right][u]\right]$, is a diagonal matrix called the spectral matrix: the elements on the diagonal are the eigenvalues ω_n^2.

To get the particular solution for an exciting-force vector $\{F(t)\}$, we transform it to principal coordinates by multiplying it with the matrix $g_c \left[[u]^{-1}[M]^{-1}\right]$, and write the particular solution of each mode, using the one-degree-of-freedom solutions developed for periodic and transient excitations in Part I: "Simple Systems."

To get the complete solution for Initial Conditions $\{x_{(0)}\}$ and $\{\dot{x}_{(0)}\}$, we obtain the Initial Conditions in the Principal Coordinates by from the transformations $\{p_{(0)}\} = [u]^{-1}\{x_{(0)}\}$ and $\{\dot{p}_{(0)}\} = [u]^{-1}\{\dot{x}_{(0)}\}$. We develop the complementary solution in each mode, from the one-degree-of-freedom solutions in Chapter 4.

Once we have the modal solution, we can express it in the original coordinates again by writing

$$\{x(t)\} = [u]\{p(t)\} \tag{24.8}$$

Example continued: Our simple symmetrical two-mass example has the characteristic matrix (Equation 24.4)

$$\frac{kg_c}{m}\begin{bmatrix} 2-\lambda & -1 \\ -1 & 2-\lambda \end{bmatrix}\begin{Bmatrix} x_1 \\ x_2 \end{Bmatrix} = \frac{g_c}{m}\begin{Bmatrix} F_o \sin\omega_{ex}t \\ 0 \end{Bmatrix}$$

with the natural frequencies

$$\lambda_{I,II} \triangleq \omega_{I,II}^2 \frac{m}{kg_c} = 1,\ 3, \text{ respectively}$$

and the modal matrix (Equation 24.5)

$$\begin{Bmatrix} x_1 \\ x_2 \end{Bmatrix} = \begin{bmatrix} 0.707 & -0.707 \\ 0.707 & 0.707 \end{bmatrix}\begin{Bmatrix} p_I \\ p_{II} \end{Bmatrix}$$
$$\begin{Bmatrix} p_I \\ p_{II} \end{Bmatrix} = \begin{bmatrix} 0.707 & 0.707 \\ -0.707 & 0.707 \end{bmatrix}\begin{Bmatrix} x_1 \\ x_2 \end{Bmatrix}$$

so that the governing equation transformed to principal coordinates is

$$\frac{m}{g_c}\begin{bmatrix} 0.707 & -0.707 \\ 0.707 & 0.707 \end{bmatrix}\begin{Bmatrix} \ddot{p}_I \\ \ddot{p}_{II} \end{Bmatrix}$$

$$+k\begin{bmatrix} 2 & -1 \\ -1 & 2 \end{bmatrix}\begin{bmatrix} 0.707 & -0.707 \\ 0.707 & 0.707 \end{bmatrix}\begin{Bmatrix} p_I \\ p_{II} \end{Bmatrix} = \begin{Bmatrix} F_o \sin\omega_{ex}t \\ 0 \end{Bmatrix}$$

which we pre-multiply by the inverse of the first matrix to get Equation 24.7

$$\frac{m}{g_c}\left\{\begin{array}{c}\ddot{p}_I\\ \ddot{p}_{II}\end{array}\right\}+k\left[\begin{array}{cc}1&0\\0&3\end{array}\right]\left\{\begin{array}{c}p_I\\ p_{II}\end{array}\right\}=\left[\begin{array}{cc}0.707&0.707\\-0.707&0.707\end{array}\right]\left\{\begin{array}{c}F_o\sin\omega_{ex}t\\0\end{array}\right\}$$

which leads to two separate one-degree-of-freedom differential equations

$$\begin{aligned}\ddot{p}_I+\frac{kg_c}{m}p_I &= 0.707\frac{F_o}{k}\frac{kg_c}{m}\sin\omega_{ex}t\\ \ddot{p}_{II}+\frac{3kg_c}{m}p_{II} &= \frac{-0.707}{3}\frac{F_o}{k}\frac{3kg_c}{m}\sin\omega_{ex}t\end{aligned}$$

Exercise 24.2 *Continue to obtain the particular solution and transform it back to raw x-coordinates.*

Exercise 24.3 *Repeat the solution process for the unsymmetrical system*

$$\frac{m}{g_c}\left[\begin{array}{cc}1&0\\0&2\end{array}\right]\left\{\begin{array}{c}\ddot{x}_1\\ \ddot{x}_2\end{array}\right\}+k\left[\begin{array}{cc}5&-4\\-4&6\end{array}\right]\left\{\begin{array}{c}x_1\\ x_2\end{array}\right\}=\left\{\begin{array}{c}F_o\sin\omega_{ex}t\\0\end{array}\right\}$$

You should get the intermediate result

$$\frac{m}{g_c}\left\{\begin{array}{c}\ddot{p}_1\\ \ddot{p}_2\end{array}\right\}+k\left[\begin{array}{cc}1&0\\0&7\end{array}\right]\left\{\begin{array}{c}p_1\\ p_2\end{array}\right\}=\left[\begin{array}{cc}1/3&1/3\\-1/3&1/6\end{array}\right]\left\{\begin{array}{c}F_o\sin\omega_{ex}t\\0\end{array}\right\}$$

24.3 Elegant Transformation

The general governing equation for an undamped force-excited system is

$$\frac{1}{g_c}[M]\left\{\ddot{x}\right\}+[K]\{x\}=\{F(t)\} \tag{24.9}$$

We rescale the coordinates proportionately to the square-root of the mass associated with each, by computing the square root of the mass matrix $[M]^{1/2}$, as well as its inverse $[M]^{-1/2}$, and replacing the original coordinate system $\{x\}$ with the weighted coordinates $\{q\}$

$$\{q\} = [M]^{1/2}\{x\} \tag{24.10}$$

$$\{x\} = [M]^{-1/2}\{q\} \tag{24.11}$$

where q has the dimensions of $\sqrt{m}x$, so that the equation becomes

$$\frac{1}{g_c}[M][M]^{-1/2}\left\{\ddot{q}\right\}+[K][M]^{-1/2}\{q\}=\{F(t)\} \tag{24.12}$$

and then pre-multiplying each term by $[M]^{-1/2}$:

$$[M]^{-1/2}[M][M]^{-1/2}\left\{\ddot{q}\right\}+g_c[M]^{-1/2}[K][M]^{-1/2}\{q\}=g_c[M]^{-1/2}\{F(t)\} \tag{24.13}$$

Multiplying the matrices, we get the identity matrix from $[M]^{-1/2}[M][M]^{-1/2}$ and a symmetrical Characteristic Matrix $\left[\widetilde{K}\right]\equiv[M]^{-1/2}[K][M]^{-1/2}$:

$$[I]\left\{\ddot{q}\right\}+g_c\left[\widetilde{K}\right]\{q\}=g_c[M]^{-1/2}\{F(t)\} \tag{24.14}$$

Solving the eigenvalue problem, we obtain the same eigenvalues we would have obtained otherwise, but the eigenvectors in these coordinates are orthogonal. We normalize the eigenvectors to form the orthonormal modal matrix $[u]$

$$[u]\triangleq[\{q_{rel}\}_n] \tag{24.15}$$

We can use this modal matrix as the transformation matrix from the principal coordinates $\{p\}$ to the coordinates $\{q\}$, and substitute $\{q\}=[u]\{p\}$:

$$[u]\left\{\ddot{p}\right\}+g_c\left[\widetilde{K}\right][u]\{p\}=g_c[M]^{-1/2}\{F(t)\} \tag{24.16}$$

In order to get rid of the coupling terms and separate the equations, we pre-multiply by the transpose of the modal matrix

$$[u]^T[u]\left\{\ddot{p}\right\}+g_c[u]^T\left[\widetilde{K}\right][u]\{p\}=g_c[u]^T[M]^{-1/2}\{F(t)\}$$

Since the transpose of an orthonormal matrix is also its inverse, $[u]^T[u]$ is the identity matrix; and since the eigenvectors are also orthogonal with respect to the stiffness matrix, $\left[[u]^T\left[\widetilde{K}\right][u]\right]$ is a diagonal matrix and is called the spectral matrix:

$$[I]\left\{\ddot{p}\right\}+[\Lambda]\{p\}=g_c[u]^T[M]^{-1/2}\{F(t)\}$$

The diagonal elements in the spectral matrix are the eigenvalues ω_n^2.

To get the particular solution for an exciting function vector $\{F(t)\}$, we transform it to principal coordinates by multiplying with $g_c[u]^T[M]^{-1/2}$ and write the particular solution of each mode, using the one-degree-of-freedom solutions developed for either periodic or transient excitation.

To get the complete solution for Initial Conditions $\{x_{(0)}\}$ and $\left\{\dot{x}_{(0)}\right\}$, we obtain the Initial Conditions in the Principal Coordinates from the transformations

$$\begin{aligned}\{p_{(0)}\}&=[u]^{-1}\{q_{(0)}\}=[u]^{-1}[M]^{1/2}\{x_{(0)}\}\\ \left\{\dot{p}_{(0)}\right\}&=[u]^{-1}\left\{\dot{q}_{(0)}\right\}=[u]^{-1}[M]^{1/2}\left\{\dot{x}_{(0)}\right\}\end{aligned}$$

We develop the complementary solution in each mode, from the one-degree-of-freedom solutions developed earlier.

Once we have the complete modal solution $\{p(t)\}$, we can express it in the original coordinates again by writing

$$\{x(t)\} = [M]^{-1/2}\{q(t)\} = [M]^{-1/2}[u]\{p(t)\}$$

Exercise 24.4 *Use the elegant method to solve the unsymmetrical system*

$$\frac{m}{g_c}\begin{bmatrix} 1 & 0 \\ 0 & 2 \end{bmatrix}\begin{Bmatrix} \ddot{x}_1 \\ \ddot{x}_2 \end{Bmatrix} + k\begin{bmatrix} 5 & -4 \\ -4 & 6 \end{bmatrix}\begin{Bmatrix} x_1 \\ x_2 \end{Bmatrix} = \begin{Bmatrix} F_o \sin\omega_{ex}t \\ 0 \end{Bmatrix}$$

Comparing the methods, we note that the simple method works if we are able to do unsymmetrical eigenvalue problems, and if we don't mind inverting (rather than transposing) the modal matrix. With the advent of programs like *Matlab,* simplicity may be preferable.

The elegant method, applied to problems with symmetrical mass and flexibility matrices, can use efficient programs that are restricted to symmetrical characteristic matrices, and avoids inverting the modal matrix. It is computationally efficient *if* the mass matrix is diagonal and its square-root (and inverse) easily obtained. However, the elegant method introduces an intermediate coordinate transformation with potentially confusing dimensions.

24.4 The Harmonic Vibration Absorber

When a machine of mass M on spring supports K is subjected to an equivalent excitation force $F_o \sin\omega_{ex}t$, it will undergo motion and transmit forces to the ground, especially when $\omega_{ex} \approx \sqrt{Kg_c/M}$. One possible solution is to add a small parasitic mass m by means of a coupling spring k to detune the system away from the excitation frequency. The governing equation is similar to the quarter-car model (Section 20.5), but the relative magnitude of the masses is reversed

$$\frac{1}{g_c}\begin{bmatrix} M & 0 \\ 0 & m \end{bmatrix}\begin{Bmatrix} \ddot{x}_1 \\ \ddot{x}_2 \end{Bmatrix} + \begin{bmatrix} K+k & -k \\ -k & k \end{bmatrix}\begin{Bmatrix} x_1 \\ x_2 \end{Bmatrix} = \begin{Bmatrix} F_o \sin\omega_{ex}t \\ 0 \end{Bmatrix}$$

Using the simple method of pre-multiplying each term by the inverse of the mass matrix, we find the characteristic matrix

$$\begin{Bmatrix} \ddot{x}_1 \\ \ddot{x}_2 \end{Bmatrix} + \frac{Kg_c}{M}\begin{bmatrix} 1+\frac{k}{K} & \frac{-k}{K} \\ -\frac{Mk}{Km} & \frac{Mk}{Km} \end{bmatrix}\begin{Bmatrix} x_1 \\ x_2 \end{Bmatrix} = \begin{Bmatrix} \frac{F_o g_c}{M}\sin\omega_{ex}t \\ 0 \end{Bmatrix}$$

with the eigenvalues

$$\lambda_{I,II} = \left(\frac{Kg_c}{M}\right)\frac{\left(1+\frac{k}{K}+\frac{M}{m}\frac{k}{K}\right) \mp \sqrt{\left(1+\frac{k}{K}+\frac{M}{m}\frac{k}{K}\right)^2 - 4\frac{M}{m}\frac{k}{K}}}{2}$$

We see that the addition of the parasitic spring-mass assembly leads to two natural frequencies. If we selected our coupling spring so that $k/m = K/M$ and selected our mass so that $m \ll M$ (and therefore $k \ll K$), this simplifies to

$$\begin{aligned}
\lambda_{I,II} &= \frac{Kg_c}{M}\left(\left(1+\frac{k}{2K}\right)\mp\sqrt{\frac{k}{K}+\left(\frac{k}{2K}\right)^2}\right) \cong \frac{Kg_c}{M}\left(1\mp\sqrt{\frac{k}{K}}\right) \\
\frac{\lambda_I+\lambda_{II}}{2} &= \frac{Kg_c}{M}\left(1+\frac{k}{2K}\right) \cong \frac{Kg_c}{M} \\
\lambda_{II}-\lambda_I &= \frac{Kg_c}{M}\left(2\sqrt{\frac{k}{K}+\left(\frac{k}{2K}\right)^2}\right) \cong \frac{Kg_c}{M}\left(2\sqrt{\frac{k}{K}}\right) = \frac{Kg_c}{M}\left(2\sqrt{\frac{m}{M}}\right)
\end{aligned}$$

so that the two new natural frequencies span the natural frequency that we had before we added the parasitic mass. Therefore we can use the added mass and spring to shift the natural frequencies away from the troublesome excitation. However, we have to be sure that our excitation will not drift, because there are now other resonances nearby—how near depends on the size of the parasitic assembly.

The eigenvectors are

$$\left\{\begin{array}{c} x_1/x_2 \\ x_2/x_2 \end{array}\right\}_{I,II} = \left\{\begin{array}{c} 1-\frac{1}{2}\left(\left(1+\frac{m}{M}+\frac{K}{k}\frac{m}{M}\right)\mp\sqrt{\left(1+\frac{m}{M}+\frac{K}{k}\frac{m}{M}\right)^2-4\frac{m}{M}\frac{K}{k}}\right) \\ 1 \end{array}\right\}$$

which for $k/m = K/M$, $m \ll M$, and $k \ll K$ simplifies to

$$\left\{\begin{array}{c} x_1/x_2 \\ x_2/x_2 \end{array}\right\}_{I,II} = \left\{\begin{array}{c} \frac{-m}{2M}\pm\sqrt{\frac{m}{M}+\left(\frac{m}{2M}\right)^2} \\ 1 \end{array}\right\} \cong \left\{\begin{array}{c} \pm\sqrt{\frac{m}{M}} \\ 1 \end{array}\right\} = \left\{\begin{array}{c} \pm\sqrt{\frac{k}{K}} \\ 1 \end{array}\right\}$$

showing that the first mode represents motion of both masses in the same direction, and the second represents motion in opposite directions.

As we saw in the first example in this chapter, there is one frequency of excitation where the response of the first mass is zero. Solving for X_1 and setting it equal to zero, we find that this occurs when $\omega_{ex}^2 = kg_c/m$. At that point, the main mass M does not move at all, but the small mass m vibrates.

At equilibrium amplitude X_2, the forces from the coupling spring must exactly balance the excitation force (so that the main mass does not move), and we can conclude that $X_2 k = F_o$, and the force and displacement are exactly out-of-phase.

The question is: how does the undamped small mass know to vibrate if the main mass does not move? The answer is that whenever the amplitude or timing of the oscillation drifts away from the equilibrium, the large mass *will* move just enough to restore synchronization.

Now that we understand how the harmonic balancer works, we can design one for any given situation:

- The resonant frequency $\sqrt{kg_c/m}$ of the parasitic mass-spring system must be as close as possible to the excitation frequency. Although common examples often show this as the same as the frequency of the main system before adding the parasitic mass (because excitation near resonance is particularly troublesome), that is *not* the correct design criterion: only the actual excitation frequency matters.

- The size of the mass (and spring stiffness) depends on what the uncertainty about the system parameters and excitation frequency is; a larger parasitic mass has a wider range of near-ideal operation. On should check what the two (new) natural frequencies are, and be sure that there is no long-term resonant excitement of the combined system.

- The length of travel $2X_2$ required in the design can be obtained from the force and the coupling-spring constant—lighter parasitic systems require more travel.

- The damping in the coupling spring should be low. If we including damping in the analysis, we find that complete cancellation is not possible; because the main mass must move enough so that the "base-excitation" of the small mass overcomes the energy lost in the damper.

Harmonic balancers are used in piston engines to oppose torsional vibrations in the shaft. They involve either a torsion-spring-driven flywheel within a flywheel (for constant speed engines), or pendulum-like elements operated by centrifugal force—since the centrifugal force is proportional to the square of the rotational speed, the natural frequency stays in tune with the number of power strokes per revolution.

Problem 24.5 *Find the particular solution for x_1 in the two-mass system described by*

$$\frac{m}{g_c}\begin{bmatrix} 1 & 0 \\ 0 & 1 \end{bmatrix}\begin{Bmatrix} \ddot{x}_1 \\ \ddot{x}_2 \end{Bmatrix} + k\begin{bmatrix} 5 & -4 \\ -4 & 5 \end{bmatrix}\begin{Bmatrix} x_1 \\ x_2 \end{Bmatrix} = \begin{Bmatrix} F_o sin \omega_{ex} t \\ 0 \end{Bmatrix}$$

Intermediate result:

$$\ddot{p}_1 + \left(\frac{kg_c}{m}\right) p_1 = \left(\frac{kg_c}{m}\right) \cdot \frac{1}{k} \cdot \frac{1}{2}\sqrt{2} \cdot F_o sin\left(\omega_{ex} t\right)$$

Problem 24.6 *Find the zero-initial-state solution for the two-mass system described by*

$$\frac{m}{g_c}\begin{bmatrix} 1 & 0 \\ 0 & 1 \end{bmatrix}\begin{Bmatrix} \ddot{x}_1 \\ \ddot{x}_2 \end{Bmatrix} + k\begin{bmatrix} 5 & -4 \\ -4 & 5 \end{bmatrix}\begin{Bmatrix} x_1 \\ x_2 \end{Bmatrix} = \begin{Bmatrix} F_o sin \omega_{ex} t \\ 0 \end{Bmatrix}$$

for $\omega_{ex}/\omega_1 = 2$ *and* $\omega_{ex}/\omega_2 = 2/3$. *Partial Answer:*

$$p_1 = \left(\frac{-\sqrt{2}}{6}\right)\frac{F_o}{k} sin\left(\omega_{ex}t\right) + \left(\frac{\sqrt{2}}{3}\right)\frac{F_o}{k} sin\left(\omega_1 t\right)$$

$$\begin{aligned} x_1 &= \left(\frac{-1}{6} + \frac{1}{10}\right)\frac{F_o}{k} sin\left(\omega_{ex}t\right) \\ &\quad + \frac{1}{3}\cdot\frac{F_o}{k} sin\left(\omega_1 t\right) - \frac{1}{15}\cdot\frac{F_o}{k} sin\left(\omega_2 t\right) \end{aligned}$$

Chapter 25

DAMPED MULTI-DEGREE-OF-FREEDOM SYSTEMS and state-variable formulations

Traditional modal analyses deals with undamped multi-degree-of-freedom systems. We will now extend our analysis to systems with damping.

25.1 Problem Statement

The governing equation of a damped system is

$$\frac{1}{g_c}[M]\left\{\ddot{x}\right\}+[C]\left\{\dot{x}\right\}+[K]\{x\}=0 \tag{25.1}$$

where the matrices $[M]$, $[C]$, and $[K]$ represent mass, damping, and stiffness coefficients. In undamped systems, our approach was to find principal coordinates $\{p\}$ which diagonalize these coefficient matrices. We achieved this for the mass matrix by pre-multiplying each term by the inverse of $[M]$

$$[I]\left\{\ddot{x}\right\}+g_c\left[M^{-1}C\right]\left\{\dot{x}\right\}+g_c\left[M^{-1}K\right]\{x\}=0 \tag{25.2}$$

We can diagonalize one additional coefficient matrix, the stiffness matrix, by solving the eigenvalue problem of the characteristic matrix $\left[M^{-1}K\right]$ and assembling a transformation matrix from the eigenvectors. That shows that we can always diagonalize two of the coefficient matrices. But we usually cannot find a transformation matrix composed of real numbers which will allow us to diagonalize a third coefficient matrix, the damping matrix, at the same time. The next section will point out some exceptions.

25.2 Rayleigh Damping

There are some systems in which real-number modal analysis is possible because the damping matrix happens to be filled with a fortuitous set of values. For example, if we place a damper from each mass to ground, and make the damping for each coordinate proportional to the mass associated with that coordinate, $[C] = \alpha\,[M]$, where α is the proportionality constant, then any coordinate system that diagonalizes the mass matrix also diagonalizes the damping matrix. The dimensionless damping ratio ζ for each mode will be inversely proportional to the natural frequency for that mode, so that all modes will have the same half-life.

Similarly, if we place a damper next to each spring, and make the damping for each coordinate proportional to the stiffness associated with that coordinate, $[C] = \beta\,[K]$, then any coordinate system that diagonalizes the stiffness matrix also diagonalizes the damping matrix. The damping ratio of each mode will be proportional to its frequency, so that higher-frequencies modes decay more rapidly.

Finally, if the damping matrix is the sum of a matrix proportional to mass and another matrix proportional to stiffness, $[C] = \alpha\,[M] + \beta\,[K]$, then the principal coordinate system that diagonalizes both mass and stiffness matrices also diagonalizes the damping matrix.

Inspecting a particular problem, it is not always easy to tell whether $[C] = \alpha\,[M] + \beta\,[K]$, since α and β will be unknown constants. A simple test is Caughey's criterion: *iff*

$$\left[M^{-1}C\right]\left[M^{-1}K\right] = \left[M^{-1}K\right]\left[M^{-1}C\right]$$

that is, if the multiplication of the mass-normalized damping matrix with the mass-normalized stiffness matrix is commutative, *then* real-number modal analysis is possible. If not, we will need the additional flexibility of complex numbers to describe the response of the system.

We can test this by adding a shock-absorber to our quarter-car model (Section 20.5). Damping must only be added between the sprung and unsprung masses: energy absorption in the sidewalls of the tire would add rolling resistance and convert forward motion into heat. Expressing our coefficients in terms of reference magnitudes m, c, and k leads to the terms

$$\frac{m}{g_c}\begin{bmatrix} 1 & 0 \\ 0 & 10 \end{bmatrix}\begin{Bmatrix} \ddot{x}_1 \\ \ddot{x}_2 \end{Bmatrix} + c\begin{bmatrix} 1 & -1 \\ -1 & 1 \end{bmatrix}\begin{Bmatrix} \dot{x}_1 \\ \dot{x}_2 \end{Bmatrix} + k\begin{bmatrix} 11 & -1 \\ -1 & 1 \end{bmatrix}\begin{Bmatrix} x_1 \\ x_2 \end{Bmatrix}$$

Normalizing with the inverse of the mass matrix, we obtain

$$\begin{bmatrix} 1 & 0 \\ 0 & 1 \end{bmatrix} \begin{Bmatrix} \ddot{x}_1 \\ \ddot{x}_2 \end{Bmatrix} + \frac{cg_c}{m} \begin{bmatrix} 1 & -1 \\ -0.1 & 0.1 \end{bmatrix} \begin{Bmatrix} \dot{x}_1 \\ \dot{x}_2 \end{Bmatrix} + \frac{kg_c}{m} \begin{bmatrix} 11 & -1 \\ -0.1 & 0.1 \end{bmatrix} \begin{Bmatrix} x_1 \\ x_2 \end{Bmatrix}$$

Trying Caughey's criterion, we find that one order of multiplication

$$\begin{bmatrix} 1 & -1 \\ -0.1 & 0.1 \end{bmatrix} \begin{bmatrix} 11 & -1 \\ -0.1 & 0.1 \end{bmatrix} = \begin{bmatrix} 11.1 & -1.1 \\ -1.11 & 0.11 \end{bmatrix}$$

gives us a different product than the other order of multiplication

$$\begin{bmatrix} 11 & -1 \\ -0.1 & 0.1 \end{bmatrix} \begin{bmatrix} 1 & -1 \\ -0.1 & 0.1 \end{bmatrix} = \begin{bmatrix} 11.1 & -11.1 \\ -0.11 & 0.11 \end{bmatrix}$$

The difference tells us that diagonalization of the damping matrix is not going to be possible with real-number modes.

Exercise 25.1 *Repeat using the elegant normalization of Section 20.6. The two products should still be unequal, though they may be transposes of each other.*

We can check our conclusion by solving the standard eigenvalue problem:

$$\det \begin{bmatrix} 11-\lambda & -1 \\ -0.1 & 0.1-\lambda \end{bmatrix} = 0$$

leading to $\left(\frac{m}{kg_c}\right)\omega^2_{I,II} \triangleq \lambda_{I,II} = 0.090\,833,\ 11.009$, with the eigenvectors

$$[u] = \left[\begin{Bmatrix} 0.091\,593 \\ 0.999\,202 \end{Bmatrix}_I, \begin{Bmatrix} 0.999\,958 \\ -0.009\,166 \end{Bmatrix}_{II} \right]$$

If we use this modal matrix to transform to the principal coordinates of the undamped problem, we obtain:

$$\begin{bmatrix} 1 & 0 \\ 0 & 1 \end{bmatrix} \begin{Bmatrix} \ddot{q}_1 \\ \ddot{q}_2 \end{Bmatrix} + \frac{cg_c}{m} \begin{bmatrix} 0.082 & -0.092 \\ -0.916 & 1.018 \end{bmatrix} \begin{Bmatrix} \dot{q}_1 \\ \dot{q}_2 \end{Bmatrix}$$
$$+ \frac{kg_c}{m} \begin{bmatrix} 0.091 & 0 \\ 0 & 11.009 \end{bmatrix} \begin{Bmatrix} q_1 \\ q_2 \end{Bmatrix} = \begin{Bmatrix} 0 \\ 0 \end{Bmatrix}$$

As Caughey's criterion predicted, the damping matrix is *not* diagonal.

Exercise 25.2 *Repeat using the elegant procedure of Section 20.6. It should still lead to off-diagonal values in the damping matrix, although they should both have the same value of* -0.290*, because of better relative scaling of the principal coordinates.*

25.3 Damped Modes

As we have been hinting, we could solve damped linear systems if we allowed complex numbers. Because we anticipated sinusoidal-type solutions, we would not only set $\left\{\ddot{x}\right\} = -\omega^2 \{x\}$ as we did in the chapters on undamped problems, but also $\left\{\dot{x}\right\} = i\omega \{x\}$. Substituting these in the set of differential Equations 25.1, we would obtain the set of algebraic characteristic equations

$$\left[-\omega^2 \frac{1}{g_c} [M] + i\omega [C] + [K]\right] \{x\} = 0$$

and set the determinant of the combined matrix equal to zero. For example, in our automotive problem from the previous section we would end up with the eigenvalue problem

$$\det \begin{bmatrix} \left(11k - \frac{m}{g_c}\omega^2 + ic\omega\right) & (-k - ic\omega) \\ (-k - ic\omega) & \left(k - 10\frac{m}{g_c}\omega^2 + ic\omega\right) \end{bmatrix} = 0$$

so that this two-degree-of-freedom problem gives us a fourth-order characteristic equation for ω. Unlike the undamped case, where ω turned out to have only real values, the roots in this damped case may be real, complex, or imaginary. As in one-degree-of-freedom problems, we will interpret the real components as sinusoidal response to initial conditions, and positive imaginary components as exponential decay. This is confusing and cumbersome, so let us switch to using $s = i\omega$. This returns us to the notation of Chapter 9, where negative real roots s represent decaying exponential functions, conjugate pairs of imaginary roots stand for sinusoidal functions, and complex-conjugate roots describe exponentially decaying sinusoidal functions. Using s also avoids the inconvenience of inputting complex numbers into the characteristic matrix; we now have

$$\left[s^2 \frac{1}{g_c} [M] + s [C] + [K]\right] \{x\} = 0$$

In our quarter-car with damping (Section 25.2) this becomes:

$$\det \begin{bmatrix} \left(11k + cs + \frac{m}{g_c}s^2\right) & (-k - cs) \\ (-k - cs) & \left(k + cs + 10\frac{m}{g_c}s^2\right) \end{bmatrix} = 0$$

Although we can obtain our characteristic equation from this, the process is tedious for large problems. We can reduce it to a Standard Form eigenvalue problem by expressing it in terms of state variables, as described in the next section.

Exercise 25.3 *Show that the Laplace transform method lead to the same characteristic equation, and obtain the roots.*

25.4 Hamilton's Canonical Form

We can express each of the second-order differential Equation 25.1 as two first-order differential equations. For an n-degree-of-freedom problem this leads to $2n$ equations. The first set of n equations arises from the definition of the velocity state variables $\{v\} \triangleq \{\dot{x}\}$

$$\frac{d}{dt}\{x\} = \{v\}$$

The second set of n equations is the normalized set of Equations 25.2 rewritten in terms of the state variables $\{x\}$ and $\{v\}$

$$\frac{d}{dt}\{v\} = -g_c\left[M^{-1}C\right]\{v\} - g_c\left[M^{-1}K\right]\{x\}$$

Writing s wherever we find the operator d/dt, and organizing the two sets of equations into one set of $2n$ equations, we obtain:

$$s\begin{bmatrix} [I] & [0] \\ [0] & [I] \end{bmatrix}\left\{\begin{matrix} \{x\} \\ \{v\} \end{matrix}\right\} = \begin{bmatrix} [0] & [I] \\ [-g_cM^{-1}K] & [-g_cM^{-1}C] \end{bmatrix}\left\{\begin{matrix} \{x\} \\ \{v\} \end{matrix}\right\} \tag{25.3}$$

This is a standard-form eigenvalue problem, which can be solved with an eigenvalue program which can return complex-number values. A *FORTRAN* subroutine like `EIGENP` based on Algorithm 343, *Communications of the ACM,* Vol. 11, No. 12 (1968) is suitable. If the number of degrees of freedom and the required precision are moderate, applications programs like *Matlab* are convenient.

Exercise 25.4 *Express a damped one-degree-of-freedom problem in this state-variable form, and show that it leads to the familiar solutions of Chapter 9.*

Our automotive example, in this form, would yield

$$s[I]\left\{\begin{matrix} x_1 \\ x_2 \\ \dot{x}_1 \\ \dot{x}_2 \end{matrix}\right\} = \begin{bmatrix} \begin{bmatrix} 0 & 0 \\ 0 & 0 \end{bmatrix} & \begin{bmatrix} 1 & 0 \\ 0 & 1 \end{bmatrix} \\ \frac{kg_c}{m}\begin{bmatrix} -11 & +1 \\ +0.1 & -0.1 \end{bmatrix} & \frac{cg_c}{m}\begin{bmatrix} -1 & +1 \\ +0.1 & -0.1 \end{bmatrix} \end{bmatrix}\left\{\begin{matrix} x_1 \\ x_2 \\ \dot{x}_1 \\ \dot{x}_2 \end{matrix}\right\}$$

which can be written with dimensionless matrices (see Section 4.3) by defining a time-scale $t^* \triangleq \omega_{\text{ref}}t$ in terms of the reference magnitudes $\omega_{\text{ref}} = \sqrt{kg_c/m}$. (If the reference magnitudes are merely standard S.I. or English units, then $\omega_{\text{ref}} = 1$ radian/second.)

$$\frac{s}{\omega_{\text{ref}}}[I]\left\{\begin{matrix} x_1 \\ x_2 \\ \frac{\dot{x}_1}{\omega_{\text{ref}}} \\ \frac{\dot{x}_2}{\omega_{\text{ref}}} \end{matrix}\right\} = \begin{bmatrix} \begin{bmatrix} 0 & 0 \\ 0 & 0 \end{bmatrix} & \begin{bmatrix} 1 & 0 \\ 0 & 1 \end{bmatrix} \\ \begin{bmatrix} -11 & +1 \\ +0.1 & -0.1 \end{bmatrix} & 2\zeta_{\text{ref}}\begin{bmatrix} -1 & +1 \\ +0.1 & -0.1 \end{bmatrix} \end{bmatrix}\left\{\begin{matrix} x_1 \\ x_2 \\ \frac{\dot{x}_1}{\omega_{\text{ref}}} \\ \frac{\dot{x}_2}{\omega_{\text{ref}}} \end{matrix}\right\}$$

where the similitude parameter of damping, written as $\zeta_{\text{ref}} \triangleq c/2\sqrt{km/g_c}$, typically has a value near 0.5 in automotive practice, so that our eigenvalue problem becomes

$$\det \begin{vmatrix} 0-\lambda & 0 & 1 & 0 \\ 0 & 0-\lambda & 0 & 1 \\ -11 & 1 & -1-\lambda & 1 \\ 0.1 & -0.1 & 0.1 & -0.1-\lambda \end{vmatrix} = 0$$

where $\lambda \triangleq s/\omega_{\text{ref}}$. Using these numbers, we obtain the complex conjugate eigenvalues

$$\begin{aligned} s_{I,II} &= (-0.1409 \pm i0.2813)\,\omega_{\text{ref}} \\ s_{III,IV} &= (-1.5984 + i2.7469)\,\omega_{\text{ref}} \end{aligned}$$

The imaginary parts represent damped modal frequencies; they differ somewhat from the values of $\sqrt{0.0908} = \pm 0.3014$ and $\sqrt{11.01} = \pm 3.318$ we calculated for undamped modes in Section 25.2. (In this case, the damped values are slightly lower than the undamped values, but we cannot always count on that in non-Rayleigh damping.) The negative real parts represent the decaying coefficients in the solution, which contains the forms $e^{-0.1409t^*}(\sin(0.2813t^*)$ and $e^{-1.5984t^*}(\sin(2.7469t^*)$. The corresponding eigenvectors are also contain complex numbers

$$[u] = \left[\begin{Bmatrix} +0.9888 \mp i0.0516 \\ -0.0461 \pm i0.0877 \\ -0.0450 \pm i0.0876 \\ -0.0052 \mp i0.0084 \end{Bmatrix}_{I,II} \begin{Bmatrix} +0.0871 \pm i0.0196 \\ +0.2411 \pm i0.6592 \\ -0.0478 \pm i0.0515 \\ -0.7008 \mp i0.0799 \end{Bmatrix}_{III,IV} \right]$$

In the eigenvalues, we interpreted complex numbers as the decaying and oscillating components of the motion. In the eigenvectors we interpret them as in- and out-of-phase motions composing a mode.

Exercise 25.5 *Repeat the same procedure, but with c reduced to zero; compare your result with the results obtained for the damped problem, and with the undamped modes in Section 25.2.*

Exercise 25.6 *Repeat the procedure for the same automotive example, but with damping increased to* $\varsigma_{ref} = 1.5$*; explain the absence of imaginary components in some of the roots.*

Unlike real-number eigenvectors, there is no standard way to normalize complex-number eigenvectors, and different application programs will return different-looking results. Plotting the values in polar coordinates will show that the different-looking vectors represent identical amplitude-ratios and phase-angles.

The $2n$ by $2n$ matrix in Equation 25.3, which represents the coefficients in $2n$ first-order differential equations, can now be diagonalized using the matrix of eigenvectors $[u]$ as a transformation matrix. We define $2n$ principal coordinates such that $[u]\{p\}$ equals the state vector composed of $\{x\}$ and $\left\{\dot{x}\right\}$; our differential equations then become

$$\frac{d}{dt}\left[[u]^{-1}[I][u]\right]\{p\} = \left[[u]^{-1}\left[\begin{array}{cc} [0] & [I] \\ [-g_cM^{-1}K] & [-g_cM^{-1}C] \end{array}\right][u]\right]\{p\}$$

The matrix multiplication leads to a diagonal matrix containing the roots of the characteristic equation. For example, in our automotive problem,

$$\frac{d}{dt}\begin{bmatrix} 1 & 0 & 0 & 0 \\ 0 & 1 & 0 & 0 \\ 0 & 0 & 1 & 0 \\ 0 & 0 & 0 & 1 \end{bmatrix}\{p\} = \begin{bmatrix} s_1 & 0 & 0 & 0 \\ 0 & s_2 & 0 & 0 \\ 0 & 0 & s_3 & 0 \\ 0 & 0 & 0 & s_4 \end{bmatrix}\{p\}$$

Exercise 25.7 *Verify that our automotive example leads to this result by carrying out the numerical matrix multiplication procedure*

We have now separated our n-degree-of-freedom problem into $2n$ uncoupled first-order ordinary differential equations with a complex coefficient, $\dot{p}_i - s_i p_i = 0$, with the solution $p_i(t) = p_i(0)e^{s_1 t}$. When the roots of the characteristic equations s_i come in complex-conjugate pairs, then the solutions are indeed decaying exponentials. The initial values $p_i(0)$ are obtained from the Initial Conditions by the transformation

$$\{p(0)\} = [u]^{-1}\left\{\begin{array}{c} \{x(0)\} \\ \{v(0)\} \end{array}\right\}$$

The solutions for p are then transformed back to the coordinates x by the transformation

$$\left\{\begin{array}{c} \{x(t)\} \\ \left\{\dot{x}(t)\right\} \end{array}\right\} = [u]\{p(t)\}$$

You will note the redundancy of numbers in this procedure; all along, we did not really have $(2n)^2$ different input data because of the various symmetries and conjugates. We also don't really have $2n$ different solutions, since $\{v(t)\}$ must equal $\frac{d}{dt}\{x(t)\}$.

Exercise 25.8 *Add a forcing function to Equation 25.3 and trace what happens to it when you follow the solution procedure.*

25.5 Closure

Traditionally, damped problems have been handled by assuming either negligible or Rayleigh damping. With the computer power now available at any desktop, it is possible to handle arbitrary damping.

Problem 25.9 *Solve our automotive example for the initial conditions* $x_1(0) = x_2(0) = 1$, $\dot{x}_1(0) = \dot{x}_2(0) = 0$.

Chapter 26

WHIRLING and damping

The deflection of a mass on a rotating shaft has at least two degrees of freedom. In Chapter 8 we saw that there are two special cases which can be treated as one-degree-of-freedom problems:

1. The "flat" shaft (Section 8.6.1) which can bend in only one direction, can be treated in a rotating coordinate system, leading to Equation 8.11

$$\ddot{r} + \left(\frac{kg_c}{m} - \Omega^2\right) \cdot r = \frac{kg_c}{m} \cdot r_o$$

   We can add a damping term *iff* it is inherent in the material of the shaft and therefore *rotates with the coordinate system*

$$\ddot{r} + \frac{cg_c}{m}\dot{r} + \left(\frac{kg_c}{m} - \Omega^2\right) \cdot r = \frac{kg_c}{m} \cdot r_o \tag{26.1}$$

   The rotating damping does not affect the steady-state excursion

$$\frac{\overline{r}}{r_o} = \frac{\frac{kg_c}{m}}{\frac{kg_c}{m} - \Omega^2} \tag{26.2}$$

   even though it shows up as an oscillation in the fixed horizontal and vertical direction coordinates

$$\begin{aligned} x_{ss} &= \overline{r}\cos\Omega t \\ y_{ss} &= \overline{r}\sin\Omega t \end{aligned} \tag{26.3}$$

2. The "circular" shaft (Section 8.6.2) which has equal stiffness in all directions, can be treated in uncoupled fixed coordinates, leading to Equations

8.15

$$\ddot{x} + \left(\frac{kg_c}{M}\right) \cdot x = \frac{em}{M} \cdot \Omega^2 \cos \Omega t$$
$$\ddot{y} + \left(\frac{kg_c}{M}\right) \cdot y = \frac{em}{M} \cdot \Omega^2 \sin \Omega t$$

We can add damping terms *iff* they are *stationary* in this fixed coordinate system, i.e., if the spring constants k and the damping c arise in the elasticity of the bearing supports (rather than the flexing of the shaft)

$$\ddot{x} + \left(\frac{c_x g_c}{M}\right) \cdot \dot{x} + \left(\frac{k_x g_c}{M}\right) \cdot x = \frac{em}{M} \cdot \Omega^2 \cos \Omega t$$
$$\ddot{y} + \left(\frac{c_y g_c}{M}\right) \cdot \dot{y} + \left(\frac{k_y g_c}{M}\right) \cdot y = \frac{em}{M} \cdot \Omega^2 \sin \Omega t \tag{26.4}$$

Other cases are treated as two-degree-of-freedom systems.

26.1 Whirling

26.1.1 Undamped Case

Instead of using the Cartesian coordinates of Equations 8.15, we can express the motion of a mass M with an unbalance em on a "circular" shaft in the polar coordinates r, the radial deflection of the mass from its rest state, and θ, the angular position (so that $\dot{\theta}$ is the whirl)

$$\ddot{r} + \left(\frac{kg_c}{M} - \dot{\theta}^2\right) r = \frac{em}{M}\Omega^2 \cos(\Omega t - \theta)$$
$$r\ddot{\theta} + 2\dot{r}\dot{\theta} = \frac{em}{M}\Omega^2 \sin(\Omega t - \theta) \tag{26.5}$$

where k represents the effective spring constant, and Ω the angular velocity of the shaft.

Exercise 26.1 *Derive Equations 26.5 from Newton's law. Hint: express the acceleration as the vector sum of the acceleration of the nominal center of the mass and the acceleration of the c.g. relative to it.*

Exercise 26.2 *Derive Equations 26.5 from Lagrange's equation. Hint: start with*

$$U = \frac{1}{2}k\left[(r\cos\theta + e\cos\Omega t)^2 + (r\sin\theta + e\sin\Omega t)^2\right]$$

For steady-state synchronous whirl, $\ddot{r}$, $\dot{r}$, and $\ddot{\theta}$ are zero, the θ of the response is identical to the excitation Ω, and the phase angle $\varphi \triangleq (\Omega t - \theta)$ is a constant, so that the equations simplify to

$$\begin{aligned} \left(\frac{kg_c}{M} - \Omega^2\right) r &= \frac{em}{M}\Omega^2 \cos\varphi \\ 0 &= \sin\varphi \end{aligned} \tag{26.6}$$

and the solution is

$$\begin{aligned} \varphi &= 0 \text{ or } \pi \\ \left|\frac{rM}{em}\right| &= \frac{\frac{\Omega^2}{\omega_n^2}}{1 - \frac{\Omega^2}{\omega_n^2}} \end{aligned} \tag{26.7}$$

This is identical to the solution obtained from Equations 8.15.

26.1.2 Bearing Damping

If the shaft is rigid, but the support bearings have uniform elasticity and damping, we can add this non-rotating damping to Equations 26.5

$$\begin{aligned} \ddot{r} + \left(\frac{cg_c}{M}\right)\dot{r} + \left(\frac{kg_c}{M} - \dot{\theta}^2\right) r &= \frac{em}{M}\Omega^2 \cos(\Omega t - \theta) \\ r\,\ddot{\theta} + \left(\frac{cg_c}{M} r + 2\,\dot{r}\right)\dot{\theta} &= \frac{em}{M}\Omega^2 \sin(\Omega t - \theta) \end{aligned} \tag{26.8}$$

For steady-state synchronous whirl, $\dot{\theta} = \Omega$, and the phase angle $\varphi \triangleq (\Omega t - \theta)$ is a constant, so that the equations simplify to

$$\begin{aligned} \left(\frac{kg_c}{M} - \Omega^2\right) r &= \frac{em}{M}\Omega^2 \cos\varphi \\ \left(\frac{cg_c}{M} r\right)\Omega &= \frac{em}{M}\Omega^2 \sin\varphi \end{aligned} \tag{26.9}$$

and the solution is

$$\begin{aligned} \tan\varphi &= \frac{2\zeta\frac{\Omega}{\omega_n}}{1 - \frac{\Omega^2}{\omega_n^2}} \\ \left|\frac{rM}{em}\right| &= \frac{\frac{\Omega^2}{\omega_n^2}}{\sqrt{\left(1 - \frac{\Omega^2}{\omega_n^2}\right)^2 - \left(2\zeta\frac{\Omega}{\omega_n}\right)^2}} \end{aligned} \tag{26.10}$$

This is identical to the solution obtained from Equations 26.4.

26.1.3 Hysteresis Damping

The situation changes when the bearings are rigid and the shaft is elastic, because any material damping in the shaft will turn with it, so that we have to describe it in a rotating coordinate system.

Exercise 26.3 *Transform the governing equations to a coordinate system rotating at angular velocity Ω.*

When we look at steady-state synchronous whirl, the damping drops out and the equations simplify to

$$\left|\frac{rM}{em}\right| = \frac{\frac{\Omega^2}{\omega_n^2}}{1 - \frac{\Omega^2}{\omega_n^2}} \tag{26.11}$$

The damping does not enter in, because the shaft does not flex—it merely rotates in a flexed position.

We find it disturbing that this expression is not the same as Equation 26.2, contrary to our expectation that for synchronous whirl the "flat" shaft and the "circular" shaft should turn out the same. The answer is that we chose different definitions for the radius r for the two cases. In whirling, it is customary to define r as the measure of the location of the nominal center of the shaft, relative to the rest position where the spring force is zero. For the flat shaft, we had defined r as the measure of the location of the center-of-gravity of the mass; the measure of the displacement relative to the rest position is $r - r_o$. If we transform Equation 26.2 accordingly, it becomes

$$\frac{|\overline{r} - r_o|}{r_o} = \frac{\Omega^2}{\frac{kg_c}{m} - \Omega^2} \tag{26.12}$$

which reduces to Equation 26.11 if we let the m within the unbalance em be equal to M.

26.2 Dynamic Unbalance

So far we have looked at unbalances em which are in the plane of a flywheel, whether they are due to a small mass added or subtracted from the edge of the flywheel, or a small error in drilling the center of the flywheel. This kind of unbalance is traditionally called a *static* unbalance, because it can be detected in a static test. Simple wheel balancers place the wheel horizontally on a central pivot; if the center of gravity is not at the center of the hub, the wheel will tilt. Balancing weights can then be added to the rim until the wheel rests level.

However, it is possible for a wheel to be statically balanced and yet vibrate when it spins. This is called *dynamic* unbalance. One way to generate it is to add a small mass m to a wheel rim on the inner side, and another one 180 degrees around the wheel, but on the outer side. If the rim width is $2a$, the two weights will generate a force-couple $2\,(am)\,\Omega^2$, analogous to the simple force $(em)\,\Omega^2$ observed in static unbalance. (A single weight attached to the outer side of the wheel would, of course, generate both a force *and* a couple.)

The same result can be achieved by drilling the central hole in a flywheel at a slight angle to the plane of the flywheel, causing a wobble. One way to look at this is that the axis of rotation does not concide with a principal axis of the flywheel, so that the mixed-subscript second moments I_{xy} etc. are not equal to zero. The cure is to add small weights to the sides of the rim on opposing sides, until the rotational axis is a principal axis.

Unlike static unbalance, which can lead to unbounded resonant amplitudes at critical speeds, isolated dynamic unbalance is self-limiting—at worst, the flywheel tilts enough to approach rotation around one of its principal axes—so that careful balancing limits the stresses in the shaft and bearings at all speeds.

26.3 Stodola's Gyroscopic Effects

So far, we have assumed that the flywheel is symmetrically supported at the center of a shaft, so that lateral deflection (due to static unbalance) is not spring-coupled to tilt deflection, and vice versa. In general, this is not true.

Let us consider an overhanging flywheel. For simplicity, let us describe the flywheel as a thin disk, and the shaft as a cantilever firmly mounted on a bearing (an unrealistic assumption). The flywheel has a mass m and a rotational moment of inertia $J_{zz} = 0.5mR^2$; for tilting the moment of inertia is $I_{xx} = I_{yy} = 0.25mR^2$. The deflection at the end of the cantilever due to a force F is $y = FL^3/3EI$ and $\phi = dy/dx = FL^2/2EI$; a moment load M leads to $\phi = dy/dx = ML/EI$ and $y = ML^2/2EI$; therefore the flexibility matrix is

$$\left\{ \begin{array}{c} y \\ \phi \end{array} \right\}_{\text{end}} = \left[\begin{array}{cc} \frac{L^3}{3EI} & \frac{L^2}{2EI} \\ \frac{L^2}{2EI} & \frac{L}{EI} \end{array} \right] \left\{ \begin{array}{c} F \\ M \end{array} \right\}_{\text{end}}$$

$$\left\{ \begin{array}{c} y \\ \phi L \end{array} \right\}_{\text{end}} = \frac{L^3}{EI} \left[\begin{array}{cc} 1/3 & 1/2 \\ 1/2 & 1 \end{array} \right] \left\{ \begin{array}{c} F \\ M/L \end{array} \right\}_{\text{end}}$$

and the stiffness matrix

$$\left\{ \begin{array}{c} F \\ M \end{array} \right\}_{\text{end}} = \left[\begin{array}{cc} \frac{12EI}{L^3} & \frac{-6EI}{L^2} \\ \frac{-6EI}{L^2} & \frac{4EI}{L} \end{array} \right] \left\{ \begin{array}{c} y \\ \phi \end{array} \right\}_{\text{end}}$$

$$\left\{ \begin{array}{c} F \\ M/L \end{array} \right\}_{\text{end}} = \frac{EI}{L^3} \left[\begin{array}{cc} 12 & -6 \\ -6 & 4 \end{array} \right] \left\{ \begin{array}{c} y \\ \phi L \end{array} \right\}_{\text{end}}$$

Therefore our **non-rotating** natural frequencies would be obtained from

$$\frac{m}{g_c} \left[\begin{array}{cc} 1 & 0 \\ 0 & D \end{array} \right] \left\{ \begin{array}{c} \ddot{y} \\ \ddot{\phi}\, L \end{array} \right\} + \frac{EI}{L^3} \left[\begin{array}{cc} 12 & -6 \\ -6 & 4 \end{array} \right] \left\{ \begin{array}{c} y \\ \phi L \end{array} \right\} = \left\{ \begin{array}{c} 0 \\ 0 \end{array} \right\}$$

$$\text{where } D \triangleq 0.25\frac{R^2}{L^2}$$

with computer algebra giving us the eigenvalues

$$\omega^2_{I,II} = \frac{EIg_c}{mL^3}\left(6 + \frac{2}{D} \mp 2\sqrt{9 + \frac{3}{D} + \frac{1}{D^2}}\right)$$

$$\omega^2_I < 3\frac{EIg_c}{mL^3}$$

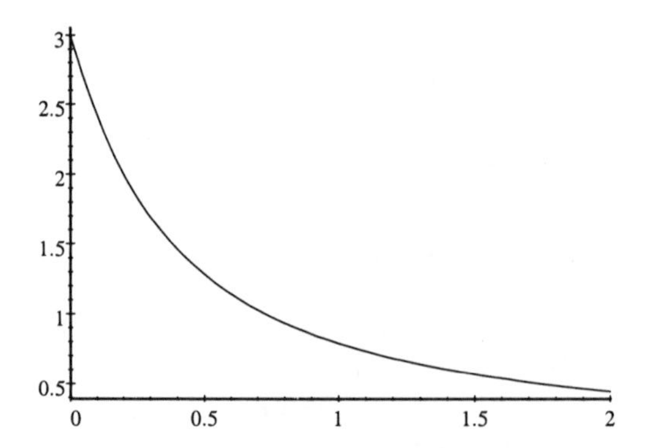

so that the lowest natural frequency is less than it would have been for a concentrated mass.

Exercise 26.4 *Plot both natural frequency as a function of* $0.5 < D < 2$. *Hint: This is most easily done using a computer plotting package.*

When the shaft is **rotating**, centrifugal forces enter in two different ways: on the one hand, any deflection of the center of the mass causes forces which *reduce* the restoring spring force; on the other hand, any tilting of the plane of the disk causes forces which tend to *restore* it to rotation around the principal axis with the largest moment of inertia. (I've chosen my words carefully—if the rotational body is a very thick disk, for which $J_{zz} < I_{xx} = I_{yy}$ the tilting of the flywheel may cause forces which push it further away from its normal orientation.[1])

If we restrict ourselves to the special case of **synchronous whirl**—that is, the situation where the deflection moves around at the (as yet unknown) rotational rate Ω of the shaft—we can use a simple rotating coordinate system. The centrifugal force due to the displacement y of the mass is $\Omega^2 my/g_c$ away from $y = 0$. The moment on the disk due to the tilt ϕ is $\Omega^2 I_{yy}\phi/g_c$ towards $\phi = 0$. We add these centrifugal forces to the restoring forces due to the elasticity

[1] *see* J.P. Den Hartog, *Mechanical Vibrations*, reissued in 1985 by Dover Publications, New York, ISBN 0-486-64785-4.

to obtain an expanded stiffness matrix for our thin-disk problem

$$\begin{aligned}
\left\{ \begin{array}{c} F \\ M \end{array} \right\}_{\text{end}} &= \left[\begin{array}{cc} \frac{12EI}{L^3} & \frac{-6EI}{L^2} \\ \frac{-6EI}{L^2} & \frac{4EI}{L} \end{array} \right] \left\{ \begin{array}{c} y \\ \phi \end{array} \right\}_{\text{end}} + \left\{ \begin{array}{c} -\Omega^2 \frac{m}{g_c} y \\ \Omega^2 \frac{0.25mR^2}{g_c} \phi \end{array} \right\} \\
&= \left[\begin{array}{cc} \left(\frac{12EI}{L^3} - \Omega^2 \frac{m}{g_c} \right) & \left(\frac{-6EI}{L^2} \right) \\ \left(\frac{-6EI}{L^2} \right) & \left(\frac{4EI}{L} + \Omega^2 \frac{0.25mR^2}{g_c} \right) \end{array} \right] \left\{ \begin{array}{c} y \\ \phi \end{array} \right\}_{\text{end}} \\
\left\{ \begin{array}{c} F \\ M/L \end{array} \right\}_{\text{end}} &= \left[\frac{EI}{L^3} \left[\begin{array}{cc} 12 & -6 \\ -6 & 4 \end{array} \right] + \Omega^2 \frac{m}{g_c} \left[\begin{array}{cc} -1 & 0 \\ 0 & 0.25\frac{R^2}{L^2} \end{array} \right] \right] \left\{ \begin{array}{c} y \\ \phi L \end{array} \right\}_{\text{end}} \\
&= \frac{EI}{L^3} \left[\begin{array}{cc} 12 - K^2 & -6 \\ -6 & 4 + DK^2 \end{array} \right] \left\{ \begin{array}{c} y \\ \phi L \end{array} \right\}_{\text{end}} \\
\text{where } K &\triangleq \Omega^2 \frac{mL^3}{EIg_c} \text{ and } D \triangleq \frac{I_{xx}}{mL^2} = 0.25\frac{R^2}{L^2}
\end{aligned}$$

The critical speed Ω_{crit} occurs when the restoring forces go to zero, $(F, M) = (0, 0)$, so that we can set the determinant of the expanded stiffness matrix equal to zero and obtain the characteristic equation

$$\begin{aligned}
\left(12 - K^2\right)\left(4 + DK^2\right) - (-6)^2 &= 0 \\
K^4 + \left(\frac{4}{D} - 12\right) K^2 - \frac{12}{D} &= 0
\end{aligned}$$

with the positive root of the unknown K^2 as a function of the system parameter D

$$K^2 = \left(6 - \frac{2}{D}\right) + \sqrt{\left(6 - \frac{2}{D}\right)^2 + \frac{12}{D}}$$

and the critical speed

$$\Omega^2_{crit} = \frac{EIg_c}{mL^3} \left(\left(6 - \frac{2}{D}\right) + \sqrt{\left(6 - \frac{2}{D}\right)^2 + \frac{12}{D}} \right)$$

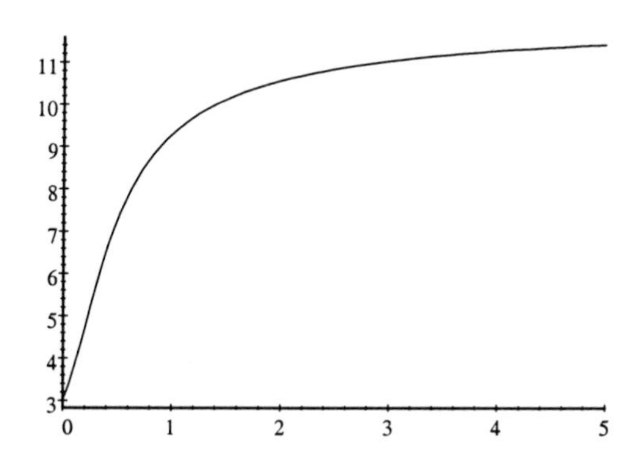

We see that for a concentrated mass ($D = 0$) the critical speed is the expected $\Omega_{crit}^2 = 3EIg_c/mL^3$; but that for our disk-flywheel example ($D > 0$) the stiffening due to the gyroscopic forces raises the first critical speed, contrary to our expectation from non-rotating analysis.

Exercise 26.5 *What would the natural frequency be if the end of the cantilever were constrained (by gyroscopic forces) to stay parallel to the base ($\theta_{end} = 0$)?*

For determining this critical speed, we assumed synchronous whirl, with the deflection in our rotating reference plane; for the general vibrational solution of a rotating disk, we would need to consider deflections in two directions, x and y (or r and θ), and angular deflections in two directions, ϕ_x and ϕ_y. Therefore, this becomes a four-degrees-of-freedom problem. If there are several disks, the number of degrees-of-freedom increases accordingly.

For a "circular" shaft, the resulting natural frequencies are commonly interpreted as *forward whirl,* where the shaft deflection rotates in the same direction as the shaft—generally more slowly than the shaft rotation, except for the synchronous whirl at important critical speeds—although *reverse whirl* is also observed occasionally.

Problem 26.6 *Study the stationary-shaft natural frequencies, and the first critical speed, of a flywheel mounted on a simply-supported weightless shaft, if the flywheel is a thin circular disk.*

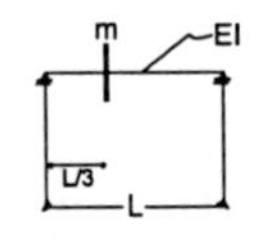

Chapter 27

TRANSFER MATRICES and finite elements

A different perspective on large systems can be obtained by looking at them element by element, breaking the formulation up into simple repeated steps.

27.1 Transfer Matrices

The crankshafts of large piston engines can be represented as a series of torsion-bars and flywheels. The "boxcar" model of Chapter 20 can be used to analyze them, but since we are mostly interested in the lowest natural frequency, other methods may be quicker. One traditional approach is Holzer's method, and the transfer-matrix formulation accomplishes the same thing in a systematic procedure. (Figure 27.1)

27.1.1 Torsional and Translational Systems

Let us consider a "boxcar" system with a great number of elements. Following a historical convention, we will number the springs and masses in pairs from left to right; the subscript n designates an "element" consisting of a spring k_n and the mass m_n adjoining it on the right.

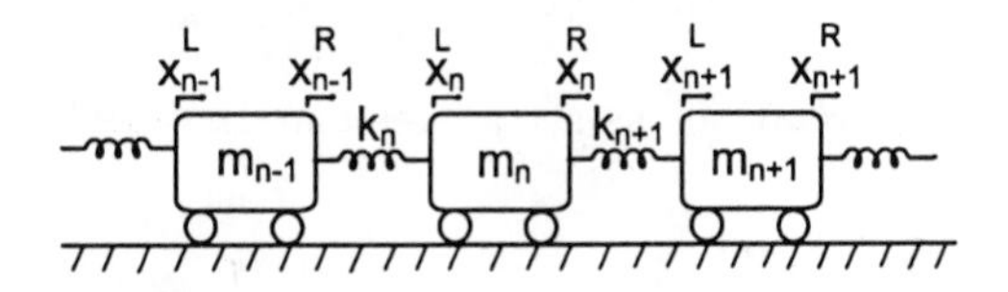

Figure 27.1: Transfer Variables Notation

The state at any point between a spring and a mass is described by the axial displacement x and the tension T at the connection. Because of our numbering convention, there are twice as many connections as "elements," and we need to add a superscript showing the location relative to the *mass*. We will designate the state to the *right* of mass m_{n-1} (and to the left of spring k_n and element n) by x^R_{n-1} and T^R_{n-1}; the state to the *left* of the of mass m_n (and to the right of spring k_n, i.e., in the middle of element n) is x^L_n and T^L_n of element n; the state to the *right* of the of mass m_n (and to the left of spring k_{n+1} and element $n+1$) is x^R_n and T^R_n; etc.

A massless spring has the same tension T (or compression $-T$) at both ends (we use a tension-compression convention rather than a directional-force convention)

$$T^L_n = T^R_{n-1} \tag{27.1}$$

and a change-in-length proportional to the tension

$$x^L_n = x^R_{n-1} + \frac{1}{k_n} T^R_{n-1} \tag{27.2}$$

We can express the two equations as a transfer matrix called the "field matrix" for element n

$$\left\{ \begin{array}{c} x \\ T \end{array} \right\}^L_n = \left[\begin{array}{cc} 1 & 1/k \\ 0 & 1 \end{array} \right]_n \left\{ \begin{array}{c} x \\ T \end{array} \right\}^R_{n-1} \tag{27.3}$$

showing that the state to the right of the spring $\left(x^L_n, T^L_n\right)$ is a function of k_n as well as the state to the left of the spring $\left(x^R_{n-1}, T^R_{n-1}\right)$.

A rigid mass has the same deflection x on both sides (we use a consistent directional convention for all x's)

$$x^R_n = x^L_n$$

and a change-in-tension proportional to the acceleration

$$T^R_n = T^L_n + \frac{m_n}{g_c} \ddot{x}^L_n$$

If we are looking at sinusoidal motions $x = \sin \omega t$, we can replace $\ddot{x}$ with $-\omega^2 x$

$$T^R_n = T^L_n - \frac{\omega^2 m_n}{g_c} x^L_n$$

We can express these equations as a transfer matrix called the "point matrix" for element n

$$\left\{ \begin{array}{c} x \\ T \end{array} \right\}^R_n = \left[\begin{array}{cc} 1 & 0 \\ -\omega^2 m/g_c & 1 \end{array} \right]_n \left\{ \begin{array}{c} x \\ T \end{array} \right\}^L_n \tag{27.4}$$

showing that the state to the right of the spring $\left(x_n^R, T_n^R\right)$ is a function of $\omega^2 m_n$ as well as the state to the left of the spring $\left(x_n^L, T_n^L\right)$.

The beauty of transfer matrices is that we can combine them; by substituting the expression from Equation 27.3 for the last column vector in Equation 27.4, we obtain

$$\left\{ \begin{array}{c} x \\ T \end{array} \right\}_n^R = \left[\begin{array}{cc} 1 & 0 \\ -\omega^2 m/g_c & 1 \end{array} \right]_n \left[\begin{array}{cc} 1 & 1/k \\ 0 & 1 \end{array} \right]_n \left\{ \begin{array}{c} x \\ T \end{array} \right\}_{n-1}^R$$

which we can multiply out, using computer algebra like *Maple,* to obtain the transfer matrix for the combined element n

$$\left\{ \begin{array}{c} x \\ T \end{array} \right\}_n^R = \left[\begin{array}{cc} 1 & 1/k \\ -\omega^2 m/g_c & 1-\omega^2 m/k g_c \end{array} \right]_n \left\{ \begin{array}{c} x \\ T \end{array} \right\}_{n-1}^R \tag{27.5}$$

The drawback of the historical convention is that the mathematical expression is reversed relative to the left-to-right schema. The vector on the left side of the equation, represents the right end of the element, and vice-versa. The mass' matrix appears to the right of the spring's matrix, opposite to their location on the schematic drawing. It is important to watch out for errors in the sequence of the matrices: changing the order of multiplication will change the answer!

Expanding to our complete system of many masses and springs, we can write out a string of spring and mass matrices and multiply them together to get a transfer matrix for the whole system. The overall transfer matrix will consist of four polynomial functions of ω^2 incorporating powers up to ω^{2n}

$$\left\{ \begin{array}{c} x \\ T \end{array} \right\}_{right-end} = \left[\begin{array}{cc} A\left(\omega^2\right) & B\left(\omega^2\right) \\ C\left(\omega^2\right) & D\left(\omega^2\right) \end{array} \right]_{overall} \left\{ \begin{array}{c} x \\ T \end{array} \right\}_{left-end} \tag{27.6}$$

The system need not start with a spring at the left end, nor does it need to end with a mass at the right end.

We can study **free vibrations** by looking at the constraints at the ends. Where the system ends with a spring *fixed* to an immovable foundation, the end state vector is $(0, T_{end})$; where it ends with a *free* mass, there is no tension and the end state vector is $(x_{end}, 0)$. By writing out Equation 27.6 we can deduce that the polynomials that are multiplied with a non-zero left-end state variable determine whether the right-end state-variables are indeed zero. We conclude that

$$A\left(\omega_n^2\right) = 0 \text{ makes right end } \textit{fixed,} \text{ if left end is } \textit{free} \tag{27.7}$$
$$B\left(\omega_n^2\right) = 0 \text{ makes right end } \textit{fixed,} \text{ if left end is } \textit{fixed} \tag{27.8}$$
$$C\left(\omega_n^2\right) = 0 \text{ makes right end } \textit{free,} \text{ if left end is } \textit{free} \tag{27.9}$$
$$D\left(\omega_n^2\right) = 0 \text{ makes right end } \textit{free,} \text{ if left end is } \textit{fixed} \tag{27.10}$$

Therefore each pair of end conditions leads to a characteristic equation of order n or less. (It may look like $2n$, but only even terms appear in an undamped problem.)

We can study periodic **forced vibrations** by substituting a forcing function for the left end condition in Equation 27.6 and solving for the right-end response. For a fixed-spring right end, and a force excitation at a free mass on the left end,

$$\left\{ \begin{array}{c} 0 \\ T \end{array} \right\}_{right-end} = \left[\begin{array}{cc} A\left(\omega_{ex}^2\right) & B\left(\omega_{ex}^2\right) \\ C\left(\omega_{ex}^2\right) & D\left(\omega_{ex}^2\right) \end{array} \right]_{overall} \left\{ \begin{array}{c} x \\ T_o \sin \omega_{ex} \end{array} \right\}_{left-end}$$

which gives us two equations in two unknowns: the force response T at the right end and the displacement response x at the left end.

Exercise 27.1 *Write the forced-vibration equations for the other three possible pairs of end conditions.*

We can add damping to the transfer matrices. If the damper c_k is parallel to a spring, the field matrix Equation 27.3 becomes

$$\left\{ \begin{array}{c} x \\ T \end{array} \right\}_n^L = \left[\begin{array}{cc} 1 & 1/\left(k + i\omega c_k\right) \\ 0 & 1 \end{array} \right]_n \left\{ \begin{array}{c} x \\ T \end{array} \right\}_{n-1}^R \tag{27.11}$$

but if the damper c_m couples a mass to the ground, the point matrix Equation 27.4 becomes

$$\left\{ \begin{array}{c} x \\ T \end{array} \right\}_n^R = \left[\begin{array}{cc} 1 & 0 \\ \left(-\omega^2 m/g_c + i\omega c_m\right) & 1 \end{array} \right]_n \left\{ \begin{array}{c} x \\ T \end{array} \right\}_n^L \tag{27.12}$$

which multiplies out to

$$\begin{aligned} \left\{ \begin{array}{c} x \\ T \end{array} \right\}_n^R &= \left[\begin{array}{cc} 1 & 0 \\ \frac{-\omega^2 m}{g_c + i\omega c_m} & 1 \end{array} \right]_n \left[\begin{array}{cc} 1 & \frac{1}{k+i\omega c_k} \\ 0 & 1 \end{array} \right]_n \left\{ \begin{array}{c} x \\ T \end{array} \right\}_{n-1}^R \\ &= \left[\begin{array}{cc} 1 & \frac{1}{k+i\omega c_k} \\ \frac{-\omega^2 m}{g_c + i\omega c_m}{}_m & 1 + \frac{-\omega^2 m/g_c + i\omega c_m}{k + i\omega c_k} \end{array} \right]_n \left\{ \begin{array}{c} x \\ T \end{array} \right\}_{n-1}^R \end{aligned} \tag{27.13}$$

Evidently, damping leads to complex-number characteristic equations. In free-vibration analysis, it usually results in complex-number roots for the natural frequencies ω_n, which we can interpret as decay by setting $s = i\omega$ as we did in Chapter 25. In forced-vibration analysis, it leads to complex-number responses, which we can interpret as phase-angles.

We can easily derive additional transfer matrices for leveraged connectors (and for gears in rotational systems). For example, a stiff and weightless lever of ratio r can be described by

$$\left\{ \begin{array}{c} x \\ T \end{array} \right\}_{long} = \left[\begin{array}{cc} r & 0 \\ 0 & 1/r \end{array} \right]_{lever} \left\{ \begin{array}{c} x \\ T \end{array} \right\}_{short}$$

The strength of the transfer-matrix method is the ease of formulating any system which consist of a string of elements; we need only insert the proper values of

m, k, and c for each element in a string of field, point, or leverage matrices. The matrix multiplication, and the solution of characteristic equations, can then be handled by computer algebra.

Periodically repeating structures with many elements become "transmission lines" with wave-propagation characteristics similar to the continuous media discussed in Chapter 28. Conversely, continuous media can be approximated by a series of non-infinitesimal springs and masses. In general, there is little difference between a chain of repeated elements and a continuous medium, if there are at least five to seven elements per "wave" in the mode shapes.

27.1.2 Beam Elements

Myklestadt's method for wing spars in airplanes, and Prohl's method for critical speeds in rotors, can be cast into the form of transfer matrices expanded to four state variables for stations along a beam: normal displacement y, slope $\phi = y'$, moment $M = y''/EI$, and shear $V = \pm y'''/EI$ (with the sign depending on the sign convention for positive shear).

If we have a series of elastic beam elements, each with a concentrated mass at the right end, the field transfer matrix Equation 27.3 becomes

$$\left\{ \begin{array}{c} y \\ y' \\ M \\ \pm V \end{array} \right\}_n^L = \left[\begin{array}{cccc} 1 & L & \frac{L^2}{2EI} & \frac{L^3}{6EI} \\ 0 & 1 & \frac{L}{EI} & \frac{L^2}{2EI} \\ 0 & 0 & 1 & L \\ 0 & 0 & 0 & 1 \end{array} \right]_n \left\{ \begin{array}{c} y \\ y' \\ M \\ \pm V \end{array} \right\}_{n-1}^R \tag{27.14}$$

and the point transfer matrix Equation 27.4 becomes

$$\left\{ \begin{array}{c} y \\ y' \\ M \\ \pm V \end{array} \right\}_n^R = \left[\begin{array}{cccc} 1 & 0 & 0 & 0 \\ 0 & 1 & 0 & 0 \\ 0 & 0 & 1 & 0 \\ \frac{\omega^2 m}{g_c} & 0 & 0 & 1 \end{array} \right]_n \left\{ \begin{array}{c} y \\ y' \\ M \\ \pm V \end{array} \right\}_n^L \tag{27.15}$$

(Additional terms may be added for rotational inertia, etc.)

Exercise 27.2 *Find the element matrix for a section of beam. Hint, multiply field and point matrices, as we did to obtain Equation 27.5.*

When we multiply a string of these four-by-four matrices, we obtain 16 characteristic functions $A\left(\omega^2\right)$ for the overall system. For a wing, we would write a fixed built-in condition for the left-end base, and look for a free right-end tip

$$\left\{ \begin{array}{c} y_{tip} \\ y'_{tip} \\ 0 \\ 0 \end{array} \right\}_{free} = \left[\begin{array}{cccc} A_{11} & A_{12} & A_{13} & A_{14} \\ A_{21} & A_{22} & A_{23} & A_{24} \\ A_{31} & A_{32} & A_{33} & A_{34} \\ A_{41} & A_{42} & A_{43} & A_{44} \end{array} \right]_{wing} \left\{ \begin{array}{c} 0 \\ 0 \\ M_{base} \\ \pm V_{base} \end{array} \right\}_{fixed} \tag{27.16}$$

We conclude that the two simultaneous characteristic equations for ω_n^2 are

$$\begin{bmatrix} A_{33} & A_{34} \\ A_{43} & A_{44} \end{bmatrix} \begin{Bmatrix} M_{base} \\ \pm V_{base} \end{Bmatrix} = \begin{Bmatrix} 0 \\ 0 \end{Bmatrix}$$

Exercise 27.3 *For a rotor, we might have a simple-support condition at each end. Find the characteristic equations. Hint: the end conditions for a pinned support are* $(0, y', 0, \pm V)$.

The transfer-matrix approach can combine bending effects in beams from this section, with tension effects from the previous section, expanding to a six-state-variable approach. The field transfer matrix for a beam element becomes

$$\begin{Bmatrix} y \\ y' \\ M \\ \pm V \\ x \\ T \end{Bmatrix}_n^L = \begin{bmatrix} 1 & L & \frac{L^2}{2EI} & L^3 & 0 & 0 \\ 0 & 1 & \frac{L}{EI} & \frac{L^2}{2EI} & 0 & 0 \\ 0 & 0 & 1 & L & 0 & 0 \\ 0 & 0 & 0 & 1 & 0 & 0 \\ 0 & 0 & 0 & 0 & 1 & \frac{1}{k} \\ 0 & 0 & 0 & 0 & 0 & 1 \end{bmatrix}_n \begin{Bmatrix} y \\ y' \\ M \\ \pm V \\ x \\ T \end{Bmatrix}_{n-1}^R$$

and the point transfer matrix

$$\begin{Bmatrix} y \\ y' \\ M \\ \mp V \\ x \\ T \end{Bmatrix}_n^R = \begin{bmatrix} 1 & 0 & 0 & 0 & 0 & 0 \\ 0 & 1 & 0 & 0 & 0 & 0 \\ 0 & 0 & 1 & 0 & 0 & 0 \\ \frac{\omega^2 m}{g_c} & 0 & 0 & 1 & 0 & 0 \\ 0 & 0 & 0 & 0 & 1 & 0 \\ 0 & 0 & 0 & 0 & -\omega^2 m/g_c & 1 \end{bmatrix}_n \begin{Bmatrix} y \\ y' \\ M \\ \pm V \\ x \\ T \end{Bmatrix}_n^L$$

Other matrices can take us around corners; for example, a ninety-degree-bend can be accomplished by a transfer matrix that exchanges the x- and y-directions

$$\begin{Bmatrix} y \\ y' \\ M \\ \pm V \\ x \\ T \end{Bmatrix}_{vert.} = \begin{bmatrix} 0 & 0 & 0 & 0 & 1 & 0 \\ 0 & 1 & 0 & 0 & 0 & 0 \\ 0 & 0 & 1 & 0 & 0 & 0 \\ 0 & 0 & 0 & 0 & 0 & \pm 1 \\ -1 & 0 & 0 & 0 & 0 & 0 \\ 0 & 0 & 0 & \mp 1 & 0 & 0 \end{bmatrix}_{bend} \begin{Bmatrix} y \\ y' \\ M \\ \mp V \\ x \\ T \end{Bmatrix}_{horiz.}$$

This allows us to analyze the natural frequencies of bent beams.

Exercise 27.4 *Specify the state vectors for a built-in fixed end, for a free end, and for a pinned support.*

27.2 Finite Elements

Although our analysis may be applied to a periodic structure such as a series of rotors on a shaft, it can also be applied to non-infinitesimal elements of a continuous structure, approximated by a series of springs and masses. If we do that, however, we would prefer to incorporate the mass into the element itself, rather than segregate it into a lumped mass off to one side. We can do this by assigning a local mode shape to the deflection *within* the element; if the element is tiny, a static-deflection mode-shape is an excellent assumption for the internal deflections.

27.2.1 Static Analysis

For *axial* deflection of an element of a uniform rod of length ℓ, static deflection under tension T is

$$\begin{aligned} \frac{du}{dx} &= \frac{T}{EA} = \text{constant} \\ u(x) &= \frac{T}{EA}x + u_1 \\ u_2 - u_1 &= \frac{\ell}{EA}T = \frac{1}{k}T \end{aligned} \tag{27.17}$$

where $k \triangleq EA/\ell$. Note that we have switched to calling the x-direction deflection u, in order to avoid confusion between the location x along the element and the elastic deflection u which varies as a function of x. Except for the change in notation, this is the same result as Equation 27.2 for a weightless spring.

Using this static deflection and expressing the potential energy in terms of the relative deflections u_1 at the left end and u_2 at the right end, and the forces applied to each end as $F_1 = -T$ and $F_2 = T$

$$\begin{aligned} \text{for } \frac{du}{dx} &= \left(\frac{u_2 - u_1}{\ell}\right) \\ \text{P.E.} &= \frac{1}{2}\int_0^\ell EA\left(\frac{du}{dx}\right)^2 dx \\ &= \frac{1}{2}ku_1^2 - \frac{2}{2}ku_1u_2 + \frac{1}{2}ku_1^2 \end{aligned}$$

Therefore, for this assumed deflection, the equivalent-stiffness matrix is

$$\begin{bmatrix} k & -k \\ -k & k \end{bmatrix} \begin{Bmatrix} u_1 \\ u_2 \end{Bmatrix} = \begin{Bmatrix} F_1 \\ F_2 \end{Bmatrix} \tag{27.18}$$

This is enough information to solve the equilibrium problem.

If we have a stepped rod under compression, we can divide it up into uniform finite elements, match the forces F at the stations where the elements join, and write the equations for the displacements

$$\begin{bmatrix} k_a & -k_a & 0 & 0 & 0 & 0 \\ -k_a & k_a+k_b & -k_b & 0 & 0 & 0 \\ 0 & -k_b & k_b+k_c & -k_c & 0 & 0 \\ 0 & 0 & -k_c & k_c+k_d & -k_d & 0 \\ 0 & 0 & 0 & -k_d & k_d+k_e & \cdots \\ 0 & 0 & 0 & 0 & \vdots & \ddots \end{bmatrix} \begin{Bmatrix} u_1 \\ u_2 \\ u_3 \\ u_4 \\ u_5 \\ \vdots \end{Bmatrix} = \begin{Bmatrix} F_1 \\ 0 \\ 0 \\ 0 \\ \vdots \\ F_{end} \end{Bmatrix}$$

The matrix is a tri-diagonal matrix, which can be inverted efficiently by computers, even for large numbers of elements.

27.2.2 Dynamic Analysis

To solve dynamic problems, we need to develop a consistent mass matrix, based on the same assumed deflection, from the kinetic energy

$$\begin{aligned} \text{for } \dot{u}(x) &= \dot{u}_1 + \frac{x}{\ell}\left(\dot{u}_2 - \dot{u}_1\right) = \left(1 - \frac{x}{\ell}\right)\dot{u}_1 + \frac{x}{\ell}\dot{u}_2 \\ \text{K.E.} &= g\frac{1}{2}\int_0^\ell \frac{\mu}{g_c}\left(\dot{u}_{(x)}\right)^2 dx \\ &= \frac{1}{2}\left(\frac{\mu\ell}{3g_c}\right)\dot{u}_1^2 - \frac{2}{2}\left(\frac{\mu\ell}{6g_c}\right)\dot{u}_1\dot{u}_2 + \frac{1}{2}\left(\frac{\mu\ell}{3g_c}\right)\dot{u}_1^2 \end{aligned}$$

where we define the element's mass-per-unit-length $\mu \triangleq M/\ell$. Therefore the equivalent-mass matrix is

$$\frac{1}{g_c}\begin{bmatrix} \frac{M}{3} & \frac{M}{6} \\ \frac{M}{6} & \frac{M}{3} \end{bmatrix}\begin{Bmatrix} \ddot{u}_1 \\ \ddot{u}_2 \end{Bmatrix} + \begin{bmatrix} k & -k \\ -k & k \end{bmatrix}\begin{Bmatrix} u_1 \\ u_2 \end{Bmatrix} = \begin{Bmatrix} F_1 \\ F_2 \end{Bmatrix} \tag{27.19}$$

as we already found in our study of coupling springs in Chapter 22.

We can add this to the equilibrium study of axial deflection of a rod

$$\frac{1}{g_c}\begin{bmatrix} \frac{M_a}{3} & \frac{M_a}{6} & 0 & 0 \\ \frac{M_a}{6} & \frac{M_a+M_b}{3} & \frac{M_b}{6} & 0 \\ 0 & \frac{M_b}{6} & \frac{M_b+M_c}{3} & \frac{M_c}{6} \\ 0 & 0 & \frac{M_c}{6} & \ddots \end{bmatrix}\begin{Bmatrix} \ddot{u}_1 \\ \ddot{u}_2 \\ \ddot{u}_3 \\ \vdots \end{Bmatrix}$$

$$+ \begin{bmatrix} k_a & -k_a & 0 & 0 \\ -k_a & k_a+k_b & -k_b & 0 \\ 0 & -k_b & k_b+k_c & -k_c \\ 0 & 0 & -k_c & \ddots \end{bmatrix}\begin{Bmatrix} u_1 \\ u_2 \\ u_3 \\ \vdots \end{Bmatrix} = \begin{Bmatrix} 0 \\ 0 \\ 0 \\ \vdots \end{Bmatrix}$$

leading to a pair of tri-diagonal matrices.

Since breaking the problem up into a greater number of smaller elements obviously increases the time and cost of computation, the burning question is "how many is enough?" This question is usually answered by starting with a coarse "grid" and then repeat with ever finer grids until the answer no longer changes.

Exercise 27.5 *Find the axial natural frequency of a uniform rod, using a single element, two elements, and three elements. Compare the results.*

27.2.3 Beams

In addition to longitudinal deflection, we can allow elements to deflect transversely in shear to obtain

$$\frac{1}{g_c}\begin{bmatrix} \frac{M}{3} & \frac{M}{6} \\ \frac{M}{6} & \frac{M}{3} \end{bmatrix}\begin{Bmatrix} \ddot{v}_1 \\ \ddot{v}_2 \end{Bmatrix} + \begin{bmatrix} k & -k \\ -k & k \end{bmatrix}\begin{Bmatrix} v_1 \\ v_2 \end{Bmatrix} = \begin{Bmatrix} P_1 \\ P_2 \end{Bmatrix}$$

where $k = GA/\ell$; or—if the elements are slender—we can look at them as beams in bending. Using static deflection (due to transverse loads P and moments M) as the mode-shape, we can obtain equivalent stiffness and consistent equivalent mass in terms of transverse deflections v and rotations θ at the ends

$$\frac{M}{420 g_c}\begin{bmatrix} 156 & 22 & 54 & -13 \\ 22 & 4 & 13 & -3 \\ 54 & 13 & 156 & -22 \\ -13 & -3 & -22 & 4 \end{bmatrix}\begin{Bmatrix} \ddot{v}_1 \\ \ell\,\ddot{\theta}_1 \\ \ddot{v}_2 \\ \ell\,\ddot{\theta}_2 \end{Bmatrix}$$

$$+\frac{EI}{\ell^3}\begin{bmatrix} 12 & 6 & -12 & 6 \\ 6 & 4 & -6 & 2 \\ -12 & -6 & 12 & -6 \\ 6 & 2 & -6 & 4 \end{bmatrix}\begin{Bmatrix} v_1 \\ \ell\theta_1 \\ v_2 \\ \ell\theta_2 \end{Bmatrix} = \begin{Bmatrix} P_1 \\ M_1/\ell \\ P_2 \\ M_2/\ell \end{Bmatrix}$$

27.2.4 Global Coordinates

In order to move beyond linear systems, we need to be able to connect elements at arbitrary angles to each other. To do this, we transform the local deflection coordinates into a global coordinate system. Each deflection u (in the local x-direction, which may be different for each element) can be expressed as a deflection $\overline{u}$, $\overline{v}$, and $\overline{w}$ (in the global $\overline{x}$-, $\overline{y}$-, and $\overline{z}$-directions, which are the same for all elements).

For example, if an element is vertical, the transformation is simply an interchange of x- and y-directions, and the transformation at each end is

$$\begin{Bmatrix} u \\ v \end{Bmatrix} = \begin{bmatrix} 0 & 1 \\ -1 & 0 \end{bmatrix}\begin{Bmatrix} \overline{u} \\ \overline{v} \end{Bmatrix}$$

or, more generally, if the element makes an angle of α with the global coordinate system, the two-dimensional transformation at each end is

$$\left\{ \begin{array}{c} u \\ v \end{array} \right\} = \left[\begin{array}{cc} \cos\alpha & \sin\alpha \\ -\sin\alpha & \cos\alpha \end{array} \right] \left\{ \begin{array}{c} \overline{u} \\ \overline{v} \end{array} \right\} \tag{27.20}$$

while angular deflections θ are unchanged.

27.3 Closure

In the decades around 1970, the growth in computing power made possible the development of the finite-element analysis methods (FEA or FEM) for the study of complex structures, and revolutionized the way engineers do their work. Since most engineers do not develop finite-element programs from scratch, but use comprehensive "codes" like *NASTRAN, ANSYS,* etc., we have restricted the coverage in this textbook to simple demonstrations of the key concepts needed to apply these programs, which are now the standard means of verifying the static and dynamic performance of the final design of structures.

In the early design stages, a numerical solution is not as useful, because it only gives the results for a particular configuration, without showing the trends resulting from different modifications and improvements. Algebraic solutions, even when simplified or truncated, are more valuable for finding optimized—or at least feasible—design configurations. In the recent past, the continuing growth in computing power has made possible symbolic-algebra programs to support mathematical analysis, initiating a second revolution in the way engineers work.

In the next few chapters, we will study analytical solutions of continuous systems, and of structures represented by equivalent continuous systems.

Problem 27.6 *Find the axial natural frequency of a rod which is fixed at one end, stepped-down to one-half the area at the midpoint, and free at the other.*

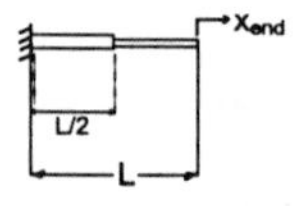

Problem 27.7 *Find the transverse natural frequency of a uniform bending beam, built-in at one end and simply supported at the other.*

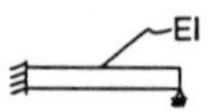

Part IV

Continuous Systems

Chapter 28

TENSIONED STRINGS and threadlines

The simplest continuous systems, such as the strings in a musical instrument, obey the classic wave equation.

28.1 The Wave Equation

If we put a control volume around an element Δx of a uniform string (Figure 28.1) of constant mass-per-unit-length $\mu = \rho A = m/L$, and under tension $T = \sigma A$, we can apply Newton's law and balance the forces for deflections y at any location x

$$\frac{\mu}{g_c}\Delta x\frac{\partial^2 y}{\partial t^2} = -T\left(\frac{\partial y}{\partial x}\right) + T\left(\left(\frac{\partial y}{\partial x}\right) + \Delta x\frac{\partial}{\partial x}\left(\frac{\partial y}{\partial x}\right) + \frac{(\Delta x)^2}{2}\frac{\partial^2}{\partial x^2}\left(\frac{\partial y}{\partial x}\right) + \ldots\right)$$

Note that deflection y is a function of both time t and location x; the left side of the equation represents mass times acceleration; the right side, the y-direction force components due to the tension at each end of the element, placing the left

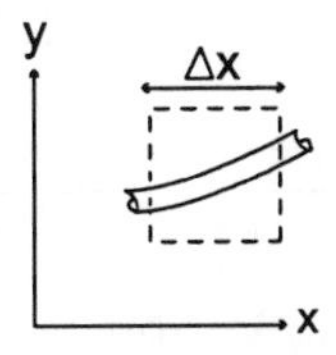

Figure 28.1: Element of a String

end at the x and using a Taylor series expansion for the right end at $x + \Delta x$. Crossing off canceling terms and dividing by Δx, the equation simplifies to

$$\frac{\mu}{g_c} \cdot \frac{\partial^2 y}{\partial t^2} = T \cdot \left(\frac{\partial^2 y}{\partial x^2} + \frac{\Delta x}{2} \cdot \frac{\partial^3 y}{\partial x^3} + \ldots \right)$$

Decreasing Δx to an infinitesimal value, the higher-order terms become negligible, and we can normalize the remaining terms

$$\frac{\partial^2 y}{\partial t^2} = \left(\frac{T g_c}{\mu} \right) \cdot \frac{\partial^2 y}{\partial x^2} \tag{28.1}$$

This is the wave equation. It is a linear, second-order Partial Differential Equation with constant coefficients. Because its form visually resembles the equation for a hyperbola, $t^2 = a x^2$, it is called a "hyperbolic P.D.E." Its Standard Form is

$$\boxed{\ddot{y} = c^2 \cdot y''} \tag{28.2}$$

where dots continue to represent derivatives with respect to time, primes represent derivatives with respect to spacial location, and c^2 replaces the parameter group $(T g_c / \mu)$ or $(\sigma g_c / \rho)$, which has the dimensions $\mathrm{m}^2/\mathrm{s}^2$ or $\mathrm{ft}^2/\mathrm{sec}^2$.

In developing this equation, we have made a number of simplifying assumptions:

- Deflections y must be small and smoothly continuous, so that $\partial y / \partial x \equiv \tan\theta \cong \theta \cong \sin\theta$ and $\cos\theta \cong 1$.
- Tension T is constant and unaffected by the deflection.
- There are no other restoring forces: the stiffness of the string is negligible.

28.2 d'Alembert's Solution

The first approach to solving the wave equation is to investigate traveling-wave solutions. If c is a constant, any differentiable function F_1 of the single argument $(x - ct)$ will satisfy the wave equation, because

$$\begin{aligned} \frac{\partial^2 F_1(x - ct)}{\partial t^2} &= c^2 \frac{d^2 F_1(x - ct)}{d^2\,(x - ct)} \\ \frac{\partial^2 F_1(x - ct)}{\partial x^2} &= \frac{d^2 F_1(x - ct)}{d^2\,(x - ct)} \end{aligned}$$

so that the terms in the governing equation automatically cancel when we have a function of this type. The function F_1 can represent any continuous shape; this shape can be called a "wave." As time progresses to increasing t, the crest of this wave travels along in space in the direction of increasing x, since the argument is $(x - ct)$. The speed of the wave is the parameter c from the governing equation.

Similarly, we can show that any twice-differentiable function F_2 of the single argument $(x + ct)$ will also satisfy the wave equation. This function moves along in space in the direction of decreasing x, at the velocity c. Furthermore, any combination

$$y(x,t) = F_1(x - ct) + F_2(x + ct) \tag{28.3}$$

is also a solution.

Traveling-wave-type solutions are valuable when we are well away from the supports; at the ends, inverted reflections occur. For example, if one end constraint is that $y = 0$ for all time at point $x = 0$, then a function $F(ct + x)$ approaching the origin from the right must create an equal and opposite function $-F(ct - x)$ so that the two functions cancel out at $x = 0$ for all times t. If F is a periodic function like $\sin(ct + x)$, the superposition of the reflected waves adds up to standing waves, somewhat like the surging of the surf when an ocean wave reflects off a seawall.

Exercise 28.1 *Find the wave-length and the frequency of the standing wave resulting from the superposition* $\sin(ct + x) - \sin(ct - x)$.

If we have to consider reflection from both ends of a string of length L, we note that the fundamental period of repetition is the time it takes for the wave to travel back and forth, a distance of two times the length of the string

$$\tau = 2L/c \tag{28.4}$$

Hence, only certain lengths and frequencies of waves that fit into L and τ are possible solutions to the homogeneous equation. In Section 28.3, we investigate stationary-wave solutions for problems where we must consider the end conditions.

28.3 Separation of Variables

The second approach to solving the wave equation is to separate variables by trying out the heuristic assumption that the deflection is the product of a function in space and a function in time

$$y(x,t) = X(x) \cdot T(t) \tag{28.5}$$

so that

$$\begin{aligned}
\frac{\partial^2 y}{\partial t^2} &= X(x) \cdot \frac{d^2 T}{dt^2} \\
\frac{\partial^2 y}{\partial x^2} &= \frac{d^2 X}{dx^2} \cdot T(t)
\end{aligned}$$

Substituting this in the governing equation,

$$X(x) \cdot \frac{d^2T}{dt^2} = c^2 \cdot \frac{d^2X}{dx^2} \cdot T(t)$$

If c is a constant with respect to time, we can rearrange the terms in order to separate X and T:

$$\frac{1}{T(t)} \cdot \frac{d^2T}{dt^2} = c^2 \cdot \frac{1}{X(x)} \cdot \frac{d^2X}{dx^2}$$

Because one side of the equation is a function only of time, and the other is a function only of space, each side must be equal to a constant if they are to be equal for all values of time and space. We will later find it convenient if we call this unknown constant $-\omega_n^2$

$$\frac{1}{T(t)} \cdot \frac{d^2T}{dt^2} = -\omega_n^2 = c^2 \cdot \frac{1}{X(x)} \cdot \frac{d^2X}{dx^2} \tag{28.6}$$

and separate the sides into two Ordinary Differential Equations

$$\frac{d^2T}{dt^2} + \omega_n^2 \cdot T(t) = 0 \tag{28.7}$$

$$\frac{d^2X}{dx^2} + \frac{\omega_n^2}{c^2} \cdot X(x) = 0 \tag{28.8}$$

If c is a constant with respect to space also, the form of the equations is familiar: the solutions follow Section 4.5

$$T = A \sin \omega_n t + B \cos \omega_n t \tag{28.9}$$

$$X = C \sin \frac{\omega_n}{c} x + D \cos \frac{\omega_n}{c} x \tag{28.10}$$

where the unknown constant ω_n^2 will be determined from the end conditions. We know that both at $x = 0$ and at $x = L$, y must be equal to zero at all times. Therefore, $X = 0$ at $x = 0$

$$0 = X(0) = C \sin 0 + D \cos 0 = D$$

so we can set D equal to zero from now on, and $X = 0$ at $x = L$

$$0 = X(L) = C \sin \frac{\omega_n}{c} L + 0 \cos \frac{\omega_n}{c} L = C \sin \frac{\omega_n}{c} L$$

which lets us conclude that either $C = 0$ (the trivial solution) or else

$$\frac{\omega_n}{c} L = n\pi \tag{28.11}$$

Thus we have found the separation-of-variables constant $\omega_n = n\pi c/L$. It can have a whole series of discrete values: When $n = 1$, the function X describes a half-sinewave between the end supports, and $\omega_1 = \pi c/L$; when $n = 2$, the X describes a full sinewave, and $\omega_2 = 2\pi c/L$; etc.

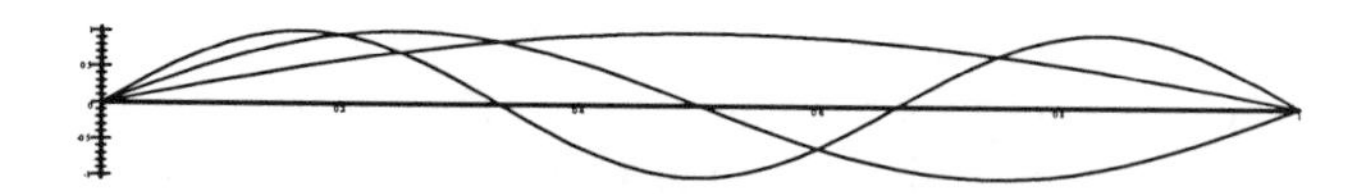

The fundamental period is $\tau_1 = 2L/c$, as we already know from d'Alembert's solution (Equation 28.4).

The complete solution is the sum of all of the discrete responses

$$y(x,t) = X(x) \cdot T(t) = \sum_n \left(C_n \sin \frac{n\pi x}{L}\right)\left(A_n \sin \frac{n\pi ct}{L} + B_n \cos \frac{n\pi ct}{L}\right) \tag{28.12}$$

C_n is redundant and can be set equal to one (effectively absorbing it into A_n and B_n). As demonstrated in an example below, values for A_n and B_n must be obtained from the initial conditions. B_n are the coefficients b_n of a Fourier analysis of the initial-displacement shape; A_n come from the coefficients b_n of a Fourier analysis of the initial-velocity shape, with each b_n divided by its ω_n.

When we look at our solution, we recognize why strings make good musical instruments:[1]

- The multiple frequencies of a string from Equation 28.11

$$\boxed{f_n = n \cdot \tfrac{c}{2L} = \tfrac{1}{2\pi} \cdot (n\pi) \cdot \tfrac{1}{L}\sqrt{\tfrac{Tg_c}{\mu}}} \tag{28.13}$$

  are perceived as harmonious by the ear, because they are integer multiples of the fundamental frequency.

- Each string in an instrument can be conveniently tuned to any desired fundamental frequency

$$f_1 = \frac{c}{2L} = \frac{1}{2L}\sqrt{\frac{Tg_c}{\mu}} = \frac{1}{2L}\sqrt{\frac{\sigma g_c}{\rho}} \tag{28.14}$$

  by adjusting the tensile stress σ, limited only by the tensile strength $\sigma_{\max}$ of the material.

- One string on a guitar, violin, or clavichord can be used for several separate notes by stopping it at different lengths L.

- A rich set of overtones, involving many harmonics, can be achieved by exciting the string at about one-seventh of the length from one end; guitar players can change sound quality by strumming nearer or farther from the center of the strings.

[1] Hermann L.F. Helmholtz, *On the Sensations of Tone, as a Physiological Basis for the Theory of Music,* translated by Alexander J. Ellis, re-issued by Dover Publications, New York, 1954, ISBN 0-48660753-4.

- The brightness (i.e., the amount of higher harmonics) can be adjusted by choosing different initial conditions. On the one hand, a pop guitarist's pick or a harpsichord's plectrum pluck a triangle-shaped initial displacement, creating many higher-order terms in the Fourier series for a penetrating sound, ideal for polyphonic music. On the other hand, a classical guitarist's fingertips or a piano's padded hammers create a rounder initial shape for the displacement or velocity distribution, yielding fewer overtones and a more mellow sound, ideal for harmonious chords. Either compacting or fluffing-up the felts on the hammers of a piano changes the "voicing" of the piano.

Example: Suppose you have deflected the center of a string by a unit amount with a plectrum made of a small quill, and then let it slip off. This constitutes a displacement initial condition shaped like a symmetrical triangle. We fit this initial conditions to the known solution (Equation 28.12)

$$\begin{aligned} y(x,t) &= \sum_n \left(A_n \sin\frac{n\pi x}{L} \sin\frac{n\pi ct}{L} + B_n \sin\frac{n\pi x}{L} \cos\frac{n\pi ct}{L} \right) \\ \left[\frac{\partial y}{\partial t}\right]_{x=\text{const}} &= \sum_n \left(A_n \frac{n\pi c}{L} \sin\frac{n\pi x}{L} \cos\frac{n\pi ct}{L} - B_n \frac{n\pi c}{L} \sin\frac{n\pi x}{L} \sin\frac{n\pi ct}{L} \right) \end{aligned}$$

With the initial conditions

$$\begin{aligned} y(x,0) &= \sum_n B_n \sin\frac{n\pi x}{L} \\ \frac{\partial y}{\partial t} &= \sum_n A_n \frac{n\pi c}{L} \sin\frac{n\pi x}{L} \text{ at } t = 0 \end{aligned}$$

By comparing these expressions to the Fourier series of Section 7.7, we see that the$\left(A_n \frac{n\pi c}{L}\right)$ are identical to the Fourier-series sine-coefficients b_n of the initial-velocity distribution; because the initial velocities are zero, all $A_n = 0$. Similarly, the B_n are identical to the Fourier-series sine-coefficients b_n of the initial-displacement distribution, which is a symmetrical triangle of width L and unit height, so that Equation 7.20 becomes

$$\begin{aligned} B_n &= \frac{\int_0^{L/2} \frac{2x}{L} \sin\frac{n\pi x}{L} dx + \int_{L/2}^{L} \frac{2(L-x)}{L} \sin\frac{n\pi x}{L} dx}{\int_0^L \sin^2 \frac{n\pi x}{L} dx} \\ B_{n-\text{odd}} &= \frac{2\int_0^{(n\pi/2)} \frac{2}{n\pi}\left(\frac{n\pi x}{L}\right) \sin\frac{n\pi x}{L} d\left(\frac{n\pi x}{L}\right) \frac{L}{n\pi}}{\frac{L}{2}} \\ B_{n-\text{even}} &= 0 \end{aligned}$$

which simplifies to

$$B_{n-\text{odd}} = \frac{8}{n^2\pi^2} \int_0^{(n\pi/2)} \left(\frac{n\pi x}{L}\right) \sin\left(\frac{n\pi x}{L}\right) d\left(\frac{n\pi x}{L}\right) = \frac{8}{n^2\pi^2} \int_0^{n\pi/2} \xi \sin\xi d\xi$$

From integral tables

$$B_{n-\text{odd}} = \frac{8}{n^2\pi^2}\left[\sin\xi - \xi\cos\xi\right]_0^{n\pi/2} = \frac{8}{n^2\pi^2}\left[\sin\frac{n\pi}{2} - \frac{n\pi}{2}\cos\frac{n\pi}{2}\right] = \frac{8}{n^2\pi^2}$$

The first eight values are:

$$\begin{aligned} B_1 &= 8/\pi^2 \\ B_{2,4,6,8\ldots} &= 0 \\ B_3 &= 8/9\pi^2 \\ B_5 &= 8/25\pi^2 \\ B_7 &= 8/49\pi^2 \end{aligned}$$

and the solution is

$$y = \frac{8}{\pi^2}\sin\frac{\pi x}{L}\cos\frac{\pi ct}{L} + \frac{8}{9\pi^2}\sin\frac{3\pi x}{L}\cos\frac{3\pi ct}{L} + \frac{8}{25\pi^2}\sin\frac{5\pi x}{L}\cos\frac{5\pi ct}{L} + \ldots$$

28.4 Equivalence of Solutions

It is important to realize that traveling-wave solutions and standing-wave solutions can be different ways of expressing the same steady oscillation. For the string supported at $x = 0$ and $x = L$, the longest-wavelength solution starts with the wave $F_1 = A\sin\pi(x - ct)/L$ traveling to the right. The reflection at $x = L$ requires the superposition of an opposite-sign wave $F_2 = -A\sin\pi(x + ct)/L$ traveling to the left; the sum of the two at $x = L$ always adds up to $y = 0$. This wave F_2 requires a reflection at $x = 0$; this reflection in turn accounts for F_1 and the two add up to $y = 0$ at that point. This same fortunate coincidence of matching reflections holds for all integral fractions of this wavelength. Therefore, acceptable traveling-wave solutions for stationary oscillation have the form

$$y(x,t) = A\sin\frac{n\pi(x - ct)}{L} - A\sin\frac{n\pi(x + ct)}{L} \tag{28.15}$$

By using trigonometric formulae, we can reduce this to the standing-wave solutions.

Exercise 28.2 *Convert the traveling-wave solution above, to a standing-wave solution.*

For pure standing waves, the separation-of-variables solution is convenient. However, there are many wave motions, such as the fluttering of a flag, which also have a traveling-wave component, so that d'Alembert solution terms are needed.

Traveling waves are important because they carry energy along the string. The average power transmitted by a sinusoidal wave is

$$\Pi_{avg} = \frac{1}{2}A^2\omega^2\frac{T}{c} = \frac{1}{2}A^2\omega^2\sqrt{\frac{T\mu}{g_c}} \tag{28.16}$$

28.5 Distributed Damping

From our experience with damping in discrete multi-degree-of-freedom systems (Section 25.2), we suspect that damping can be handled if it is either proportional to the mass loading or proportional to the restoring force. The former case would be, for example, uniform displacement damping along the uniform string due to the surrounding fluid

$$\ddot{y} + \alpha \dot{y} = c^2 y'' \tag{28.17}$$

Separation of variables leads to

$$\begin{aligned} \frac{d^2T}{dt^2} + \alpha\frac{dT}{dt} + \omega_n^2 \cdot T(t) &= 0 \\ \frac{d^2X}{dx^2} + \frac{\omega_n^2}{c^2} \cdot X(x) &= 0 \end{aligned}$$

The latter equation tells us that the classic mode-shapes are preserved. There is modal damping of $\zeta = \alpha/2\omega_n$; that is to say, the higher-frequency modes have a lower damping ratio (but damp out in about the same time).

The other case would be material damping within the string

$$\ddot{y} = c^2 y'' + \beta \dot{y}'' \tag{28.18}$$

and separation of variables leads to

$$\begin{aligned} \frac{d^2T}{dt^2} + \left(\frac{\beta\omega_n^2}{c^2}\right)\frac{dT}{dt} + \omega_n^2 \cdot T(t) &= 0 \\ \frac{d^2X}{dx^2} + \frac{\omega_n^2}{c^2} \cdot X(x) &= 0 \end{aligned}$$

so that classical mode-shapes are still preserved. There is modal damping of $\zeta = \beta\omega_n/2c^2$; that is to say, the higher-frequency modes have a higher damping ratio and damp out faster.

It is possible, of course, to have both kinds of damping

$$\ddot{y} + \alpha \dot{y} = c^2 y'' + \beta \dot{y}'' \tag{28.19}$$

and modal $\zeta = \alpha/2\omega_n + \beta\omega_n/2c^2$. This would give us a minimum damping ratio for the ω_n nearest to $c\sqrt{\alpha/\beta}$.

Any other kind of damping would take us away from classical modes, and require description of the modes by complex numbers to account for phase differences in the motion at different locations along the string.

In general, higher frequencies damp out faster. In a harpsichord, this means that the sharp initial sound, which does not blend well into chords, rapidly mellows out into tones which can be used in harmonious chords; if the chords are played as arpeggios, the initial clash is avoided but the enduring harmonies are still heard.

28.6 Support Damping

Damping may also occur at the ends of the string, because the termination is not ideal. In the traveling-wave formulation, this can be expressed as a reduction of amplitude in the reflected wave.

Pianos and violins extract energy from one end of the string, and transmit it to an elastic soundboard by means of the bridge. The soundboard then radiates the music into the air. The historical development of instruments has lead to a balance between extracting a lot of sound (but having the tone die out quickly) on the one hand, and maintaining notes for a long time (but extracting only a modest volume) on the other hand. Both loudness and persistence can be achieved at the same time if there is a lot of energy in the string; this requires the string to have an adequate mass, especially for low notes, and explains why bass strings need to be heavy.

Additional damping may occur at the other end of the string. In a strummed instrument like a guitar, this is minimized by having rigid frets which provide a nearly ideal termination; the finger is placed on the far side of the fret. In a bowed instrument like a violin, most notes are played by stopping the strings directly with a finger—the additional damping is not a problem since the bow keeps adding energy, and the greater damping of higher harmonics results in a more mellow sound—and the difference between finger-stopped notes and open notes is noticeable. (Some viols have adjustable frets, but the finger should be placed not beside the fret, but directly over it, providing a result intermediate between open strings and finger-stopped strings.) In general, instruments for the polyphonic music of the Baroque era are constructed to produce more overtones for more easily identified multiple melody lines; instruments for the orchestral harmonies of the romantic era produce a more "mellow" sound which blends better.

28.7 Numerical Solution

The wave equation can be solved numerically by substituting finite-difference expressions for the differential expressions. We will record displacement y at discrete locations along the string, and at discrete time intervals; for $y(x,t)$ we substitute $y_{n,m}$, where the first integer subscript n counts spacial increments along the string, and the second subscript m counts many time intervals have elapsed. For second derivatives we substitute second-difference expressions as we did in Sections 6.7 and 16.3.1:

$$\begin{aligned} \frac{\partial^2 y}{\partial t^2} &= c^2 \cdot \frac{\partial^2 y}{\partial x^2} \\ \frac{y_{n,m+1} - 2y_{n,m} + y_{n,m-1}}{(\Delta t)^2} &\cong c^2 \cdot \frac{y_{n+1,m} - 2y_{n,m} + y_{n-1,m}}{(\Delta x)^2} \end{aligned}$$

Assuming that we have the older-time values, we solve for the newest value

$$y_{n,m+1} \cong \left(\frac{c\Delta t}{\Delta x}\right)^2 y_{n+1,m} + 2\left(1 - \left(\frac{c\Delta t}{\Delta x}\right)^2\right) y_{n,m} + \left(\frac{c\Delta t}{\Delta x}\right)^2 y_{n-1,m} - y_{n,m-1} \tag{28.20}$$

28.7.1 Finite-Difference Procedure

A particularly simple algorithm results when we chose Δt or Δx such that $c\Delta t/\Delta x = 1$

$$x_{n,m+1} \cong y_{n+1,m} + y_{n-1,m} - y_{n,m-1} \tag{28.21}$$

If we have the values $y_{n,0}$ to define the displacements at $t = 0$, and also $y_{n,1}$ to establish the change to $t = \Delta t$ and therefore the initial velocity, we can calculate the values $y_{n,2}$ one time increment Δt later, and so on. This is easily done in a tabular format or a spreadsheet.

Exercise 28.3 *With pencil and paper, set up a table describing a string by the displacement at* 20 *evenly spaced locations. Start with an initial displacements* 1.0 *at time zero at the seventh and ninth location, and the displacement* 2.0 *at the eighth location one time interval later; all other initial displacements are zero. Calculate the displacement for five time intervals.*

The locations at the end supports of the string must be kept at $y = 0$; we can explain this to ourselves by imagining that each value located at one Δx from the boundary is matched by an equal but opposite-sign value located one Δx beyond the boundary. We discover that the procedure gives us a negative wave reflected from the boundary.

Exercise 28.4 *Continue the previous Exercise for another forty time-steps. How long before the pattern repeats itself? What is the displacement half-way between repeated patterns?*

Note that the displacement at any one time is not enough to tell us what is going to happen: our wave might be going to the right or to the left. We also need to know a second state variable, the velocity, to tell whether a wave will be going to the right or to the left.

Exercise 28.5 *For the final displacements of the previous exercise, find the velocity distribution*

$$\dot{y}_{n,45\frac{1}{2}} = \frac{y_{n,46} - y_{n,45}}{\Delta t}$$

28.7.2 Stability

Unlike Southwell's relaxation procedure for the elliptical Partial Differential Equation (P.D.E.) in Section 6.7, this finite-difference procedure for a hyperbolic P.D.E. Equation 28.20 is not robust. If we try to cover more time by increasing Δt relative to Δx, it goes unstable—just like the Ordinary Differential Equation procedure of Section 16.3.1 did when we chose too large a time-step.

Exercise 28.6 *Repeat the first few steps of the preceding exercises using* $c\Delta t/\Delta x = 1.4$*, so that the algorithm becomes*

$$y_{n,m+1} \cong 2y_{n+1,m} - 2y_{n,m} + 2y_{n-1,m} - y_{n,m-1}$$

What appears to happen to the traveling waves?

28.7.3 Numerical Diffusion

Clearly, our first choice of $c\Delta t/\Delta x = 1.0$ is just marginally stable: it did not blow up, but neither would it damp out round-off errors. Perhaps we should go the other way, and decrease Δt, as we did to get more accurate results with the Central Difference Method (Section 16.3.1). Unfortunately, that tends to smear out our initial conditions, as a trial with a small traveling wave will show.

Exercise 28.7 *Repeat the first few steps of the preceding exercises using* $c\Delta t/\Delta x = 0.7$*, so that the algorithm becomes*

$$y_{n,m+1} \cong 0.5y_{n+1,m} + y_{n,m} + 0.5y_{n-1,m} - y_{n,m-1}$$

What appears to happen to the traveling waves?

If we had started with a large standing wave as an initial condition, the effect of this smearing would not be as evident as it is on a small traveling wave: it would show up as an apparent "numerical damping" of the wave. To get accurate results, more sophisticated procedures are needed.

Exercise 28.8 *Repeat the previous problem; as an initial displacement, use a half-sinewave distributed over the entire length of the string for the two initial time-steps. What is the apparent logarithmic decrement of this undamped system?*

Real physical damping in the string must not be handled in this way, but should be addressed by programming the equations given in Section 28.5.

Exercise 28.9 *Develop a finite-difference algorithm from one of the modal-damping P.D.E.'s above.*

28.8 Threadlines

A common situation in engineering is a translating string-like element: a belt or chain in power transmissions, a band-saw in machine tools, a thread in weaving looms, a paper web in printing plants, or a plastic web in film manufacturing.

If a string moves at velocity v, we can look at the vibration in two ways. We can analyze it in a coordinate system that steadily moves along in the "machine direction" (MD), and measure ξ from some point on the moving web, applying Equation 28.1

$$\frac{\mu}{g_c}\left[\frac{\partial^2 y}{\partial t^2}\right]_{\xi=const} - T\left[\frac{\partial^2 y}{\partial \xi^2}\right]_{t=const.} = 0 \tag{28.22}$$

which is a simple equation, but is difficult to fit to stationary boundary conditions. If we change to fixed coordinates, with the MD coordinate $x = \xi + vt$, the end conditions occur at fixed locations $x = 0$ and $x = L$, but additional terms show up in the equation to account for the velocity of the web relative to the coordinate system

$$\frac{\mu}{g_c}\left[\frac{\partial^2 y}{\partial t^2}\right]_{x=const} + 2\frac{\mu}{g_c}v\left[\frac{\partial^2 y}{\partial t \partial x}\right]_{x,t=const} + \left(\frac{\mu}{g_c}v^2 - T\right)\left[\frac{\partial^2 y}{\partial x^2}\right]_{t=const.} = 0 \tag{28.23}$$

It is important to realize that the first term is *not* the same in the two equations! The added terms are loosely called "gyroscopic" terms, because the middle one has the appearance a Coriolis-force term, and the third one contains a centrifugal-force term. Because of the centrifugal-force term, tension measured by the reaction force on an idler wheel or by the contact pressure on a roller is not the actual tension experienced by the thread or belt material, but an *apparent* tension $\left(T - \mu v^2/g_c\right)$. This apparent tension provides the restoring force in the equation: when $v^2 > T g_c/\mu$ the threadline becomes unstable and deflects to one side or the other.

Exercise 28.10 *Using a fixed, open control volume, derive Equation 28.23 with the gyroscopic terms.*

We can apply d'Alembert's solution to the threadline, writing

$$\begin{aligned} y(\xi,t) &= F_1(\xi - ct) + F_2(\xi + ct) \\ y(x,t) &= F_1(x - vt - ct) + F_2(x - vt + ct) \end{aligned}$$

where $c \triangleq \sqrt{T g_c/\mu}$. If the threadline is stable, $v^2 < T g_c/\mu$, and waves travel downstream with the absolute velocity $c + v$ and upstream with the absolute velocity $c - v$.

The difference in the speed of wave propagation creates a problem if we try numerical simulation using the simple procedure of Section 28.7, which requires us to choose a Δt which is just right for the wave speed. A Δt which is stable for the downstream direction, will be over-damped in the opposite direction, and give inaccurate results.

Because of the Coriolis-like mixed-derivative term, separation of variables is not possible in the fixed coordinate system. Young-Bae Chang's solution[2] has the form

$$y = A \sin\left(\frac{n\pi x}{L}\right) \cos\left(\frac{v}{c}\frac{n\pi x}{L} + \left(1 - \frac{v^2}{c^2}\right)\frac{n\pi ct}{L}\right)$$

where $c^2 \triangleq Tg_c/\mu$. The sinusoidal coefficient assures us that the end conditions are met. If we sketch out the motion of the fundamental mode, we find that each point on the web has a sinusoidal motion, just like a classic mode, but that there is a phase-difference between the motions at different stations along the web span. To express this in the form of a classic mode, we would have to use complex numbers for the amplitude distribution.

Exercise 28.11 *Check this solution by plugging it back into the partial differential equation. This is quite laborious, and the use of computer algebra is recommended.*

Exercise 28.12 *Check this solution by first showing that it can be converted to a traveling-wave form*

$$y(x,t) = F_1(x - vt - ct) + F_2(x - vt + ct)$$

and then showing that all solutions of this form satisfy the partial differential equation

$$\ddot{y} + 2v\,\dot{y}' + \left(v^2 - c^2\right) y'' = 0$$

This is a much quicker and easier way of proving that the solution is correct.

Fluid traveling along the outside of a string or a transmission belt, or through the inside of a hose under tension, creates similar centrifugal and Coriolis terms as the motion of the belt itself.[3]

Exercise 28.13 *What are the equations for a string hanging like a pendulum? Hint: Put a control volume around an element Δx of the string under variable tension, and apply Newton's law*

$$\frac{\mu}{g_c} \cdot \Delta x \cdot \frac{\partial^2 y}{\partial t^2} = -T \cdot \frac{\partial y}{\partial x} + \left(T\frac{\partial y}{\partial x} + \Delta x \cdot \frac{\partial}{\partial x}\left(T\frac{\partial y}{\partial x}\right) + \frac{(\Delta x)^2}{2} \cdot \frac{\partial^2}{\partial x^2}\left(T\frac{\partial y}{\partial x}\right) + \ldots\right)$$

---

[2] Y.B. Chang and P.M. Moretti, "Interaction of Fluttering Webs with Surrounding Air," *TAPPI Journal,* Vol. 74, No. 3 (March 1991), pp. 231–236.

[3] Y.B. Chang, S.J. Fox, D.G. Lilley, and P.M. Moretti, "Aerodynamics of Moving Belts, Tapes, and Webs," DE-Vol. 36, *Machinery Dynamics and Element Vibrations,* pages 33–40, ASME Design Conference, Miami, Florida, Sept. 22–25, 1991, ISBN No. 0-7918-0627-8.

Crossing off canceling terms, dividing by Δx, and decreasing Δx to an infinitesimal value, the equation simplifies to

$$\frac{\mu}{g_c} \cdot \frac{\partial^2 y}{\partial t^2} = \frac{\partial}{\partial x}\left(T\frac{\partial y}{\partial x}\right) = T\frac{\partial^2 y}{\partial x^2} + \frac{\partial T}{\partial x} \cdot \frac{\partial y}{\partial x}$$

where $\partial T/\partial x = \mp \mu g/g_c$ (the sign depends on whether you placed the origin at the top or the bottom of the string)

$$\ddot{y} = \left(\frac{T(x)\, g_c}{\mu}\right) \cdot y'' \mp g \cdot y'$$

Proceed with separation of variables. The Ordinary Differential Equation for time should look familiar, but the equation for space is unusual. Is there any rope-thickness distribution $\mu(x)$ which would make this problem easy to solve?

Exercise 28.14 *Let's keep a heavy, limp cord roughly horizontal by fixing one end to a support, but maintaining the tension by pulling on the other end with a long, almost weightless fishing line. What are the end conditions for a modal analysis? What happens to a traveling wave approaching the end with the "weightless" support?*

Exercise 28.15 *What happens when a traveling wave, on a constant-tension string, encounters a step-change in the linear density μ of the string?*

Problem 28.16 *Suppose you have deflected one point on a string, 1/7 of the length from one end, by a unit amount with a plectrum made of a small quill, and then let it slip off. What is the distribution of the harmonics, i.e., the amplitude of the overtones relative to the fundamental?*

Problem 28.17 *Suppose you have deflected one point on a string, 1/6 of the length from one end, by a unit amount with a plectrum made of a small quill, and then let it slip off. What is the distribution of the harmonics, i.e., the amplitude of the overtones relative to the fundamental?*

Problem 28.18 *Investigate what happens when a traveling wave, on a constant tension string, encounters a concentrated mass fastened to the string.*

Problem 28.19 *Investigate what happens when a traveling wave, on a constant tension string, encounters a step-change in the linear density μ of the string.*

Chapter 29

PRESSURE AND SHEAR WAVES and special end conditions

Pressure waves in rods and pipes, and torsional waves in shaft, follow the same governing equation as the taut string. However, end conditions can be more complex.

29.1 Pressure Waves

If we look at pressure/tension waves along a rod (e.g., the sucker-rod of an oil well) or in a fluid medium (e.g., in an organ pipe), we get the governing equation for longitudinal displacement u

$$\frac{\partial^2 u}{\partial t^2} = \left(\frac{EAg_c}{\mu}\right) \cdot \frac{\partial^2 u}{\partial x^2} \tag{29.1}$$

which has the same form as the taut-string Equation 28.2; therefore, the same forms of solutions must apply. The difference is that c^2 now equals EAg_c/μ or Eg_c/ρ, a property of the solid material.

Exercise 29.1 *Calculate the speed of sound in some common metals, using handbook data for elasticity and density.*

Using Young's modulus of elasticity E implies that the rod is thin (relative to the wavelengths). If its diameter is very large—the problem of plane waves in a solid—the rod cannot expand sideways in proportion to Poisson's ratio μ_P, and we must substitute $E/\left(1 - 2\mu_P\right)$.

The same equations can be used for pressure waves inside pipes. For perfect gases, the speed of sound is a function of temperature, $c = \sqrt{\gamma \overline{R} T / M_{\mathrm{mol}}}$, where λ is the ratio of specific heats c_p/c_v, $\overline{R} = 8314$ N-m/kmol-K (1545 ft-lbf/lbmol-^{o}R), T is the absolute temperature in Kelvin (or in degrees Rankine), and M_{mol} is the molecular weight.

Exercise 29.2 *Calculate the speed of sound for room-temperature helium ($\gamma =$ 1.67), air ($\gamma \cong 1.4$ and average $M_{mol} \cong 29$ kg/kmol or lbm/lbmol), and carbon dioxide . Calculate c for steel from values for Young's modulus E and density ρ from an engineering handbook. Look up the velocity of sound for water in a* Handbook of Chemistry and Physics.

In the preceding chapter on tensioned-string vibrations, where the end conditions were simply $y = 0$, pressure waves have a choice of two boundary conditions at each end. Either the displacement is clamped and $y = 0$; or else the end is free, the force remains at zero, and therefore the gradient of the displacement $\partial y / \partial x = 0$. Fitting the modal solution from Section 28.3

$$\begin{aligned} u(x,t) &= X(x) \cdot T(t) \\ &= \sum_n \left(C_n \sin \frac{\omega_n}{c} x + D_n \cos \frac{\omega_n}{c} x \right) \left(A_n \sin \frac{n\pi ct}{L} + B_n \cos \frac{n\pi ct}{L} \right) \end{aligned} \tag{29.2}$$

to the alternative end conditions, gives us three possible cases:

- If $u = 0$ at $x = 0$ and at $x = L$,

$$u(x,t) = \sum_n \left(\sin \frac{n\pi x}{L} \right) \left(A_n \sin \frac{n\pi ct}{L} + B_n \cos \frac{n\pi ct}{L} \right) \tag{29.3}$$

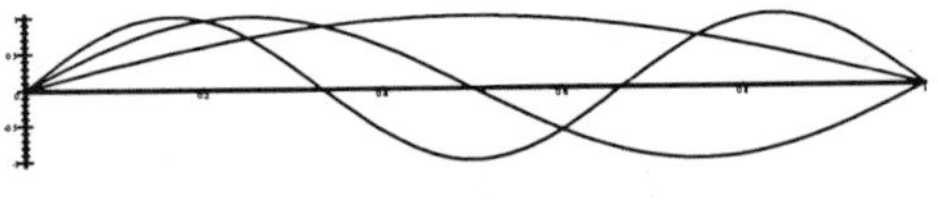

  This matched set of end conditions gives us the wavelength relationship $L = \lambda_I / 2$ and all the harmonic terms $\omega_{II} = 2\omega_I$, $\omega_{III} = 3\omega_I$, etc.

- If $\partial u / \partial x = 0$ at $x = 0$ and at $x = L$,

$$u(x,t) = \sum_n \left(\cos \frac{n\pi x}{L} \right) \left(A_n \sin \frac{n\pi ct}{L} + B_n \cos \frac{n\pi ct}{L} \right) \tag{29.4}$$

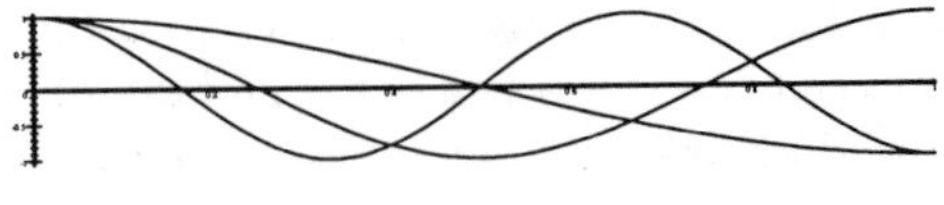

This is also a matched set of end conditions and gives us $L = \lambda_I/2$ and all the harmonic terms.

- If $u = 0$ at $x = 0$, and $\partial u/\partial x = 0$ at $x = L$,

$$u = \sum_n \left(\sin\frac{(2n-1)\,\pi x}{2L}\right)\left(A_n \sin\frac{(2n-1)\,\pi ct}{2L} + B_n \cos\frac{(2n-1)\,\pi ct}{2L}\right) \tag{29.5}$$

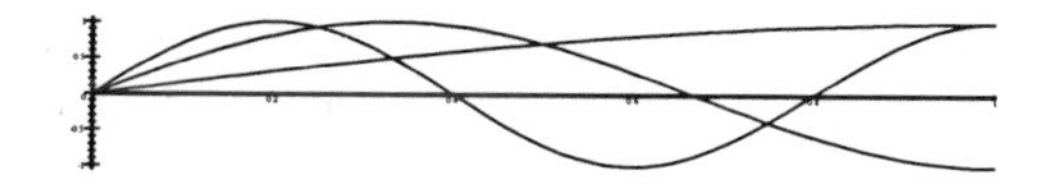

or, if the end conditions are reversed,

$$u = \sum_n \left(\cos\frac{(2n-1)\,\pi x}{2L}\right)\left(A_n \sin\frac{(2n-1)\,\pi ct}{2L} + B_n \cos\frac{(2n-1)\,\pi ct}{2L}\right) \tag{29.6}$$

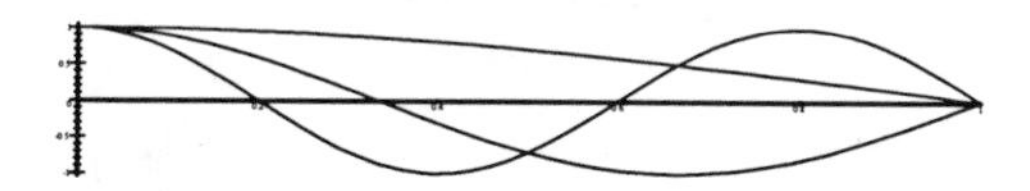

This mixed set of end conditions gives us the wavelength relationship $L = \lambda_I/4$ and only the odd harmonic terms ω_I, $3\omega_I$, $5\omega_I$, etc.

This explains why closed-end organ pipes need only be half as long as open-ended ones for a given note: all pipes have an opening at the base (where there is a fipple or a reed); if they are open-ended flutes, they have a matching open end condition at the top and their length is one-half wavelength, $\lambda/2$; if they are closed-top bourdons, their length is $\lambda/4$ and they produce notes an octave lower than flutes of the same length. Therefore, the longest pipes on an organ generally include bourdons, to provide low pedal notes. However, as the equation shows, bourdons produce only odd series of harmonics, and flutes of the same length are also necessary. When ranks of each are played together, the bourdons provide the low fundamental, and the flutes fill in the missing even harmonics.[1]

29.2 Shear and Torsional Waves

Similar waves can be identified in shear and torsional problems; for a compact cross-section, $c^2 = Gg_c/\rho$. Here, as in pressure waves, both fixed and free terminations are possible.

[1] Hermann L. F. Helmholtz, *On the Sensations of Tone as a Physiological Basis for the Theory of Music,* translated and expanded by Alexander J. Ellis, reprinted by Dover Publications, New York, 1954, ISBN 0-48660753-4.

Exercise 29.3 *Show that the velocity of shear waves is about 60% of the velocity of pressure waves for common structural materials.*

The wave equation also has application to electrical transmission lines, but resistance-damping (Section 28.5) is generally a factor.

29.3 Other Terminations

Rigidly clamped ends ($u = 0$) and perfectly free ends ($\partial u/\partial x = 0$) are only two possible end conditions; other simple terminations include elastic clamps and lumped inertia.

29.3.1 Elastic Clamp

If there is a spring k_{end} restraining the end of a rod or a shaft, the force balance at $x = L$ is

$$
\begin{aligned}
EA\left[\frac{\partial u}{\partial x}\right]_{x=L} &= -k_{\text{end}}\,[u]_{x=L} \\
GJ_{\text{section}}\left[\frac{\partial \theta}{\partial x}\right]_{x=L} &= -\kappa_T\,[\theta]_{x=L}
\end{aligned}
$$

which establishes a fixed relationship between slope and deflection at the end.

Example: Let the left end $x = 0$ be fixed at $u = 0$, and the right end $x = L$ be attached to a spring k_{end}. Applying the left-end boundary condition to the general solution

$$
\begin{aligned}
u(x,t) &= \sum_n \left(C_n \sin\frac{\omega_n}{c}x + D_n \cos\frac{\omega_n}{c}x\right)(A_n \sin\omega_n t + B_n \cos\omega_n t) \\
u(0,t) &= \sum_n D_n (A_n \sin\omega_n t + B_n \cos\omega_n t) = 0
\end{aligned}
$$

so that $D_n = 0$ (and C_n can be absorbed into A_n and B_n), leading to

$$
\begin{aligned}
u(x,t) &= \sum_n \left(\sin\frac{\omega_n}{c}x\right)(A_n \sin\omega_n t + B_n \cos\omega_n t) \\
\frac{\partial u}{\partial x} &= \sum_n \left(\frac{\omega_n}{c}\cos\frac{\omega_n}{c}x\right)(A_n \sin\omega_n t + B_n \cos\omega_n t)
\end{aligned}
$$

Applying the right-end boundary condition to this solution

$$
\begin{aligned}
EA\left[\frac{\partial u}{\partial x}\right]_{x=L} &= -k_{\text{end}}\,[u]_{x=L} \\
EA\left[\sum_n \frac{\omega_n}{c}\cos\frac{\omega_n}{c}x\right]_{x=L} &= -k_{\text{end}}\left[\sum_n \sin\frac{\omega_n}{c}x\right]_{x=L}
\end{aligned}
$$

Therefore, for each value of n,

$$\left(\frac{\omega_n}{c}L\right)\cot\left(\frac{\omega_n}{c}L\right) = \frac{-k_{\text{end}}L}{EA}$$

If the spring constants of the rod EA/L and the end spring k_{end} are given, we can solve for the group $(\omega_n L/c)$ and therefore for ω_n. The mode shapes will still be sinewaves, terminating where the slope and displacement have the proper relationship.

Exercise 29.4 *Solve the same problem if the left end $x = 0$ is free so that $\partial u/\partial x = 0$, and the right end $x = L$ is attached to a spring k_{end}.*

29.3.2 Tip Mass

If there is a lumped mass M_{end} at the otherwise free end of a rod, or a flywheel J_{end} at the end of a shaft, the force balance at $x = L$ is

$$\begin{aligned}
EA\left[\frac{\partial u}{\partial x}\right]_{x=L} &= -\frac{M_{\text{end}}}{g_c}\left[\frac{\partial^2 u}{\partial t^2}\right]_{x=L} = +\frac{M_{\text{end}}}{g_c}\omega^2\,[u]_{x=L} \\
GJ_{\text{section}}\left[\frac{\partial \theta}{\partial x}\right]_{x=L} &= -\frac{J_{\text{end}}}{g_c}\left[\frac{\partial^2 \theta}{\partial t^2}\right]_{x=L} = +\frac{J_{\text{end}}}{g_c}\omega^2\,[\theta]_{x=L}
\end{aligned}$$

which, at any given frequency, establishes a fixed relationship between slope and deflection at the end.

Example: Let the left end $x = 0$ be fixed at $u = 0$, and the right end $x = L$ have a tip mass M_{end}. Starting with the modal solution which satisfies the left-end boundary condition

$$\begin{aligned}
u(x,t) &= \sum_n \left(\sin\frac{\omega_n}{c}x\right)(A_n \sin\omega_n t + B_n \cos\omega_n t) \\
\frac{\partial u}{\partial x} &= \sum_n \left(\frac{\omega_n}{c}\cos\omega_n t\right)(A_n \sin\omega_n t + B_n \cos\omega_n t) \\
\frac{\partial^2 u}{\partial t^2} &= \sum_n -\omega_n^2\left(\sin\frac{\omega_n}{c}x\right)(A_n \sin\omega_n t + B_n \cos\omega_n t)
\end{aligned}$$

Applying the right-end boundary condition to this solution

$$\begin{aligned}
EA\left[\frac{\partial u}{\partial x}\right]_{x=L} &= -\frac{M_{\text{end}}}{g_c}\left[\frac{\partial^2 u}{\partial t^2}\right]_{x=L} = +\frac{M_{\text{end}}}{g_c}\omega^2\,[u]_{x=L} \\
EA\left[\sum_n \left(\frac{\omega_n}{c}\cos\frac{\omega_n}{c}x\right)\right]_{x=L} &= \frac{M_{\text{end}}}{g_c}\left[\sum_n \omega_n^2\left(\sin\frac{\omega_n}{c}x\right)\right]_{x=L}
\end{aligned}$$

which leads us to

$$\begin{aligned}
\frac{EA}{Mc^2} &= \frac{w}{c}\tan \\
\left(\frac{\omega_n}{c}L\right)\tan\left(\frac{\omega_n}{c}L\right) &= \frac{EAg_c}{M_{\text{end}}}\frac{L}{c^2} = \frac{\mu L}{M_{\text{end}}}
\end{aligned}$$

If the mass of the rod μL and the end mass M_{end} are given, we can solve for the group $(\omega_n L/c)$ and therefore for ω_n. The mode shapes will still be sinewaves, terminated where the slope and displacement have the proper relationship for that frequency.

Exercise 29.5 *Solve the same problem if the right end $x = 0$ is free so that $\partial u/\partial x = 0$, and the left end $x = L$ has a tip mass M_{end}.*

Exercise 29.6 *Let the left end $x = 0$ be fixed at $u = 0$, and the right end $x = L$ have both a tip mass M_{end} and a restraining spring k_{end}. What is the characteristic equation for the natural frequencies?*

29.3.3 Damped End

Especially in rotating systems, there may be a damped termination

$$
\begin{aligned}
EA\left[\frac{\partial u}{\partial x}\right]_{x=L} &= -C_{\text{end}}\left[\frac{\partial u}{\partial t}\right]_{x=L} \\
GJ_{\text{section}}\left[\frac{\partial \theta}{\partial x}\right]_{x=L} &= -\Sigma_{\text{end}}\left[\frac{\partial \theta}{\partial t}\right]_{x=L}
\end{aligned}
$$

As we might expect, this damping will make classic modes inapplicable, unless we introduce complex-number modes.

Example: Let the left end $x = 0$ be fixed at $u = 0$, and the right end $x = L$ have a tip damper C_{end}. Starting with the modal solution which satisfies the left-end boundary condition

$$
\begin{aligned}
u(x,t) &= \sum_n \left(\sin\frac{\omega_n}{c}x\right)(A_n \sin\omega_n t + B_n \cos\omega_n t) \\
\frac{\partial u}{\partial x} &= \sum_n \left(\frac{\omega_n}{c}\cos\frac{\omega_n}{c}x\right)(A_n \sin\omega_n t + B_n \cos\omega_n t) \\
\frac{\partial u}{\partial t} &= \sum_n \left(\sin\frac{\omega_n}{c}x\right)(A_n\omega_n \cos\omega_n t - B_n\omega_n \sin\omega_n t)
\end{aligned}
$$

Applying the right-end boundary condition to this solution

$$
EA\left[\frac{\partial u}{\partial x}\right]_{x=L} = -C_{\text{end}}\left[\frac{\partial u}{\partial t}\right]_{x=L}
$$

where

$$
\begin{aligned}
\left[\frac{\partial u}{\partial x}\right]_{x=L} &= \frac{\omega_n}{c}\cos\left(\frac{\omega_n}{c}L\right)(A_n \sin\omega_n t + B_n \cos\omega_n t) \\
\left[\frac{\partial u}{\partial t}\right]_{x=L} &= \sin\left(\frac{\omega_n}{c}L\right)(A_n\omega_n \cos\omega_n t - B_n\omega_n \sin\omega_n t)
\end{aligned}
$$

We can split this equation into $\sin \omega_n t$ and $\cos \omega_n t$ components

$$
\begin{aligned}
EA\frac{\omega_n}{c}\cos\left(\frac{\omega_n}{c}L\right)(A_n) &= -C_{\text{end}}\sin\left(\frac{\omega_n}{c}L\right)(-B_n\omega_n) \\
EA\frac{\omega_n}{c}\cos\left(\frac{\omega_n}{c}L\right)(B_n) &= -C_{\text{end}}\sin\left(\frac{\omega_n}{c}L\right)(A_n\omega_n)
\end{aligned}
$$

and obtain the characteristic equations

$$
\begin{aligned}
\cot\left(\frac{\omega_n}{c}L\right) &= \frac{C_{\text{end}}}{\sqrt{EA\mu/g_c}}\left(\frac{B_n}{A_n}\right) = \frac{-C_{\text{end}}}{\sqrt{EA\mu/g_c}}\left(\frac{A_n}{B_n}\right) \\
B_n^2 &= -A_n^2
\end{aligned}
$$

which clearly requires imaginary or complex numbers in the coefficients A and B; classical modes no longer apply.

Problem 29.7 *Calculate the pitch produced by a sixteen-foot open-ended organ pipe. (This is the nominal bottom note of many organs.) If the "middle C" of a piano has a fundamental frequency of 256 Hz, where does that place the note you calculated?*

Problem 29.8 *If pressure against the free end of a uniform rod is suddenly released, what modes of longitudinal vibration are activated?*

Chapter 30

CONTINUOUS MEDIA and acoustic measurements

We can extend the study of one-directional waves to two or three dimensions.[1] Here we will give only a few examples to illustrate the principle.

30.1 Drumheads

If we look at thin, flat membranes under uniform tension, we can derive a governing equation which is the two-dimensional expansion of the taut-string Equation 28.1

$$\frac{\mu}{g_c}\frac{\partial^2 z}{\partial t^2} - T\left(\frac{\partial^2 z}{\partial x^2} + \frac{\partial^2 z}{\partial y^2}\right) = p\,(x, y, t) \tag{30.1}$$

where μ is the mass-per-unit-area, T is the tension-per-unit-width, and p the excitation-force-per-unit-area.

Exercise 30.1 *Check the units in this equation for consistency. Note that μ and T are dimensionally different than in Chapter 28.*

In coordinate-less notation the second term is the Laplacian operator

$$\frac{\mu}{g_c}\frac{\partial^2 z}{\partial t^2} - T\left(\nabla^2 z\right) = p \tag{30.2}$$

which we can expand into any other coordinate system. For example, if we had a line-like "plane" wave on the membrane, we could rotate our Cartesian coordinate system so that all the variation in z would occur along the x-axis, and

[1] Karl F. Graff, *Wave Motion in Elastic Solids,* reprinted by Dover Publications, New York 1991, ISBN 0-486-66745-6.

the y-derivative of z would disappear, leaving only the one-directional tensioned-string equation. Therefore, we can conclude that "plane" waves propagate with the velocity $c = \sqrt{Tg_c/\mu}$ independent of orientation.

We could also expand this equation into polar coordinates if that will fit our boundary conditions better—for example for a circular drumhead for which $z = 0$ at $r = r_{\max}$

$$\begin{aligned}\frac{\mu}{g_c}\frac{\partial^2 z}{\partial t^2} - T\left(\frac{1}{r}\frac{\partial}{\partial r}\left(r\frac{\partial z}{\partial r}\right) + \frac{1}{r^2}\frac{\partial^2 z}{\partial \theta^2}\right) &= p(r,\theta,t) \\ \frac{\mu}{g_c}\frac{\partial^2 z}{\partial t^2} - T\left(\frac{\partial^2 z}{\partial r^2} + \frac{1}{r}\frac{\partial z}{\partial r} + \frac{1}{r^2}\frac{\partial^2 z}{\partial \theta^2}\right) &= p(r,\theta,t) \qquad (30.3)\end{aligned}$$

If we strike this drumhead at the center so that we expect radial symmetry

$$\begin{aligned}\frac{\mu}{g_c}\frac{\partial^2 z}{\partial t^2} - T\left(\frac{1}{r}\frac{\partial}{\partial r}\left(r\frac{\partial z}{\partial r}\right)\right) &= p(r,t) \\ \frac{\mu}{g_c}\frac{\partial^2 z}{\partial t^2} - T\left(\frac{\partial^2 z}{\partial r^2} + \frac{1}{r}\frac{\partial z}{\partial r}\right) &= p(r,t) \qquad (30.4)\end{aligned}$$

where $p = 0$ for the free-vibration solution.

We are unable to find a self-similar traveling-wave solution for this equation, which does not entirely surprise us: on the one hand, conservation of energy dictates that the amplitude of the wave must decrease at it moves outward; on the other hand, the restoring force is weakened behind the curved wave and fails to fully restore the neutral position, leaving a "tail" behind the wave which changes its shape. We can, however, separate variables (Section 28.3), leading to two Ordinary Differential Equations

$$\begin{aligned}z(r,t) &= R(r)\cdot\Theta(t) \\ \frac{d^2\Theta}{dt^2} + \omega_n^2\Theta(t) &= 0 \\ \frac{\partial^2 R}{\partial r^2} + \frac{1}{r}\frac{\partial R}{\partial r} + \frac{\omega_n^2}{c^2}R(r) &= 0 \qquad (30.5)\end{aligned}$$

leading to modal solutions. The time-variable Θ has the familiar form leading to sinusoidal results; the space-variable R appears in a Bessel's equation of the first kind of order 0, which is tabulated in standard mathematics references. The frequencies follow the form

$$\begin{aligned}f_n &= \frac{1}{2\pi}(a_n)\frac{1}{r_{\max}}\sqrt{\frac{Tg_c}{\mu}} \qquad (30.6) \\ \text{where } a_n &= 2.405,\ 5.520,\ 8.654,\ 11.792,\ 14.931,\ \text{etc.}\end{aligned}$$

so that the overtones are not harmonics of the fundamental.

30.2 Acoustic Waves

If we look at the propagation of sound in a fluid,[2] we derive a governing equation for particle displacements which resembles the compression–in-rods Equation 29.1

$$\frac{\partial^2 \overrightarrow{u}}{\partial t^2} - c^2 \nabla^2 \overrightarrow{u} = 0 \tag{30.7}$$

Because displacements in a gas are difficult to track, it is customary to write these equations either in terms of components of particle velocities $\dot{u}_x$, $\dot{u}_y$, and $\dot{u}_z$; or else in terms of the scalar quantity called "condensation"

$$s \triangleq \frac{\rho - \rho_o}{\rho_o}$$

which is the percentage deviation of the density from average density; or else in terms of the gage pressure p

$$\frac{\partial^2 p}{\partial t^2} - c^2 \nabla^2 p = 0 \tag{30.8}$$

In **Cartesian coordinates** this expands to

$$\frac{\partial^2 p}{\partial t^2} - c^2 \left(\frac{\partial^2 p}{\partial x^2} + \frac{\partial^2 p}{\partial y^2} + \frac{\partial^2 p}{\partial z^2} \right) = 0$$

If we line up the coordinate system with a *plane wave* having only x-direction variation of p and motion u_x, this reduces to the familiar

$$\frac{\partial^2 p}{\partial t^2} - c^2 \frac{\partial^2 p}{\partial^2 x} = 0$$

which we know to have traveling-wave solutions with velocity c, which is about $\frac{1}{3}$ km/s ($\frac{1}{5}$ mile/sec) in air.

In **spherical coordinates** the pressure equation expands to

$$\frac{\partial^2 p}{\partial t^2} - c^2 \left(\frac{\partial^2 p}{\partial r^2} + \frac{2}{r}\frac{\partial p}{\partial r} + \frac{1}{r^2 \sin\theta}\frac{\partial}{\partial \theta}\left(\sin\theta \frac{\partial p}{\partial \theta} \right) + \frac{1}{r^2 \sin^2\theta}\frac{\partial p}{\partial \phi} \right) = 0$$

For a *spherically symmetrical wave* this simplifies to

$$\frac{\partial^2 p}{\partial t^2} - c^2 \left(\frac{\partial^2 p}{\partial r^2} + \frac{2}{r}\frac{\partial p}{\partial r} \right) = 0$$

or

$$\frac{\partial^2 (rp)}{\partial t^2} - c^2 \frac{\partial^2 (rp)}{\partial r^2} = 0$$

[2] John William Strutt Baron Rayleigh, *The Theory of Sound,* in two volumes, reissued by Dover Publications, New York, 1976, ISBN 0-48660292-3 and ISBN 0-48660293-1.

which has a traveling-wave solution with velocity c. As the spherical wave propagates outward from a source, the function of (rp) remains the same, so that the pressure-amplitude of the wave falls off inversely with the radius.

The wavelength λ is related to the frequency f by

$$c = f \cdot \lambda \tag{30.9}$$

Because acoustic frequencies range from 20 to 20,000 Hertz, wavelengths vary from 17 meters (55 feet) to 17 mm (0.7 inches).

30.2.1 Sound Pressure

The amplitude of sound at a particular location is measured by a pressure microphone, and reported as a root-mean-square average

$$p_{eff} \triangleq \sqrt{\frac{1}{t}\int_0^t p^2 dt}$$

For a harmonic wave, $p_{eff} = p_{\max}/\sqrt{2} = 0.707 p_{\max}$.

Because the human ear's response is approximately logarithmic, it is customary to use a logarithmic scale for Sound Pressure Level (SPL), defined as

$$\mathrm{SPL} \triangleq 20 \log_{10}\left(\frac{p_{eff}}{p_{ref}}\right) \ \mathrm{dB} \tag{30.10}$$

In air, a common reference pressure is $p_{ref} = 20 \times 10^{-6}\mathrm{N/m}^2$; the logarithmic units are called "decibels (dB) referenced to 20 μPa." (In water, a common reference pressure is 1 μPa.)

Because an effective pressure of 20 μPa in a mid-range ($\sim$ 1000 Hz) note is just barely audible to an average young person, 0 dB is approximately the threshold of hearing. A sound-level change of 3 dB is the smallest step increase easily detected by the human ear. Environmental sound levels in the range of 40 to 80 dB are common in the modern world. At about 90 dB the possibility of hearing damage begins to arise; permanent hearing impairment due to industrial noise exposure, over-amplified music, or target shooting without ear protection, is widespread in the population.

Sound level measurements are generally intended to cover the audible range. In order to correspond to the sensitivity of the average ear, which rolls-off at the low and high ends of the spectrum, sound-level meters incorporate several different frequency weighting schemes.[3]

[3] Leo L. Beranek, *Noise and Vibration Control,* revised 1988, Institute of Noise Control Engineering of the USA, Poughkeepsie, NY, ISBN 0-96220720-9.

30.2.2 Intensity

Acoustic intensity is the rate of flow of energy through a unit area normal to the direction of propagation, averaged either over a long time or (for a periodic wave) over one cycle

$$I \triangleq \frac{1}{\tau}\int_0^{\tau} p\,\dot{u}\,dt$$

For a plane or spherical harmonic wave, this integral leads to the result

$$I = \pm\frac{p_{\max}\,\dot{u}_{\max}}{2\rho_o c} = \pm\frac{p_{eff}^2}{\rho_o c}$$

with the sign depending on the direction of propagation. The product of average density and speed-of-sound ($\rho_o c$) is called "specific acoustic impedance" because it represents the ratio of pressure to particle velocity. Typical values are 415 Pa-s/m for air, and 1.48 MPa-s/m for water.

The logarithmic scale for Intensity Level (IL) is defined as

$$\mathrm{IL} \triangleq 10\log_{10}\left(\frac{I}{I_{ref}}\right)\ \mathrm{dB} \qquad (30.11)$$

A common reference Intensity in air is $I_{ref} = 10^{-12}\mathrm{W/m^2}$; the logarithmic units are called "decibels (dB) referenced to $10^{-12}\mathrm{W/m^2}$;" for plane and spherical waves in air, this is approximately equivalent to a reference pressure of 20 μPa. In water, a reference pressure of 20 μPa would correspond to an Intensity of $2.7\times 10^{-16}\mathrm{W/m^2}$; 1 μPa corresponds to $6.76\times 10^{-19}\mathrm{W/m^2}$—changing the reference pressure by a factor of 20 changes the reference Intensity by a factor of 400.

The fact that Intensity is proportional to the *square* of the effective pressure of a traveling wave explains why the coefficient of 10 in Equation 30.11 for IL changes to 20 in Equation 30.10 for SPL. Originally, the logarithm of the ratio of power fluxes was named after Alexander Graham Bell; then the factor of 10 was added to give us the "deci-Bell." An additional factor of two was introduced for the log of the ratio of pressures, to make it compatible with the log of the ratio of powers. An increase of 10 dB (or one "Bell") represents a ten-fold power levels (and hundred-fold pressure level); 3 dB is a doubling of power—buying twice as powerful an amplifier makes the music only just noticeably louder.

30.2.3 Standing Waves

In an enclosed space, acoustic waves will be reflected back from walls, and standing waves occur. A fundamental problem of acoustics is that a scalar measurement (pressure) at a single point (the microphone) cannot distinguish between standing and traveling waves. A particular SPL might be caused by a traveling wave of the corresponding IL; on the other hand, it might be caused by a standing wave. How do we tell the difference?

A standing wave can be decomposed into matching traveling waves going in opposite directions, but the value of the SPL will vary depending on whether we place the microphone near a node or near a maximum. If the sound field is stationary, we can move the microphone around to find nodes; but, in general, a single-point measurement is unsatisfactory for separating sound sources and reflections.

The problem can be tackled in several different ways:

- The source to be studied—either a loudspeaker or else a noisy machine—can be placed in a large open field in the country, so that no reflections can occur (except for the ground acting as a baffle). In this semi-infinite space we know that all waves are traveling outward; there are no standing waves. This approach is limited by the availability of quiet outdoor locations.

- The source to be studied can be placed on a wire-mesh floor in an anechoic chamber which is lined with porous sound-absorbing material such as wedges of foam. If the reflections are substantially attenuated, the outward-bound waves dominate and the returning waves can be neglected. In addition, the incursion of environmental noise can be controlled by heavy walls and doors (behind the sound-absorbing material).

- The opposite approach is to place the source into a large echo chamber with hard-surfaced reflective walls. The quality of the chamber is measured by the length of time it takes the echoing of a transient sound to dissipate. If that time is long enough, the sound field due to a continuous source is dominated by standing waves, and the total source power can be calculated from the SPL and the chamber's parameters. Misleading results due to placing the microphone on a node (for a particular frequency) can be avoided by making the chamber irregular in shape, modulating the shape with a rotating "stirrer" paddle, sampling with several microphones, and/or traversing the microphone at the end of a boom.

- Measurement can be made on-site, using a pair of microphones spaced Δx, by cross-correlating the two signals

$$\begin{aligned} I &\triangleq \left\langle p\,\dot{u}\right\rangle_{avg} \\ \text{where } \dot{u} &= \frac{-1}{\rho_o}\int_0^t \frac{\partial p}{\partial x}dt \\ \text{so } I &= \frac{-1}{\rho_o}\left\langle p\int_0^t \frac{\partial p}{\partial x}dt\right\rangle_{avg} \end{aligned}$$

If $\Delta x \ll \lambda$, we can use the two microphone pressures to obtain $p \cong (p_1+p_2)/2$ and $\partial p/\partial x \cong (p_2-p_1)/\Delta x$

$$I \cong \frac{-1}{\rho_o}\left\langle \frac{(p_1+p_2)}{2}\int_0^t \frac{(p_2-p_1)}{\Delta x}dt\right\rangle_{avg} \tag{30.12}$$

which gives us the net intensity of a traveling wave along the line running through the microphones. Therefore, we can measure the net outward radiation along a spherical control surface all around the source, and integrate it for total source power.

Similar techniques can be used to measure traveling waves along strings with known tensile stress and density, from simultaneous laser-Doppler vibrometer measurements at two locations;[4] and for traveling waves along beams and panels, from four or more simultaneous vibrometer or straingage measurements.

Problem 30.2 *Search the library for information on "panel flutter" in high-speed aircraft, and write a short descriptive bibliography.*

[4] Peter M. Moretti, Young-Bae Chang, and Krishna Vedula, "Vibration intensity measurement of fluttering webs by laser-Doppler sensors," *SPIE, Proc. First Int. Conf. on Vibration Measurements by Laser Techniques: Advances and Applications,* E.P. Tomasino, Editor, Ancona, Italy, 3–5 October 1994.

Chapter 31

BEAM VIBRATION and approximate methods

The governing equation for a slender beam is obtained by differentiating the displacement to obtain slope, differentiating slope to obtain curvature, multiplying by beam stiffness to obtain bending moment, differentiating to obtain shear, and differentiating again to obtain load. This is inserted into a force balance

$$\frac{\mu(x)}{g_c}\frac{\partial^2 y}{\partial t^2} + \frac{\partial^2}{\partial x^2}\left(EI(x)\frac{\partial^2 y}{\partial x^2}\right) = F(x,t) \tag{31.1}$$

which, in the case of a uniform beam, simplifies to the Euler-Bernoulli beam equation

$$\boxed{\tfrac{\mu}{g_c}\ddot{y} + EIy'''' = F(x,t)} \tag{31.2}$$

"Slender" means that the deflection due to bending is much larger than the additional deflection due to shear, a condition which is met by most structural beams. For example, if we look at the static deflection y of an uniform cantilever loaded by a force F at the tip

$$\begin{aligned} y_{\text{total}} &= y_{\text{bend}} + y_{\text{shear}} \\ y_{\text{bend}} &= \frac{F_{\text{tip}}}{EI}\frac{(3L-x)x^2}{6} \leq \frac{F_{\text{tip}}L^3}{3EI} \\ y_{\text{shear}} &= \frac{F_{\text{tip}}}{GA}x \leq \frac{F_{\text{tip}}L}{GA} \end{aligned}$$

we see that neglecting the shear deflection in a cantilever requires

$$L^2 \gg 3\frac{E}{G}\frac{I}{A} \cong 3 \times 2.6 \times \frac{I}{A}$$

For an I-beam, $\sqrt{I/A}$ approaches one-half of the height of the beam; for a circular cylinder, it is only one-fourth of the diameter.

(On the opposite extreme, extremely stubby beams may permit us to neglect bending and consider only shear deflection, governed by the wave equations of Chapter 20 with $c^2 \triangleq Gg_c/\mu$.)

Exercise 31.1 *Compare bending deflection and shear deflection in a simply supported uniform beam.*

"Slender" in dynamic problems also implies that inertial effects due to translational inertia are much greater than those due to rotation of the beam's cross-section. Timoshenko[1] developed a beam equation which includes the sum of bending and shear deflection, and both translational inertia and rotational inertia. To assume that both shear and rotation effects can be neglected requires that the wavelength λ which we are considering must be long relative to the height of the beam

$$\lambda \gg \sqrt{\frac{I}{A}} \tag{31.3}$$

At short wavelengths, Timoshenko's equation shows more inertia and less restoring force, and therefore lower natural frequencies, than the Euler-Bernoulli equation.

31.1 Modal Solution

Free vibrations of a uniform slender beam are governed by Equation 31.2 without a forcing term

$$\ddot{y} + \left(\frac{EIg_c}{\mu}\right) y'''' = 0 \tag{31.4}$$

31.1.1 Separation of Variables

We can follow the procedure of Section 28.3 to separate variables. Heuristically writing

$$y(x,t) = X(x)\, T(t) \tag{31.5}$$

we substitute into the governing equation

$$X(x)\frac{d^2T}{dt^2} + \frac{EIg_c}{\mu}\frac{d^4x}{dt^4}T(t) = 0$$

[1] William Weaver, Jr., Stephen P. Timoshenko, and Donovan H. Young, *Vibration Problems in Engineering,* Fifth Edition, Wiley, New York, 1990, ISBN 0-471-63228-7.

and sort out the functions of x and t

$$\frac{1}{T(t)} \cdot \frac{d^2T}{dt^2} = \text{constant} = -\left(\frac{EIg_c}{\mu}\right) \frac{1}{X(x)} \cdot \frac{d^4X}{dt^4}$$

Naming the unknown constant $-\omega_n^2$, we obtain two ordinary differential equations

$$\frac{d^2T}{dt^2} + \omega_n^2 T = 0 \tag{31.6}$$

$$\frac{d^4X}{dt^4} - \left(\frac{\mu\omega_n^2}{EIg_c}\right) T = 0 \tag{31.7}$$

We already know from Chapter 4 that the first one has the solution

$$T = A\sin\omega_n t + B\cos\omega_n t \tag{31.8}$$

Because the second equation is also linear with constant coefficients, we can try the homogeneous solution

$$\begin{aligned} X &= A_n e^{st} \\ X'''' &= A_n s^4 e^{st} \end{aligned}$$

which we plug back into the ordinary differential equation to obtain

$$A_n \left(s^4 - \frac{\mu\omega_n^2}{EIg_c}\right) e^{sx} = 0$$

giving us the characteristic equation with the roots

$$s_n = \pm\sqrt{\pm\sqrt{\frac{\mu\omega_n^2}{EIg_c}}} = \pm\sqrt[4]{\frac{\mu\omega_n^2}{EIg_c}} \text{ and } \pm i\sqrt[4]{\frac{\mu\omega_n^2}{EIg_c}}$$

The latter, imaginary pair of roots is best converted from an exponential to a trigonometric form; the real roots can be represented by either exponentials or by hyperbolic functions

$$X = C\sin\sqrt[4]{\frac{\mu\omega_n^2}{EIg_c}}x + D\cos\sqrt[4]{\frac{\mu\omega_n^2}{EIg_c}}x + E\sinh\sqrt[4]{\frac{\mu\omega_n^2}{EIg_c}}x + F\cosh\sqrt[4]{\frac{\mu\omega_n^2}{EIg_c}}x \tag{31.9}$$

where

$$\sinh sx \triangleq \frac{e^{sx} - e^{-sx}}{2} \tag{31.10}$$

$$\cosh sx \triangleq \frac{e^{sx} + e^{-sx}}{2} \tag{31.11}$$

31.1.2 End Conditions

The solution for y has to be fitted to the end constraints. A beam can have a variety of support conditions:

- A **simple**, **hinged**, or **pinned** support resists deflection y, but does not have the clamping ability to introduce moments EIy'' at that end

$$y = 0 \tag{31.12}$$
$$y'' = 0 \tag{31.13}$$

- A **clamped** or **built-in** support resists deflection and rotation

$$y = 0 \tag{31.14}$$
$$y' = 0 \tag{31.15}$$

- A **free** end has no moments EIy'' or shear EIy'''

$$y'' = 0 \tag{31.16}$$
$$y''' = 0 \tag{31.17}$$

- Theoretically, we can postulate supports that fix any two out of y, y', y'', and y'''.

Because a beam has two ends, the three common end conditions can be combined into nine kinds of beams, three of which are merely mirror images of three others. **Example**: A simply supported beam has a pinned support at each end. Starting with the modal solution Equation 31.9, condensed by substituting

$$\beta_n^4 \triangleq \frac{\mu\omega_n^2}{EIg_c} \tag{31.18}$$

$$\begin{aligned}
X &= C\sin\beta_n x + D\cos\beta_n x + E\sinh\beta_n x + F\cosh\beta_n x \\
X' &= C\beta_n\cos\beta_n x - D\beta_n\sin\beta_n x + E\beta_n\cosh\beta_n x + F\beta_n\sinh\beta_n x \\
X'' &= -C\beta_n^2\sin\beta_n x - D\beta_n^2\cos\beta_n x + E\beta_n^2\sinh\beta_n x + F\beta_n^2\cosh\beta_n x \\
X''' &= -C\beta_n^3\cos\beta_n x + D\beta_n^3\sin\beta_n x + E\beta_n^3\cosh\beta_n x + F\beta_n^3\sinh\beta_n x
\end{aligned}$$

At the left end, $x = 0$, we apply the hinged boundary condition $y = 0$ and $y'' = 0$; if they are to remain zero for all time, then $X = 0$ and $X'' = 0$ as well

$$\begin{aligned}
0 &= X_{(0)} = D + F \\
0 &= X''_{(0)} = \beta_n^2(-D + F)
\end{aligned}$$

which tells us that D and F must both be zero. To solve for C and E, we apply the hinged boundary condition $y = 0 = X$ and $y'' = 0 = X''$ to the remaining solution terms at the right end $x = L$

$$\begin{aligned} 0 &= X_{(L)} = C \sin \beta_n L + E \sinh \beta_n L \\ 0 &= X''_{(L)} = \beta_n^2 \left(-C \sin \beta_n L + E \sinh \beta_n L\right) \end{aligned}$$

This condition can only be satisfied by $E = 0$, and either $C = 0$ or else

$$\begin{aligned} \sqrt[4]{\frac{\mu \omega_n^2}{EIg_c}} L &= \beta_n L = n\pi \\ \omega_n &= \left(\beta_n L\right)^2 \frac{1}{L^2} \sqrt[2]{\frac{EIg_c}{\mu}} = (n\pi)^2 \frac{1}{L^2} \sqrt[2]{\frac{EIg_c}{\mu}} \end{aligned} \tag{31.19}$$

Therefore, the resonant frequencies have the form

$$\boxed{f_n = \tfrac{1}{2\pi} \cdot \alpha_n \cdot \tfrac{1}{L^2} \sqrt{\tfrac{EIg_c}{\mu}}} \tag{31.20}$$

where the $1/2\pi$ converts from radians to cycles, the α is a dimensionless number depending only on the geometry (such as "uniform beam on pinned end supports"), the $1/L^2$ provides for similitude scaling, and the square root contains the beam properties. Using these frequencies, the solution for a simply supported beam is

$$\begin{aligned} y\,(x,t) &= T\,(t)\, X\,(x) \\ &= \left(A_n \sin \omega_n t + B_n \cos \omega_n t\right) \sin\left(\sqrt[4]{\frac{\mu \omega_n^2}{EIg_c}}\, x\right) \\ &= \left(A_n \sin \omega_n t + B_n \cos \omega_n t\right) \sin \frac{n\pi x}{L} \end{aligned} \tag{31.21}$$

where we have absorbed the extra constant C into A and B, which must be found from initial conditions.

31.1.3 Modes of Uniform Beams

The same procedure can be used for all six geometries of uniform beams on standard support conditions to obtain the mode shapes, characteristic equation, and dimensionless frequencies α (as defined by Equation 31.20):

- The **simply-supported** beam in the example above has a pinned support at each end

$$\begin{aligned} X &= \sin \frac{n\pi x}{L} \\ \beta_n L &= n\pi = 3.1416,\ 6.2832,\ \text{etc.} \\ \alpha_n &= (n\pi)^2 = 9.8696,\ 39.478,\ 88.826,\ \text{etc.} \\ \alpha_n / \alpha_1 &= 1.00,\ 4.00,\ 9.00,\ 16.00,\ \text{etc.} \end{aligned} \tag{31.22}$$

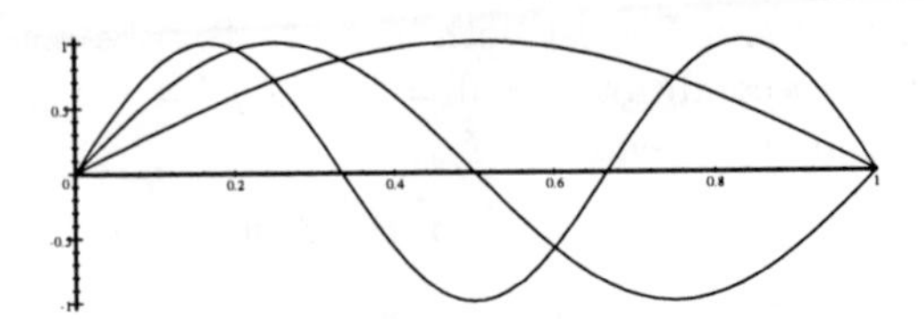

- The **cantilever** beam is clamped at one end, and free at the other

$$
\begin{aligned}
X &= (\sin\beta_n x - \sinh\beta_n x) - D_n(\cos\beta_n x - \cosh\beta_n x) \\
\cos\beta_n L &= \frac{-1}{\cosh\beta_n L} \Longrightarrow \beta_n L = 1.875104,\ 4.694091, \text{ etc.} \\
D_n &= \frac{\sin\beta_n L + \sinh\beta_n L}{\cos\beta_n L + \cosh\beta_n L} = 1.3622,\ 0.98187,\ 1.0008, \text{ etc.} \\
\alpha_n &= (\beta_n L)^2 = 3.5160,\ 22.0345,\ 61.6972, \text{ etc.} \approx \left(n - \frac{1}{2}\right)^2 \pi^2 \\
\alpha_n/\alpha_1 &= 1.00,\ 6.27,\ 17.5,\ 34.4, \text{ etc.}
\end{aligned}
\qquad (31.23)
$$

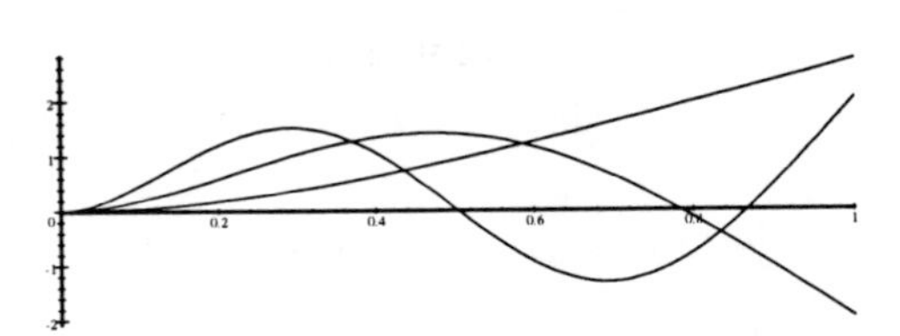

- The **clamped-clamped** beam is built-in at both ends

$$
\begin{aligned}
X &= (\sin\beta_n x - \sinh\beta_n x) - D_n(\cos\beta_n x - \cosh\beta_n x) \\
\cos\beta_n L &= \frac{1}{\cosh\beta_n L} \\
D_n &= \frac{\sin\beta_n L - \sinh\beta_n L}{\cos\beta_n L - \cosh\beta_n L} \\
\alpha_n &= 22.3733,\ 61.6728,\ 120.9034, \text{ etc.} \approx \left(n + \frac{1}{2}\right)^2 \pi^2 \\
\alpha_n/\alpha_1 &= 1.00,\ 2.76,\ 5.40,\ 8.93, \text{ etc.}
\end{aligned}
\qquad (31.24)
$$

- The **free-free** beam is unsupported, floating in space

$$\begin{aligned}
X &= (\sin\beta_n x + \sinh\beta_n x) - D_n(\cos\beta_n x + \cosh\beta_n x)\\
\cos\beta_n L &= \frac{1}{\cosh\beta_n L}\\
D_n &= \frac{\sin\beta_n L - \sinh\beta_n L}{\cos\beta_n L - \cosh\beta_n L}\\
\alpha_n &= 0,\ 22.3733,\ 61.6728,\ 120.9034,\ \text{etc.} \approx \left(n+\frac{1}{2}\right)^2\pi^2\\
\alpha_n/\alpha_1 &= 0,\ 1.00,\ 2.76,\ 5.40,\ 8.93,\ \text{etc.}
\end{aligned} \tag{31.25}$$

- The **clamped-pinned** beam is built-in at one end, and hinged at the other

$$\begin{aligned}
X &= (\sin\beta_n x - \sinh\beta_n x) - D_n(\cos\beta_n x - \cosh\beta_n x)\\
\tan\beta_n L &= \tanh\beta_n L\\
D_n &= \frac{\sin\beta_n L - \sinh\beta_n L}{\cos\beta_n L - \cosh\beta_n L}\\
\alpha_n &= 15.4182,\ 49.9645,\ 104.2477,\ \text{etc.} \approx \left(n+\frac{1}{4}\right)^2\pi^2\\
\alpha_n/\alpha_1 &= 1.00,\ 3.24,\ 6.26,\ 11.56,\ \text{etc.}
\end{aligned} \tag{31.26}$$

- The **free-pinned** beam is free at one end, and hinged at the other, free to rotate about the pin

$$\begin{aligned}
X &= \sinh\beta_n x + D_n \sin\beta_n x\\
\tan\beta_n L &= \tanh\beta_n L\\
D_n &= \frac{\sinh\beta_n L}{\sin\beta_n L}\\
\alpha_n &= 0,\ 15.4182,\ 49.9645,\ 104.2477,\ \text{etc.}\\
\alpha_n/\alpha_1 &= 0,\ 1.00,\ 3.24,\ 6.26,\ 11.56,\ \text{etc.}
\end{aligned}$$

Only the simply supported beam produces an integer-multiple series of frequencies.

Exercise 31.2 *Obtain values for $\beta_n L$ and D_n for the free-free beam, and plot the first vibrational mode. At what values of x/L is the node located?*

Exercise 31.3 *Obtain the values for $\beta_n L$ and D_n for the free-free beam, and plot the first two vibrational mode. At what values of x/L are the nodes located? How is the second mode related to the previous Exercise?*

Real support conditions are rarely ideal. In particular, a rigid clamping condition is difficult to obtain in the field. For example, when tubes are welded into a tubesheet, the tubesheet is not a perfectly rigid foundation; the tube behaves as if it were clamped not at the face of the tubesheet, but at a location which is a fraction of a diameter back into the tubesheet.

In the laboratory, a near-ideal "free-free" support condition can be achieved by supporting a beam on knife edges at the nodes of the fundamental free-free mode, which are located 22% of the length from each end; flexibility in the supports minimizes the effect of placement error.[2] This is useful for determining the beam property $\sqrt{EIg_c/\mu}$ from a frequency measurement. It is also how the resonant bars in a glockenspiel, marimba, or vibraharp are supported for a predictable fundamental note. In Latin music, a skilled percussionist will support one of the claves at roughly the same pair of points—using forefinger and thumb near one end, and the edge of the palm near the other—for a strong sound when struck with the other stick.

31.1.4 Excitation

If initial displacements are given, they must be expressed in terms of the modes. In the case of the simply supported beam, that turns out to be a Fourier-series description of the initial shape

$$y_{(0)} = \sum_n B_n \sin \frac{n\pi x}{L}$$

and the subsequent motion is the sum of the modal terms, each at its own frequency ω_n

$$y = \sum_n B_n \cos \omega_n t \sin \frac{n\pi x}{L}$$

If an initial velocity distribution along x is given, it also must be expressed in terms of the modes; for example on a simply supported beam

$$\dot{y}_{(0)} = b_n \sin \frac{n\pi x}{L}$$

where $b_n = A_n \omega_n$ and the subsequent motion is

$$y = \frac{b_n}{\omega_n} \sin \omega_n t \sin \frac{n\pi x}{L}$$

Forced excitation should also be described in terms of the modes, used as principal coordinates, and each mode treated as a one-degree-of-freedom system.

[2]P.M. Moretti, R.L. Lowery, and J.G. Withers, "Structural Characteristics of Helical-Corrugated Heat-Exchanger Tubes," ASME Paper No. 75-WA/HT-14, Winter Annual Meeting of the American Society of Mechanical Engineers, Houston, Texas, Dec. 1, 1975.

31.2 Traveling Waves in Beams

Recall that in Chapter 28, when we looked at standing waves of wavelength λ on a string, we noted that they could also be described as the superposition of a pair of traveling waves

$$2\sin\left(\frac{\omega_n}{c}x\right)\cos\left(\omega_n t\right) = \sin\frac{\omega_n(x-ct)}{c} + \sin\frac{\omega_n(x+ct)}{c} \tag{31.27}$$

where the wavelength can be found by inspecting the argument of the first sine, $\omega_n/c = 2\pi/\lambda$; and the period and frequency can be found by inspecting the argument of the cosine, $\omega_n = 2\pi/\tau = 2\pi f_n$. The wave speed must be

$$c = \lambda_n f_n = \frac{\lambda\omega_n}{2\pi} \tag{31.28}$$

which turns out to be the same velocity for all values of n when we plug in the tensioned-string values $\lambda_n = 2L/n$ and $\omega_n = \frac{n\pi}{L}\sqrt{\frac{Tg_c}{\mu}}$. In beams, the situation is not so simple.

31.2.1 Wave Velocity

When we plug in the natural frequencies observed for the sinusoidal modes on a simply supported beam, we find that

$$c_n = \lambda_n f_n = \left(\frac{2L}{n}\right)\left(\frac{1}{2\pi}\cdot(n\pi)^2\cdot\frac{1}{L^2}\sqrt{\frac{EIg_c}{\mu}}\right) = \frac{n\pi}{L}\sqrt{\frac{EIg_c}{\mu}}$$

and eliminate n by substituting $2L/\lambda$

$$c = \frac{2\pi}{\lambda}\sqrt{\frac{EIg_c}{\mu}} \tag{31.29}$$

or eliminate n by substituting $\frac{L}{\pi}\sqrt{\omega\sqrt{\frac{\mu}{EIg_c}}}$

$$c = \sqrt{\omega\sqrt{\frac{EIg_c}{\mu}}} \tag{31.30}$$

We see that the wave velocity for sinusoidal waves is not a constant, but is inversely proportional to the wave length, and directly proportional to the square-root of the frequency. This means that complex waves, which are composed of different Fourier components, are "dispersive," because some of the components travel faster than others.

Exercise 31.4 *Obtain the frequency and wavelength relationship for steady sinusoidal standing waves*

$$\omega = \left(\frac{2\pi}{\lambda}\right)^2 \sqrt{\frac{EIg_c}{\mu}}$$

by combining Equations 31.29 and 31.30.

Exercise 31.5 *Obtain the same relationship directly from the governing Equation 31.4, by inserting the trial solution*

$$y = \sin\frac{2\pi x}{\lambda}\sin\omega_n t$$

Exercise 31.6 *Obtain Equation 31.30 for steady sinusoidal traveling waves directly from the governing Equation 31.4, by inserting the trial solution*

$$y = \sin\left(\frac{\omega_n x}{c} - \omega_n t\right)$$

31.3 Beams on Multiple Supports

Industrial applications of transversely vibrating beams include tubes in a shell-and-tube heat exchanger; the individual tubes are rolled or welded into tubesheets at each end, and simply supported wherever they pass through baffle plates.

31.3.1 Bounded Estimates

The lowest natural frequency of a such a multi-span uniform beam can be estimated from the individual spans. We can carry out the mental experiment of sawing the beam into separate spans: all intermediate spans become simply supported single spans, while the first and last spans become clamped-pinned beams if the ends are built-in. Each separated span has a fundamental natural frequency which is easily obtained from Equation 31.20 (α_1 is 9.87 for intermediate spans, or 15.4 for end spans). When we mentally reconnect the spans, the combined natural frequency must be higher than that of the "softest" span, but lower than that of the "hardest" span. Therefore, the highest and lowest individual frequencies form a bounded estimate for the fundamental frequency of the multi-span beam.[3]

Two cautions: (1) the individual-span frequencies exist only in the mental experiment, and are not observed in the physical system; and (2) the separate-and-recombine approach works here, but should not be extended to arbitrarily connected complex systems.

[3] R.L. Lowery and P.M. Moretti, "Natural Frequency and Damping of Tubes on Multiple Supports," *Heat Transfer, Research and Application*, J.C. Chen, Editor, AIChE Symposium Series No, 174, Vol. 74, 1978, pp. 1–5.

31.3.2 Ideal Distribution of Supports

If we have a fixed number of supports and a given overall length, the highest natural frequency is obtained if the supports are spaced such that each span, if sawed-off into an individual span, has the same fundamental frequency. That means that all intermediate spans are the same length, and the end spans are 25% longer. Our method then gives us the same value for the upper and lower bound, fixing the fundamental frequency for that support spacing. Therefore that spacing is the most efficient use of a given number of supports.

Exercise 31.7 *Calculate the natural frequency for a clamped-clamped beam with one, two, and three optimally spaced intermediate simple supports, and compare the results with the higher modes of a clamped-clamped beams without intermediate supports.*

Since any other spacing yields a lower natural frequency, the frequency for optimally spaced supports is also an upper bound on the overall fundamental frequency, and can be used together with the lowest individual-span frequency to form a tighter pair of bounds.

31.3.3 Effect of U-Bends

Many shell-and-tube heat exchangers have hairpin-shaped tubes, with U-bends extending into the space beyond the last baffle plate. If we mentally sawed off a U-bend by itself, it would be supported only by a pair of hinges at one side, with no restoring force, giving us a zero natural frequency—an accurate but useless lower bound. The solution is to mentally saw-off a three-span subsystems consisting of the U-bend of radius R and a pair of straight spans of length L. This three-span system looks like a short hair-pin on four simple supports. If we have the fundamental frequency of that subsystem, we can use it as a lower bound for the fundamental frequency of the combined system.

The fundamental frequencies (one for out-of-plane motion and one for in-plane motion) can be expressed in the form of Equation 31.20, but α_1 is not a constant, but a function of R/L. It has been obtained by a variety of mathematical, numerical, and experimental techniques, and published in graphical form.[4]

31.3.4 Effect of Surrounding Fluid

Experiments show what we expect from the discussion in Section 6.6: the effect of the surrounding fluid is to add slightly more mass than the fluid displaced.[5] More sophisticated analysis shows mass coupling between the tubes.

[4] P.M. Moretti, Fundamental Frequencies of U-Tubes in Tube Bundles," *Trans. ASME, J. Pressure Vessel Technology,* Vol. 107, No. 2, May 1985, pp. 207–209.

[5] P.M. Moretti and R.L. Lowery, "Hydrodynamic Inertia Coefficients for a Tube Surrounded by Rigid Tubes," *Trans. ASME, J. of Pressure Vessel Technology,* Vol. 98, Series J, No. 3, August 1976, pp. 190–193.

31.4 Rayleigh-Ritz

Up to this point, we have focused on uniform beams. Let us move on to approximate analytical methods for beams of arbitrary stiffness and mass distributions.

In lumped-mass problems, we described the system in terms of specific coordinates and found kinetic and potential energy. If the system was linear and in absolute coordinates q, we found that we got the quadratic forms

$$\begin{aligned} T &= \frac{1}{2}(m_{11})\,\dot{q}_1^2 + \frac{2}{2}(m_{12})\,\dot{q}_1\dot{q}_2 + \ldots \\ &= \frac{1}{2}(m_{11})\,q_1^2\omega^2 + \frac{2}{2}(m_{12})\,q_1q_2\omega^2 + \ldots \\ U &= \frac{1}{2}(k_{11})\,q_1^2 + \frac{2}{2}(k_{12})\,q_1q_2 + \ldots \end{aligned}$$

Let us combine this idea with the concept of using Fourier coefficients, or polynomial coefficients, or other "trial functions" as coordinates for describing the deflection-shape of a continuous system. Expressing a deflection curve in terms of an arbitrary combination of these trial functions

$$y_{\max}(x) = C_1 f_1(x) + C_2 f_2(x) + \ldots$$

we then express the kinetic and potential energy of a **beam in bending** as

$$\begin{aligned} T_{\max} &= \int_0^L \frac{\mu(x)}{g_c}\,\dot{y}_{\max}^2\,dx = \int_0^L \frac{\mu(x)}{g_c} y_{\max}^2\omega^2 dx \\ &= \int_0^L \frac{\mu(x)}{g_c}\left(C_1 f_1 + C_2 f_2 + \ldots\right)^2 \omega^2 dx \\ U_{\max} &= \int_0^L EI(x)\left(\frac{d^2 y_{\max}}{dx^2}\right)^2 dx \\ &= \int_0^L EI(x)\left(C_1\frac{d^2 f_1}{dx^2} + C_2\frac{d^2 f_2}{dx^2} + \ldots\right)^2 dx \end{aligned}$$

where we have assumed sinusoidal motion for the convenience of substituting $y_{\max}^2\omega^2$ for $\dot{y}_{\max}^2$. The integrals can be evaluated even if the linear density μ or the stiffness EI vary along the beam.

The same energy approach can be taken for a non-uniform **rod in compression** by evaluating

$$\begin{aligned} T_{\max} &= \int_0^L \frac{\mu(x)}{g_c}\,\dot{u}_{\max}^2\,dx = \int_0^L \frac{\mu(x)}{g_c}\left(C_1 f_1 + C_2 f_2 + \ldots\right)^2 \omega^2 dx \\ U_{\max} &= \int_0^L EA(x)\left(\frac{du_{\max}}{dx}\right)^2 dx = \int_0^L EA(x)\left(C_1\frac{df_1}{dx} + C_2\frac{df_2}{dx} + \ldots\right)^2 dx \end{aligned}$$

or for a non-uniform **shaft in torsion** through the integrals

$$\begin{aligned}
T_{\max} &= \int_0^L \frac{J(x)}{g_c} \dot{\theta}_{\max}^2 \, dx = \int_0^L \frac{J(x)}{g_c} \left(C_1 f_1 + C_2 f_2 + \ldots\right)^2 \omega^2 dx \\
U_{\max} &= \int_0^L GJ(x) \left(\frac{d\theta_{\max}}{dx}\right)^2 dx = \int_0^L GJ(x) \left(C_1 \frac{df_1}{dx} + C_2 \frac{df_2}{dx} + \ldots\right)^2 dx
\end{aligned}$$

or even for a non-uniform **string** or a string with variable tension due to gravity, through

$$\begin{aligned}
T_{\max} &= \int_0^L \frac{\mu(x)}{g_c} \dot{y}_{\max}^2 \, dx = \int_0^L \frac{\mu(x)}{g_c} \left(C_1 f_1 + C_2 f_2 + \ldots\right)^2 \omega^2 dx \\
U_{\max} &= \int_0^L T(x) \left(\frac{dy_{\max}}{dx}\right)^2 dx = \int_0^L T(x) \left(C_1 \frac{df_1}{dx} + C_2 \frac{df_2}{dx} + \ldots\right)^2 dx
\end{aligned}$$

After the integration, all these integrals lead to the forms

$$\begin{aligned}
T_{\max} &= \frac{1}{2}\left(\frac{m_{11}}{g_c}\right) C_1^2 \omega^2 + \frac{2}{2}\left(\frac{m_{12}}{g_c}\right) C_1 C_2 \omega^2 + \ldots \\
U_{\max} &= \frac{1}{2}\left(k_{11}\right) C_1^2 + \frac{2}{2}\left(k_{12}\right) C_1 C_2 + \ldots
\end{aligned}$$

where the equivalent masses m_{ii} and mass-couplings m_{ij}, and the equivalent springs k_{ii} and spring-couplings k_{ij}, relative to "coordinates" C_i, are defined by whatever shows up in the parentheses when the integrals are evaluated.

Assembling these equivalent coefficients into mass and stiffness matrices, we define a multi-degree-of-freedom problem

$$\frac{-\omega_n^2}{g_c} \begin{bmatrix} m_{11} & m_{12} & \cdots \\ m_{21} & m_{22} & \cdots \\ \vdots & \vdots & \ddots \end{bmatrix} \begin{Bmatrix} C_1 \\ C_2 \\ \vdots \end{Bmatrix} + \begin{bmatrix} k_{11} & k_{12} & \cdots \\ k_{21} & k_{22} & \cdots \\ \vdots & \vdots & \ddots \end{bmatrix} \begin{Bmatrix} C_1 \\ C_2 \\ \vdots \end{Bmatrix} = \begin{Bmatrix} 0 \\ 0 \\ \vdots \end{Bmatrix}$$

which has the trial functions as the coordinate system, and the weighting factors C_i as the magnitudes of the maximum deflections in the "direction" of each trial-function coordinate.

This procedure has the following implications:

- A few well-chosen trial functions describe a shape efficiently.
- Because their number is not infinite, they probably will not give an exact answer.
- The procedure gives the best possible answer for the trial functions given, as can be shown from the calculus of variations.
- The lowest natural frequency may be slightly high.

- Only the lower half of the frequencies obtained is likely to be reasonably accurate.
- **It is absolutely necessary that the trial functions meet the boundary constraints**—otherwise the solution obtained won't fit the boundary constraints either.

Historically, this approach was developed from the idea that, because Rayleigh's method gives high answers for the lowest natural frequency, the assumed shape which gives the lowest answer gives the best answer. Applying variational calculus to finding the weighted combination of the trial functions resulting in an extremum of the natural frequency, leads to the same equations, as our approach of solving a multi-degree-of-freedom problem using the trial functions as the coordinate system.

31.5 Thin Plates

The governing Equation 31.1 for a beam can be expanded to the governing equation for flexural waves in a uniform, thin, flat plate of uniform thickness h

$$\frac{\mu}{g_c}\frac{\partial^2 z}{\partial t^2}+\frac{Eh^3}{12\left(1-\nu^2\right)}\left(\frac{\partial^4 z}{\partial x^4}+2\frac{\partial^4 z}{\partial x^2 \partial y^2}+\frac{\partial^4 z}{\partial y^4}\right)=F\left(x,y,t\right) \tag{31.31}$$

We can write this equation in coordinate-independent form using the biharmonic operator ∇^4

$$\frac{\mu}{g_c}\ddot{z}+\frac{Eh^3}{12\left(1-\nu^2\right)}\nabla^4 z=F\left(x,y,t\right) \tag{31.32}$$

For a radially symmetric problem—where displacement z is a function of the radius r only, we could expand this into polar coordinates and omit the θ-terms

$$\frac{\mu}{g_c}\frac{\partial^2 z}{\partial t^2}+\frac{Eh^3}{12\left(1-\nu^2\right)}\left(\frac{1}{r}\frac{\partial}{\partial r}\left(r\frac{\partial}{\partial r}\left(\frac{1}{r}\frac{\partial}{\partial r}\left(r\frac{\partial z}{\partial r}\right)\right)\right)\right)=F\left(r,t\right)$$

This is still a forbidding equation. Rayleigh-Ritz can be applied if we specify radially symmetric trial functions which fit the edge condition at the rim $r=R$ of the circular plate

$$z_{\max}(r)=C_1 f_1(r)+C_2 f_2(r)+\ldots$$

and write the kinetic and potential energies

$$\begin{aligned}
T_{\max} &= \int_0^R \frac{\mu}{g_c} z_{\max}^2 \omega^2 2\pi r dr \\
U_{\max} &= \int_0^R \frac{Eh^3}{12\left(1-\nu^2\right)}\left(\frac{1}{r}\frac{\partial}{\partial r}\left(r\frac{\partial z_{\max}}{\partial r}\right)\right)^2 2\pi r dr
\end{aligned}$$

Problem 31.8 *Find the axial natural frequency of a rod which is fixed at one end, stepped-down to one-half the area at the midpoint, and free at the other.*

Problem 31.9 *Find the transverse natural frequency of a uniform bending beam, built-in at one end and simply supported at the other.*

Chapter 32

COLUMN VIBRATION and rails and pipes

32.1 Tensioned Beams

Uniform beams under tension obey the equation

$$\frac{\mu}{g_c}\ddot{y} - T \cdot y'' + EI \cdot y'''' = 0 \tag{32.1}$$

which combines the tension restoring-force terms from Equation 28.1 and the bending-stiffness terms from Equation 31.2.We can separate variables and obtain an ordinary differential equation for the mode-shape $Y(x)$

$$-\omega^2 \frac{\mu}{g_c} Y - T \cdot Y'' + EI \cdot Y'''' = 0 \tag{32.2}$$

We recall that a string has sinusoidal modes and a *pinned-pinned* beam also has sinusoidal modes

$$Y_n = A \sin\left(\frac{n\pi x}{L}\right)$$

For the *special case* of a pinned-pinned beam under tension, we can try sinusoidal modes as well, and plug them into Equation 32.2

$$A\left[-\omega_n^2 \frac{\mu}{g_c} + T \cdot \left(\frac{n\pi}{L}\right)^2 + EI \cdot \left(\frac{n\pi}{L}\right)^4\right] \sin\left(\frac{n\pi x}{L}\right) = 0$$

$$\omega_n^2 = \left(\frac{n\pi}{L}\right)^2 \frac{T g_c}{\mu} + \left(\frac{n\pi}{L}\right)^4 \frac{EI g_c}{\mu} \tag{32.3}$$

In the right-hand terms, we recognize the expressions for natural frequencies of mere strings and mere beams. For this special case of a combined system, simple addition of the restoring forces is possible, and

$$\omega^2_{\text{combined}} = \omega^2_{\text{string}} + \omega^2_{\text{beam}}$$

since the mode shapes are unaffected. With other boundary conditions, we will see the same approximate *trend,* though the solution of Equation 32.2 will be more complex, involving sines, cosines, hyperbolic sines, and hyperbolic cosines.

While the overtones of strings and beams are integer multiples of the fundamental, they are not the same series: the second mode of a string has twice the frequency of the first; but the second mode of a simply supported beam vibrates at four times the fundamental frequency. The second mode of a simply supported beam with tension will have a frequency which is between two and four times the fundamental—generally not an integer multiple—so that beams under tension do not produce tones which are perceived to be harmonious. Musical instruments usually work with strings which have a minimum of beam stiffness relative to the tension

$$\frac{EIg_c}{L^4\mu} \ll \frac{Tg_c}{L^2\mu}$$

Traditionally, this is accomplished by tightening each string as close to the breaking point as practical;[1] if the length of the instrument is not sufficient to produce low notes from highly stressed strings, the bass strings are wrapped with metal wire to increase their mass without adding to the stiffness of the structural core of the string.

32.2 Compressed Columns

A compression force $P = -T$ (negative tension) causes a reduction in the natural frequency of a beam. For the *special case* of a pinned-pinned column, where the mode-shapes are sine-waves, Equation 32.3 applies

$$\omega_1^2 = \left(\frac{\pi}{L}\right)^4 \frac{EIg_c}{\mu} - \left(\frac{\pi}{L}\right)^2 \frac{Pg_c}{\mu}$$

The lowest natural frequency goes to zero when

$$P_{\text{crit}} = EI\left(\frac{\pi}{L}\right)^2$$

This corresponds to the Euler buckling load.

[1] *vide* Thomas Mace, *Musick's monument, or, A remembrancer of the best practical musick, both divine and civil ...* Ratcliffe & Thompson, London, 1676, which directs lutenists to tighten the highest string just short of the breaking point.

In straight shell-and-tube heat-exchanger tube bundles, some of the tubes may be under compression from thermal stresses or because of the assembly process. This is a factor in determining natural frequencies when assessing the risk of tube vibration.

32.3 Rails on Elastic Foundations

In addition to tension/compression, we can add an elastic foundation to a beam. On a rail-bed, this is the equivalent of a stiff spring at each place where the rail passes over a tie. If we are interested in long waves spanning many ties, we can represent the rail-bed as a continuous support of stiffness-per-unit-length κ

$$\frac{\mu}{g_c} \ddot{y} + \kappa y - T y'' + EI y'''' = 0 \tag{32.4}$$

After separation of variables, with the unknown frequency ω_n, the ordinary differential equation for the mode shape $Y(x)$ is

$$Y'''' + \frac{(-T)}{EI} Y'' + \left(\frac{\kappa}{EI} - \frac{\omega_n^2 \mu}{EI g_c} \right) Y = 0 \tag{32.5}$$

If the ends are pinned (or very far away), we can assume a sinusoidal solution $Y = A \sin(2\pi x/\lambda)$, and get

$$A \left[\left(\frac{2\pi}{\lambda} \right)^4 - \frac{(-T)}{EI} \left(\frac{2\pi}{\lambda} \right)^2 + \left(\frac{\kappa}{EI} - \frac{\omega_n^2 \mu}{EI g_c} \right) \right] \sin\left(\frac{2\pi x}{\lambda} \right) = 0$$

Therefore, the natural frequency ω_n is a function of the wavelength λ.

$$\frac{\omega_n^2 \mu}{EI g_c} = \left(\frac{2\pi}{\lambda} \right)^4 - \frac{(-T)}{EI} \left(\frac{2\pi}{\lambda} \right)^2 + \frac{\kappa}{EI}$$

The system is unstable for a given λ when the frequency ω_n^2 goes to zero, for a value of compressive load $P_{\text{crit}} \equiv -T$

$$P_{\text{crit}} = EI \left(\frac{2\pi}{\lambda} \right)^2 + \kappa \left(\frac{2\pi}{\lambda} \right)^{-2} \tag{32.6}$$

In an infinite beam, where any value of λ is possible, the minimum buckling compressive load occurs when

$$\frac{d}{d\lambda} \left(EI \left(\frac{2\pi}{\lambda} \right)^2 + \kappa \left(\frac{2\pi}{\lambda} \right)^{-2} \right) = 0$$

which occurs at

$$\begin{aligned} \lambda_{\text{crit}} &= 2\pi \left(\frac{EI}{\kappa} \right)^{1/4} \\ P_{\text{crit}} &= 2\sqrt{EI\kappa} \end{aligned}$$

In a pinned-pinned beam, buckling will occur at some even division of the length into half-wavelengths; by trying out various possible values of λ in Equation 32.6 for P_{crit}, we can find the minimum buckling load.

32.4 Traveling Waves

When we consider that a standing wave on an infinite beam can be considered to be a superposition of two traveling waves traveling in opposite directions, we discover that the following is also a solution of the governing equation:

$$y = A\sin\left(\omega t \mp \frac{2\pi}{\lambda}x\right)$$

where

$$\begin{aligned}
\left(\frac{2\pi}{\lambda}\right)^2 &= \sqrt{\left(\frac{T}{2EI}\right)^2 + \frac{1}{EI}\left(\frac{\mu}{g_c}\omega_n^2 - \kappa\right)} - \frac{T}{2EI} \\
\omega_n^2 &= \frac{EIg_c}{\mu}\left(\frac{2\pi}{\lambda}\right)^4 + \frac{Tg_c}{\mu}\left(\frac{2\pi}{\lambda}\right)^2 + \frac{\kappa g_c}{\mu}
\end{aligned}$$

The phase velocity is

$$\begin{aligned}
v_{phase} &= \frac{\omega\lambda}{2\pi} \\
&= \sqrt{\frac{EIg_c}{\mu}\left(\frac{2\pi}{\lambda}\right)^2 + \frac{Tg_c}{\mu} + \frac{\kappa g_c}{\mu}\left(\frac{2\pi}{\lambda}\right)^{-2}} \\
&= \frac{\omega}{\sqrt{\sqrt{\left(\frac{T}{2EI}\right)^2 + \frac{1}{EI}\left(\frac{\mu}{g_c}\omega_n^2 - \kappa\right)} - \frac{T}{2EI}}}
\end{aligned}$$

Evidently, for an Euler beam

$$v_{phase} = \left(\frac{2\pi}{\lambda}\right)\sqrt{\frac{EIg_c}{\mu}} = \frac{\omega}{\sqrt{\omega_n\sqrt{\frac{\mu}{EIg_c}}}}$$

and for a tensioned string

$$v_{phase} = \sqrt{\frac{Tg_c}{\mu}}$$

32.5 Stokes' Group Velocity

The phase velocities above are the apparent velocities of continuous sine waves. What about a signal transmitted as an amplitude modulation of a continuous sine wave? We can generate a simple amplitude-modulated signal by superposing two nearly identical waves[2]

$$y = A \cdot sin\,(\omega_1 t - k_1 x) + A \cdot sin\,(\omega_2 t - k_2 x)$$

where

$$\begin{aligned} \omega_1 &= \omega + \Delta\omega,\ k_1 = k + \Delta k \\ \omega_2 &= \omega - \Delta\omega,\ k_2 = k - \Delta k \end{aligned}$$

Using the trigonometric relationship $sin(A \pm B) = sinA \cdot cosB \pm cosA \cdot sinB$, we can combine the superposed waves into

$$y = 2A \cdot sin\,(\omega t - kx) \cdot cos\,(\Delta\omega t - \Delta k x)$$

which describes a beating phenomenon, with a carrier of frequency ω and wave number k, and a modulation which has (the absolute value of) a wave with the frequency $\Delta\omega$ and the wave number Δk. Therefore the carrier velocity is

$$v_{phase} = \frac{\omega}{k} = \frac{\omega\lambda}{2\pi}$$

and the velocity of the wave packets or groups is

$$v_{group} = \frac{\Delta\omega}{\Delta k}$$

In the case of the tensioned string, where $\omega/k = c$ is a constant, $\Delta\omega/\Delta k$ has the same value, so that the groups move at the same velocity c as the carrier waves within them.

But in the case of the beam, where $v_{phase} = \omega/k = 2\pi B^2/\lambda$, $v_{group} = \Delta\omega/\Delta k = 2v_{phase}$: The group envelope moves faster than the carrier waves, so that carrier waves move backwards through the group!

In general,

$$\begin{aligned} \text{if } v_{phase} &\propto \lambda^n \\ \text{then } v_{group} &= (1-n)\,v_{phase} \end{aligned} \tag{32.7}$$

- For the tensioned string, $n = 0$ and $v_{group} = v_{phase} = c$.
- For an Euler beam, $n = -1$ and $v_{group} = 2v_{phase}$.
- For the tensioned beam, presumably $n = -0.x$ and $v_{group} = 1.x \cdot v_{phase}$; we will investigate this further below.

[2] Léon Brillouin, *Wave Propagation and Group Velocity,* Academic Press, New York, 1960.

- For surface waves in deep water, $n = +1/2$ and $v_{group} = v_{phase}/2$.

The governing equation for an Euler beam under tension T and also on an elastic foundation is

$$\frac{\mu}{g_c}\ddot{y} + \kappa y - Ty'' + EIy'''' = 0$$

which has a solution for sine waves traveling to the right

$$y = A \cdot sin\,(\omega t - kx)$$

in which

$$\begin{aligned}
k &\triangleq \frac{2\pi}{\lambda} \\
\omega &= \sqrt{\frac{EIg_c}{\mu}k^4 + \frac{Tg_c}{\mu}k^2 + \frac{\kappa g_c}{\mu}} \\
v_{phase} &= \frac{\omega}{k} = \sqrt{\frac{EIg_c}{\mu}k^2 + \frac{Tg_c}{\mu} + \frac{\kappa g_c}{\mu}k^{-2}} \\
v_{group} &= \frac{\partial \omega}{\partial k} = \frac{4\frac{EIg_c}{\mu}k^3 + 2\frac{Tg_c}{\mu}k}{2\sqrt{\frac{EIg_c}{\mu}k^4 + \frac{Tg_c}{\mu}k^2 + \frac{\kappa g_c}{\mu}}} \\
&= \frac{2\frac{EIg_c}{\mu}k^4 + \frac{Tg_c}{\mu}k^2}{\frac{EIg_c}{\mu}k^4 + \frac{Tg_c}{\mu}k^2 + \frac{\kappa g_c}{\mu}} \cdot v_{phase}
\end{aligned}$$

Note that this behaves as expected if only the tension T is present, or else only the stiffness EI is present.

For a compressed beam, where the tension T is negative, the results obtained in the previous paragraph still hold as long as the quantity under the square root is positive; wring $P \equiv -T$, this requires that

$$\begin{aligned}
k^2 &> \frac{P}{2EI} + \sqrt{\left(\frac{P}{2EI}\right)^2 - \frac{\kappa}{EI}} \\
\lambda &< 2\pi\sqrt{\frac{P}{2EI} + \sqrt{\left(\frac{P}{2EI}\right)^2 - \frac{\kappa}{EI}}} \\
P_{\text{crit}} &= EIk^2 + \kappa k^{-2} \\
&= \frac{4\pi^2 EI}{\lambda^2} + \frac{\kappa\lambda^2}{4\pi^2}
\end{aligned}$$

32.6 Pipes

If a beam is hollow and carries a flowing fluid of mass μ_{fluid} and velocity v, it also has gyroscopic terms (similar to the threadline equation of Chapter 32)

$$\frac{\mu_{total}}{g_c}\left[\frac{\partial^2 y}{\partial t^2}\right]_x + 2\cdot\frac{\mu_{fluid}}{g_c}\overline{v}\left[\frac{\partial^2 y}{\partial t \partial x}\right]_{x,t} + \frac{\mu_{fluid}}{g_c}\overline{v^2}\left[\frac{\partial^2 y}{\partial x^2}\right]_{t.} + EI\left[\frac{\partial^4 y}{\partial x^4}\right]_{t.} = 0$$

where the bar indicates an average velocity, respectively an average velocity-squared. Because of the centrifugal term, the flowing fluid decreases the natural frequency. This can be a factor in above-ground pipelines,[3] and needs to be considered when calculating the natural frequencies of tubes in shell-and-tube heat exchangers.

Similar equations can be written for cylinders (for example nuclear-reactor fuel elements) surrounded by flows which are parallel to their axis.[4]

Problem 32.1 *Bicycle mechanics claim that, if the spokes on a bicycle wheel are overtightened, the rim of the wheel may collapse laterally into a pretzel or "Pringle." The question arises: at what tension level would a wheel turn inherently unstable, even to an infinitesimal disturbance. Investigate the stability of the rim through its natural frequency as a function of spoke tension—at what value of tension does the frequency go to zero? You may represent the rim as a (straight) beam with a lateral stiffness EI, a compressive load arising from the radial tension-per-inch (opposite to the tensile stress in a pipe due to radial pressure of its contents), and a lateral elastic foundation obtained from the spring constant and angle of the spokes. Obtain dimensions, angles, etc., by examining an actual bicycle; estimate the lateral second moment I of the rim by examining a rim on your own bike, in a bike shop, or in a mail-order or Internet bike-component catalog. Is the critical tension for collapse greater or smaller than that required for a spoke to fail? Write a short engineering report on your conclusions.*

---

[3] G.W. Housner, "Bending Vibrations of a Pipe Containing Flowing Fluids," *ASME Journal of Applied Mechanics*, 1952, pp. 205–208.

[4] Michael P. Païdoussis, *Fluid-Structure Interaction: Slender Structures and Axial Flow,* Volume 1, Academic Press, London, 1998, ISBN 0-12-544360-9.

Chapter 33

MODAL ANALYZERS and cross-spectra

Just as high-speed computing has changed both numerical simulation and algebraic analysis, it also has changed experimental method. Vibration experiments on complex structures, such as vehicles, are carried out by modal testing, using integrated modal-analysis systems which incorporate date acquisition, multi-channel FFT analyzers, and graphic display of mode-shapes and frequencies.

33.1 Cross-Spectra

The key tool we need for measuring waves, is to be able to measure displacement, velocity, or acceleration at any two locations simultaneously, and obtain the ratio of the responses. If we could be sure that only one mode at a time is present, this would be simple; in complex structures, exciting a single mode is impractical, and we need statistical tools to separate modes.

33.1.1 Cross-Correlation

We can expand the concept of the autocorrelation (Chapter 18) which relates the values of a signal to time-shifted values of the same signal, and use it to compare signals from two different data channels. Thus Equation 18.6 becomes

$$R(\tau) = \lim_{T\to\infty} \frac{1}{T} \int_0^T x_1(t) \cdot x_2(t+\tau)\, dt \tag{33.1}$$

For sampled data at regular intervals Δt, Equation 18.7 becomes

$$R(n\Delta t) = \frac{1}{m} \sum_{i=1}^{m} x_i x_{i+n} \tag{33.2}$$

This cross-correlation is not necessarily symmetrical any more. If the responses are always in phase, then it will resemble the autocorrelation; but if there is any damping, there may be a phase-shift which moves the peak of the cross-correlation to one side or the other. Not surprisingly, traveling waves can be identified this way, and Equation 30.12 can be expressed in terms of cross-correlations.

If one of the data channels is a known quantity—an excitation input or a reference response which is analyzed by auto-correlation—then the relative response of the other channel can be obtained. Therefore the ability to carry out auto- and cross-correlations is a possible tool for obtaining modes.

33.1.2 Multi-Channel FFT

Although correlation methods are conceptually simple, we recall that they have two defects: First, the time-domain cross-correlations are difficult to interpret for frequency content; we would prefer their cosine-Fourier-transform, the cross-PSD (Power Spectral Density). Secondly, obtaining correlations and PSDs loses phase information.

Using the FFT algorithm, it is possible to build dedicated processors which operate on pairs of data channels, and obtain averaged auto- and cross-spectra in real time. These "multi-channel FFT analyzers" are one building block of experimental-modal-analysis systems.

The multiple channels of input can be obtained either by exciting the structure at one location and measuring the response, simultaneously or successively, at many different locations; or else by measuring the response at one location, and successively exciting the structure at many different locations by a "roving force."

33.2 Excitation

Excitation over a wide range of frequencies can be obtained in several different ways:

- From the impact with a hammer, which is instrumented and hooked up to one data input channel to form a reference, because successive blows and the resulting transients may not always be exactly identical. This is the easiest way to excite most structures.

- By attaching an electro-magnetic or piezo-electric mechanical exciter driven by a random-noise generator. This assures arbitrarily long data records and quantitative accuracy.

- By attaching a mechanical exciter providing a sweep of sinusoidal frequencies.

The main requirement is that all interesting modes be stimulated.

33.3 Measurement

Response data can also be obtained in several different ways:

- With accelerometers, which are easy to attach to a structure, and are available with three axes of measurement.
- With laser vibrometers, which measure surface velocity, and do not add any mass to the structure.
- With straingages, which measure flexural displacement.

The instrumentation system generally produces a low-voltage analog signal input for the A/D conversion.

33.4 Data Acquisition

As we observed in Chapter 18, the analog input data must go through an anti-aliasing filter before conversion to digital data. Because the quantity of input data involved in complex structures is overwhelming, it is not convenient to store the entire set of time-series data for later processing. Only small amounts of time-domain data are temporarily stored in input buffers, processed on-the-fly by dedicated "hard-wired" FFT analyzers, and the auto- and cross-spectra accumulated for each channel pair in averaging memory. The resulting frequency response functions (FRF) are much more compact, and can be stored on disk as a data base for later evaluation.

33.5 Modal Evaluation

Using a multi-modal-analysis computer program, the frequency-domain FRF values are then analyzed for frequency and mode-shape information. The mode shapes can be displayed graphically. The accuracy of the results is dependent on the same sampling parameters mentioned in Chapter 18.

Problem 33.1 *Obtain descriptive information about several FFT multi-channel analyzers. Compare systems. Write a short explanation of coherence and other functions which are not explained above. Incorporate into a short report on available technology. Hint: Start with ZONIC Corp. and Data Physics Corp. on the Internet.*

Part V

Parametric Excitation

Chapter 34

TIME-VARYING COEFFICIENTS and Mathieu's equation

Up to this point, we have discussed the response of linear equations to initial conditions and to excitation which can be isolated on the right side of the governing differential equation as non-homogeneous terms. In some structural systems, the excitation appears as a non-constant coefficient on the left-hand side of the equation. Although the resulting equation is still linear, it is difficult to solve.

34.1 Mathieu's Equation

Let us consider a simple pendulum, suspended from a pivot which oscillates in the *vertical* direction with a displacement $y_{\text{pivot}} = Y_o \cos \omega_{ex} t$. Applying the method of Section 5.6, we write potential and kinetic energy

$$
\begin{aligned}
U &= \frac{mg}{g_c}\left[L\left(1-\cos\theta\right)+Y_o\cos\omega_{ex}t\right] \\
T &= \frac{1}{2}\frac{m}{g_c}\left[\left(L\,\dot{\theta}\,\cos\theta\right)^2+\left(L\,\dot{\theta}\,\sin\theta-Y_o\omega_{ex}\sin\omega_{ex}t\right)^2\right] \\
\frac{d}{dt}\frac{\partial T}{\partial\,\dot{\theta}} &= \frac{m}{g_c}\left[L^2\,\ddot{\theta}\,-L\,\dot{\theta}\,\cos\theta\cdot Y_o\omega_{ex}\sin\omega_{ex}t-L\sin\theta\cdot Y_o\omega_{ex}^2\cos\omega_{ex}t\right] \\
\frac{-\partial T}{\partial\theta} &= \frac{m}{g_c}\left[\left(L\,\dot{\theta}\,\cos\theta\right)\left(Y_o\omega_{ex}\sin\omega_{ex}t\right)\right] \\
\frac{\partial U}{\partial\theta} &= \frac{mg}{g_c}\left[L\sin\theta\right]
\end{aligned}
$$

and find the governing equation, which incorporates the acceleration of the pivot $-Y_o\omega_{ex}^2 \cos\omega_{ex}t$

$$\frac{mL^2}{g_c}\ddot{\theta} + \frac{mgL}{g_c}\sin\theta = \frac{\left(Y_o\omega_{ex}^2 \cos\omega_{ex}t\right)mL}{g_c}\sin\theta$$

Note that the excitation force depends on the displacement θ, so that we can combine it with the restoring-torque term as if the acceleration of the pivot were a pulsation of gravity

$$\ddot{\theta} + \left(\frac{g}{L} - \frac{Y_o\omega_{ex}^2 \cos\omega_{ex}t}{L}\right)\sin\theta = 0 \tag{34.1}$$

Linearizing for small amplitudes θ

$$\ddot{\theta} + \left(\frac{g}{L} - \frac{Y_o\omega_{ex}^2 \cos\omega_{ex}t}{L}\right)\theta = 0 \tag{34.2}$$

This equation has the form of a linear equation with a time-varying spring $\ddot{x} + k(t)\,x = 0$.

Exercise 34.1 *Derive this equation, using the methods of Chapters 2 and 3.*

34.2 Canonical Form

The Standard Form of this equation is obtained by renaming the variables

$$y(z) \triangleq \theta(t) \tag{34.3}$$

$$z \triangleq \frac{\omega_{ex}t}{2} \tag{34.4}$$

$$\frac{4}{\omega_{ex}^2}\cdot\frac{d^2y}{dz^2} + \left(\frac{g}{L} - \frac{Y_o\omega_{ex}^2 \cos\omega_{ex}t}{L}\right)y = 0$$

and the parameters

$$a \triangleq \frac{4}{\omega_{ex}^2}\cdot\frac{g}{L} \tag{34.5}$$

$$q \triangleq \frac{1}{2}\cdot\frac{4}{\omega_{ex}^2}\cdot\frac{Y_o\omega_{ex}^2 \cos\omega_{ex}t}{L} \tag{34.6}$$

$$\boxed{y'' + (a - 2q\cos 2z)\,y = 0} \tag{34.7}$$

and has been extensively discussed by McLachlan.[1]

[1] N.W. McLachlan, *Ordinary Non-Linear Differential Equations in Engineering and Physical Sciences,* Second Edition, Oxford at Clarendon Press, London, 1956, pages 113–119.

34.3 Fractional Analysis

The form in which Mathieu's equation is generally written changes the time-scale to standardize the *excitation frequency* to the integer value of two, $2z = \omega_{ex}t$. We are more accustomed to adapting the time-scale to the *natural frequency* of the basic system, $\omega_n \triangleq \sqrt{g/L} = \sqrt{a}$, by defining $t^* \triangleq \omega_n t = t\sqrt{g/L}$ or $z^* \triangleq \omega_n z = z\sqrt{a}$. If we did that, we would have

$$\frac{d^2\theta}{dt^{*2}} + \left(1 - \frac{Y_o\omega_{ex}^2}{g}\cos\frac{\omega_{ex}t^*}{\sqrt{g/L}}\right)\theta = 0$$

$$\frac{d^2y}{dz^{*2}} + \left(1 - \frac{2q}{a}\cos 2z^*\right)y = 0$$

Examining the equations, we anticipate that:

- In the absence of excitation Y_o, the characteristic solution is sinusoidal, with the argument $t\sqrt{g/L} = z\sqrt{a}$; we suspect that large-amplitude forced solutions will occur when the characteristic frequency is excited.
- The forced response depends on the dimensionless parameter $Y_o\omega_{ex}^2/g = 2q/a$; if that parameter is small, we anticipate stable oscillation unless the characteristic frequency is excited.
- We suspect that when $\omega_{ex} = \sqrt{g/L}$ or $\sqrt{a} = 2$, there is a resonance with indefinitely increasing amplitudes for even very small values of $Y_o\omega_{ex}^2/g = 2q/a$.
- Because the forcing term depends on the product of the pivot motion and the displacement, we may expect another resonance when the excitation frequency is twice the characteristic frequency, $\omega_{ex} = 2\sqrt{g/L}$ or $\sqrt{a} = 1$. Moving your hand vertically at various frequencies while holding a weight on a string confirms that we can keep the pendulum moving by raising the pivot twice per cycle; we can feel that we are putting energy into the system by raising the pivot against the centrifugal force pulling down in mid-swing, and lowering it against the lesser force at maximum θ. A child standing on a swing can accomplish the same thing by flexing the knees at appropriate times.
- We wonder whether there are other resonances for other integer values $\sqrt{a} = n$ or

  $$\omega_{ex} = \frac{2}{n}\sqrt{\frac{g}{L}}$$

  If the motion of the pendulum with small excitation is approximately $A\sin\omega_n t$, the product $(A\sin\omega_n t)\times\left(Y_o\omega_{ex}^2\cos\omega_{ex}t\right)$ will have stable Fourier series components with the frequency ω_n for such even multiples.

If a resonance leads to constantly increasing amplitudes, it may be regarded as an instability.

34.4 Stability

The classical analyses of Mathieu's equation[2] show that it is indeed unstable for

$$\frac{2\sqrt{g/L}}{\omega_{ex}} = \sqrt{a} = 1,\ 2,\ 3,\ 4,\ \text{etc.} \tag{34.8}$$

when $Y_o\omega_{ex}^2/g = 2q/a$ is small. For increased values of $2q/a$, there is a tendency for each resonance to lock-in with the excitation, and the instability zones widen to include a whole band of values near $\sqrt{a} = 1,\ 2,\ 3,\ 4$, etc.; when $2q/a \gg 1$, almost any excitation frequency is unstable. The presence of damping increases the stable regions. In experiments, this is most commonly observed by the fact that the pendulum must be slightly perturbed laterally to get the response started.

Except for special cases, the solutions in the stable regions are not exactly periodic but show changes in amplitude and phase. For the purpose of stability analysis, we have linearized our pendulum equation; to analyze realistic cases we would need to consider both non-linearities and damping.

34.5 Strings

When a string deflects, it length changes slightly. Assuming a sinusoidal mode $y = y_{\max}\sin(2\pi x/\lambda)$, the length s for a span $\lambda/2$ is

$$\begin{aligned}
s &= \int_0^{\lambda/2}\sqrt{1+\left(\frac{dy}{dx}\right)^2}\,dx \\
&= \int_0^{\lambda/2}\sqrt{1+\left(y_{\max}\frac{2\pi}{\lambda}\cos\frac{2\pi x}{\lambda}\right)^2}\,dx \\
&= \frac{\lambda}{2\pi}\int_0^{\pi}\sqrt{1+\left(y_{\max}\frac{2\pi}{\lambda}\right)^2\cos^2 u}\,du \\
\frac{s}{\lambda/2} &= \frac{1}{2\pi}\int_0^{2\pi}\sqrt{1+a^2\cos^2 u}\,du = \frac{1}{2\pi}\int_0^{2\pi}\sqrt{1+a^2\sin^2 u}\,du
\end{aligned}$$

[2] N.W. McLachlan, *Theory and Application of Mathieu Functions,* Oxford Press, London, 1947.

where typically $a^2 \triangleq (2\pi y_{\max}/\lambda)^2 < 1$, so that we can use the binomial expansion

$$\begin{aligned}\frac{s}{\lambda/2} &= \frac{1}{2\pi}\int_0^{2\pi}\left(1+\frac{a^2}{2}\sin^2 u - \frac{a^4}{8}\sin^4 u + \frac{a^6}{16}\sin^6 u - \dots\right) du \\ &= 1+\frac{1}{4}a^2 - \frac{3}{64}a^4 + \frac{5}{256}a^6 - \dots\end{aligned}$$

which shows that when $a^2 \triangleq (2\pi y_{\max}/\lambda)^2 \ll 1$, then $2s/\lambda \cong 1+a^2/4$. Therefore we can write the deflection-induced additional strain in the string by the non-linear relationship

$$\Delta\epsilon = \frac{s}{\lambda/2} \cong 1+\frac{1}{4}a^2 = 1+\left(\frac{\pi y_{\max}}{\lambda}\right)^2 \tag{34.9}$$

This strain increases the stress and tension in the string, causing non-linearities which we will investigate in the next few chapters.

However, the immediate consequence of this stress is to transmit forces to the supports of the string. The tensile effect is the same whether the string deflects upward or downward, so the force has a fundamental frequency of twice the vibrational frequency of the string, plus all the odd harmonics.

Reciprocally, a longitudinal vibration of the end support of the string has the potential of causing vibrations at half the excitation frequency, plus all the harmonics, analogous to the Mathieu's equation stability map.

34.6 Beams

We have already observed that the natural frequency of beams changes when tension and compression are applied (Chapter 32). Such a loading occurs for example in connecting rods, whether in steam locomotives or automotive engines. The natural frequency changes once per revolution, so the parametric excitation is strongest at half that frequency. On the other hand, the direct lateral excitations is at the frequency of rotation itself, and is usually thought to be more significant.

34.7 Closure

The tools which are used for non-linear systems—graphical in Chapter 36 and numerical in Chapter 40—are also applicable to parametric excitation.

Exercise 34.2 *Find the stability plot for the Mathieu's equation in a mathematical-physics text, and interpret the plot coordinates, in terms of millimeters of vertical motion, for a simple pendulum with a length of one meter.*

Part VI

Non-Linear Vibration

Chapter 35

LINEARIZATION and error analysis

35.1 Linearization

Many governing equations have non-linear terms. Our first response to non-linear problems is to linearize them. For example, the equation for a pendulum reduces to

$$\ddot{\theta} + \omega_n^2 \sin\theta = 0$$

From the Maclaurin series expansion, we know that, near zero,

$$\sin\theta = [\sin\theta]_{\theta=0} + \theta \cdot \left[\frac{d\sin\theta}{d\theta}\right]_{\theta=0} + \frac{\theta^2}{2!} \cdot \left[\frac{d^2\sin\theta}{d\theta^2}\right]_{\theta=0} + \frac{\theta^3}{3!} \cdot \left[\frac{d^3\sin\theta}{d\theta^3}\right]_{\theta=0} + \ldots$$

so that the equation can be represented in the form

$$\ddot{\theta} + \omega_n^2 \left(0 + \theta + 0 - \frac{\theta^3}{6} + 0 + \frac{\theta^5}{120} + \ldots\right) = 0$$

If are dealing with small amplitudes, such that

$$\frac{\theta^2}{6} \ll 1$$

we can linearize the equation to

$$\ddot{\theta} + \omega_n^2 \theta \cong 0$$

Similarly, a spring may not exactly follow a linear element law, but may stiffen or soften slightly with deflection. This can lead, for example, to the equation

$$\ddot{x} + \frac{kg_c}{m}x + \frac{\varepsilon kg_c}{m}x^3 = 0$$

Where ε is positive for a hardening spring, and negative for a softening spring. The equation can be linearized by leaving off the cubic term if $x^2 \ll |1/\varepsilon|$.

However, if there is a biasing force which shifts the equilibrium, the conditions near equilibrium must be considered, not the conditions near $x = 0$. For example, in

$$\ddot{x} + \frac{kg_c}{m}x + \frac{\varepsilon kg_c}{m}x^3 = g$$

we must first solve for the equilibrium $\overline{x}$ from the steady-state equation

$$\frac{kg_c}{m}\overline{x} + \frac{\varepsilon kg_c}{m}\overline{x}^3 = g$$

which has a real root if

$$\varepsilon \geq \frac{-4}{27} \cdot \left(\frac{kg_c}{mg}\right)^2$$

Having obtained the at-rest value $\overline{x}$, we substitute the new, shifted coordinate $\widetilde{x} = x - \overline{x}$ into the governing equation to obtain

$$\ddot{\overline{x}} + \ddot{\widetilde{x}} + \frac{kg_c}{m}\left(\overline{x} + \widetilde{x}\right) + \frac{\varepsilon kg_c}{m}\left(\overline{x}^3 + 3\overline{x}^2\widetilde{x} + 3\overline{x}\widetilde{x}^2 + \widetilde{x}^3\right) = g$$

The first term, a derivative of a constant, is zero, and several others cancel because $\overline{x}$ represents a solution of the equilibrium equation, leaving

$$\ddot{\widetilde{x}} + \frac{kg_c}{m}\left(\widetilde{x}\right) + \frac{\varepsilon kg_c}{m}\left(3\overline{x}^2\widetilde{x} + 3\overline{x}\widetilde{x}^2 + \widetilde{x}^3\right) = 0$$

If $|\widetilde{x}| \ll (kg_c/3\overline{x}m + 3\varepsilon\overline{x})$, we can write

$$\ddot{\widetilde{x}} + \frac{kg_c}{m}\left(1 + 3\varepsilon\overline{x}^2\right)\left(\widetilde{x}\right) \cong 0$$

which is a linear equation. Note that the natural frequency depends on the equilibrium position $\overline{x}$, and that the equilibrium is stable if $\varepsilon > \left(-1/3\overline{x}^2\right)$.

35.2 Linearization Process

The linearization procedure consists of three steps:

1. Solve the governing equation for the equilibrium position $\overline{x}$ (often $\overline{x} = 0$, but not always!).

2. Expand the expressions in the equation through a Taylor series expansion about that equilibrium point, equivalent to expressing the equation in terms of $\widetilde{x} = (x - \overline{x})$:

$$f(x) = [f(x)]_{x=\overline{x}} + (x - \overline{x}) \cdot \left[\frac{df(x)}{dx}\right]_{x=\overline{x}} + \frac{(x-\overline{x})^2}{2!} \cdot \left[\frac{d^2 f(x)}{dx^2}\right]_{x=\overline{x}} + \ldots$$

3. Neglect higher order terms (i.e., higher powers of $\widetilde{x}$).

This replaces the actual element laws with linear laws which fit the equilibrium point and the slope at that point.

35.3 Example

A good example of a non-linear "spring" is the pendulum. We can add a clock-spring to the pivot of the pendulum, and adjust it so that the position θ_o at which the spring is unloaded is not necessarily straight down ($\theta_o = 0$). The governing equation then becomes

$$\frac{m\ell^2}{g_c}\ddot{\theta} + \kappa_T(\theta - \theta_o) + \frac{mg\ell}{g_c}\sin\theta = 0$$

which can be simplified to

$$\ddot{\theta} + \alpha\theta + \beta\sin\theta = \alpha\theta_o$$

where $\alpha = \kappa_T g_c/m\ell^2$ and $\beta = g/\ell$. The linearized form of the equation depends on the equilibrium position $\overline{\theta}$, obtained by solving the equilibrium equation (i.e., $\ddot{\theta} = 0$):

$$\alpha\overline{\theta} + \beta\sin\overline{\theta} = \alpha\theta_o$$

Four cases can be developed:

35.3.1 Case I

If either $\alpha = 0$ or $\theta_o = 0$, then $\overline{\theta} = 0$, the downward vertical position. Splitting θ into $\left(\overline{\theta} + \widetilde{\theta}\right)$, then $\widetilde{\theta} = \theta$, and the governing equation can be expressed as

$$\ddot{\widetilde{\theta}} + \alpha\widetilde{\theta} + \beta\sin\widetilde{\theta} = 0$$

or, using the Maclaurin series expansion for the sine,

$$\ddot{\widetilde{\theta}} + \alpha\widetilde{\theta} + \beta\left(\widetilde{\theta} - \frac{1}{6}\widetilde{\theta}^3 + \frac{1}{120}\widetilde{\theta}^5 - \ldots\right) = 0$$

which linearizes to

$$\ddot{\widetilde{\theta}} + (\alpha + \beta)\,\widetilde{\theta} \cong 0$$

35.3.2 Case II

If $\theta_o = (\pi/2 + \beta/\alpha)$, then $\overline{\theta} = \pi/2$, the horizontal position. We can show this by inserting this θ_o into the governing equation:

$$\begin{aligned} \ddot{\theta} + \alpha\theta + \beta \sin\theta &= \alpha(\pi/2 + \beta/\alpha) \\ \ddot{\theta} + \alpha(\theta - \pi/2) + \beta(\sin\theta - 1) &= 0 \end{aligned}$$

Splitting θ into $\left(\overline{\theta} + \widetilde{\theta}\right)$, we substitute $\widetilde{\theta} = (\theta - \pi/2)$,

$$\begin{aligned} \ddot{\widetilde{\theta}} + \alpha\widetilde{\theta} + \beta\left(\sin\left(\pi/2 + \widetilde{\theta}\right) - 1\right) &= 0 \\ \ddot{\widetilde{\theta}} + \alpha\widetilde{\theta} + \beta\left(\sin\pi/2\cos\widetilde{\theta} + \cos\pi/2\sin\widetilde{\theta} - 1\right) &= 0 \\ \ddot{\widetilde{\theta}} + \alpha\widetilde{\theta} + \beta\left(\cos\widetilde{\theta} - 1\right) &= 0 \end{aligned}$$

Expanding the cosine about $\widetilde{\theta} = (\theta - \pi/2)$, this reduces to

$$\ddot{\widetilde{\theta}} + \alpha\widetilde{\theta} + \beta\left(\frac{-1}{2}\widetilde{\theta}^2 + \frac{1}{24}\widetilde{\theta}^4 - \ldots\right) = 0$$

which linearizes to

$$\ddot{\widetilde{\theta}} + \alpha\widetilde{\theta} \cong 0$$

35.3.3 Case III

If $\theta_o = \pi$ and $\alpha \geq \beta$, then $\overline{\theta} = \pi$, the upward vertical position. Splitting θ into $\left(\overline{\theta} + \widetilde{\theta}\right)$, then $\widetilde{\theta} = (\theta - \pi)$, and the governing equation can be expressed as

$$\begin{aligned} \ddot{\widetilde{\theta}} + \alpha\left(\widetilde{\theta} + \pi\right) + \beta\sin\left(\widetilde{\theta} + \pi\right) &= \alpha\pi \\ \ddot{\widetilde{\theta}} + \alpha\widetilde{\theta} - \beta\sin\widetilde{\theta} &= 0 \end{aligned} \tag{35.1}$$

The sine can be expanded to

$$\ddot{\widetilde{\theta}} + \alpha\widetilde{\theta} - \beta\left(\widetilde{\theta} - \frac{1}{6}\widetilde{\theta}^3 + \frac{1}{120}\widetilde{\theta}^5 - \ldots\right) = 0$$

which linearizes to

$$\ddot{\widetilde{\theta}} + (\alpha - \beta)\widetilde{\theta} \cong 0$$

35.3.4 Case IV

In general, θ_o could have any value. Substituting $\theta = \left(\overline{\theta} + \widetilde{\theta}\right)$ into the governing equation,

$$\ddot{\widetilde{\theta}} + \alpha\left(\overline{\theta} + \widetilde{\theta}\right) + \beta \sin\left(\overline{\theta} + \widetilde{\theta}\right) = \alpha\theta_o$$

which expands to

$$\ddot{\widetilde{\theta}} + \alpha\left(\overline{\theta} + \widetilde{\theta}\right) + \beta\left(\sin\overline{\theta}\cos\widetilde{\theta} + \cos\overline{\theta}\sin\widetilde{\theta}\right) = \alpha\theta_o$$

The equilibrium solution, for which $\widetilde{\theta} = 0$, is obtained from $\alpha\overline{\theta} + \beta\sin\overline{\theta} = \alpha\theta_o$. We insert the resulting $\overline{\theta}$ into the preceding equation and expand it

$$\ddot{\widetilde{\theta}} + \alpha\left(\overline{\theta} + \widetilde{\theta}\right) + \beta\left(\sin\overline{\theta}\left(1 - \frac{1}{2}\widetilde{\theta}^2 + \ldots\right) + \cos\overline{\theta}\left(\widetilde{\theta} - \frac{1}{6}\widetilde{\theta}^3 + \ldots\right)\right) = \alpha\theta_o$$

Cancelling the terms which have been balanced out in obtaining $\overline{\theta}$:

$$\ddot{\widetilde{\theta}} + \alpha\widetilde{\theta} + \beta\left(\sin\overline{\theta}\left(\frac{-1}{2}\widetilde{\theta}^2 + \ldots\right) + \cos\overline{\theta}\left(\widetilde{\theta} - \frac{1}{6}\widetilde{\theta}^3 + \ldots\right)\right) = 0$$

which linearizes to

$$\ddot{\widetilde{\theta}} + \left(\alpha + \beta\cos\overline{\theta}\right)\widetilde{\theta} \cong 0$$

We can see that this reduces to the three special cases shown above, when $\overline{\theta} = 0$, $\pi/2$, or π, respectively.

35.4 Damping

Fluid damping typically consists of quadratic or other higher-order terms. Since quadratic terms disappear during linearization near an equilibrium displacement, this important type of damping is neglected in linearization.

The opposite problem occurs with friction damping, which generally includes "sticktion," a minimum initial force required to overcome the stationary state. This produces a deadband which locks up the system at small amplitudes, and must be neglected to produce any linearized dynamic analysis at all.

35.5 Closure

Linearization is restricted to small amplitudes. For larger amplitudes or for non-linear damping, we will need to go beyond linear methods.

Problem 35.1 *The restoring force generated by an extensional spring can generally be described by means of a polynomial*

$$F_{spring} = a_1x + a_2x^2 + a_3x^3 + a_4x^4 + a_5x^5 + \ldots$$

where a_2, a_3, etc., my be positive or negative coefficients. For a spring suspending a mass in the presence of a gravitational force

$$\frac{m}{g_c}\ddot{x} + a_1x + a_2x^2 + a_3x^3 + \ldots = \frac{mg}{g_c}$$

what is the natural frequency of small oscillations?

Problem 35.2 *For the same problem, if the spring is mounted in compression instead of tension, so that*

$$\frac{m}{g_c}\ddot{x} + a_1x + a_2x^2 + a_3x^3 + \ldots = \frac{-mg}{g_c}$$

what is the natural frequency of small oscillations?

Problem 35.3 *If a spring were described by $F_{spring} = ax^n$ where the coefficient a and the exponent n are fitted to experimental data, what is the natural frequency of oscillation of the system*

$$\frac{m}{g_c}\ddot{x} + ax^n = \frac{\pm mg}{g_c}$$

for different values of n? What are acceptable values of n? Does your answer depend on whether g is positive or negative?

Chapter 36

THE PHASE PLANE and graphical solutions

Most physical problems are not strictly linear; some deviate strongly from linearity. When we give up linearity, we give up superposition. That means we can no longer write general solutions that arise from the addition of several particular integrals for the components of the input and a complementary Initial-Conditions solution. To look at the patterns of the solutions, we can represent them in the phase plane.

36.1 The Phase Plane

The state of a non-hereditary one-degree-of-freedom system, linear or non-linear, can be described at any moment of time by specifying the state variables of displacement x and velocity v. Therefore we can show the state of the system as a point on the phase plane: a plot which has displacement x as the abscissa, and velocity normalized with a reference time-scale

$$y \triangleq \frac{v}{\omega_{ref}} \tag{36.1}$$

as the ordinate. Changes of the system show up as trajectories on this plot. Since $v = \dot{x}$, a trajectory must proceed to the right—increasing x—in the upper half of the plot where v is positive, and to the left in the lower half, for a clockwise motion. Every trajectory must cross the x-axis (where $v = 0$) vertically.

Undamped linear vibrations show up as clockwise ellipses. These ellipses become circles if we choose the reference time-scale $\omega_{ref} = \omega_n$. We know this because the energy in a linear system is

$$\text{constant} = (T + U) = \left(\frac{mv^2}{2g_c} + \frac{kx^2}{2}\right) = \left(\frac{v^2}{\omega_n^2} + x^2\right)\frac{k}{2} = \left(y^2 + x^2\right)\frac{k}{2} = \left(r^2\right)\frac{k}{2}$$

Time does not show up on the phase plane, but we know that the rate of travel around this circle is the angular velocity $\omega_{ref} = \omega_n$, because we know that the projection of the trajectory on the x-axis is $x_{\max} \sin(\omega t - \theta)$, and the projection on the y-axis is $\frac{v_{\max}}{\omega} \cos(\omega t - \theta)$.

We may go on to speculate that *non-linear* undamped vibrations will show up as distorted circular motion, and that it is possible to plot energy "contour lines" showing all possible trajectories, each line representing another energy level relative to when the system is at rest in the center of the plot. We expect *damped* systems to have trajectories spiraling inward towards the center of the plot.

A very general normalized equation for a non-linear system is

$$\ddot{x} + F(x, v) = 0 \tag{36.2}$$

$F(x, v)$ may encompass spring-like dependence on displacement x, damping-like dependence on v, or a combination of these. To eliminate time t from the equation, we can replace $\ddot{x}$ with

$$\frac{d^2x}{dt^2} = \frac{dv}{dt} = \frac{dv}{dx} \cdot \frac{dx}{dt} = \frac{dv}{dx} \cdot v = \frac{1}{2}\frac{d(v^2)}{dx} \tag{36.3}$$

resulting in

$$\frac{dv}{dx} \cdot v + F(x, v) = 0$$

from which we see that the slope of the trajectory dv/dx is

$$\frac{dv}{dx} = \frac{-F(x, v)}{v}$$

or in scaled coordinates $y \triangleq v/\omega_{ref}$:

$$\frac{dy}{dx} = \frac{-F(x, v)}{y \cdot \omega_{ref}^2} \tag{36.4}$$

This gives us enough information to plot trajectories, which we can map out all over the phase plane. We can take any initial state, identify the trajectory it is on, and follow the ensuing sequence of future states.

36.2 Phase-Plane-Delta Method

The easiest graphical method for plotting a trajectory is based on the fact that undamped linear systems can be plotted as circles. We can make Equation 36.2 *look* like a linear equation by arbitrarily *choosing* a nominal reference frequency ω_{ref}, and rewriting Equation 36.2 as

$$\ddot{x} + \omega_{ref}^2 x + f(x, v) = 0 \tag{36.5}$$

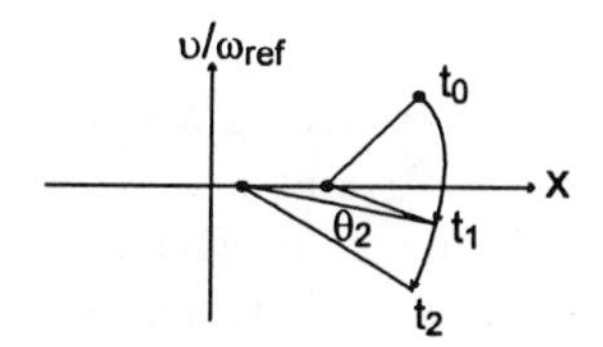

Figure 36.1: Phase-Plane Delta

where

$$f(x, v) \triangleq F(x, v) - \omega_{ref}^2 x \tag{36.6}$$

(The choice of ω_{ref} is up to each worker, but experience will show that some choices will make the upcoming tasks easier, and some will make it more tedious.) If we treat $f(x, v)$ as if it stayed *constant for a little while,* we have a linear equation with a biasing force, and can solve for a temporary pseudo-equilibrium

$$x_{center} \equiv -f(x, v)/\omega_{ref}^2 \tag{36.7}$$

If we shift the origin to this x_{center}, we have an *apparently* undamped and linear system, at least until x_{center} starts to change too much:

$$\ddot{x} + \omega_{ref}^2 (x - x_{center}) = 0 \tag{36.8}$$

The value of x_{center} will not change very fast *if* we choose ω_{ref} such that $f(x, v)$ is small or at least steady.

Therefore, the procedure is:

1. Choose ω_{ref} such that $f(x, v)$ is as weak a function of x and v as possible.

2. Use ω_{ref} to scale the ordinate of the phase plane.

3. Plot the Initial Conditions for x and v onto the phase plane, and label them $t = 0$.

4. Compute (e.g., on a hand-held calculator)

$$x_{center} \equiv \frac{\omega_{ref}^2 x - F(x, v)}{\omega_{ref}^2}$$

   for the current x and v, and plot it along the x-abscissa on the phase plane. This is the origin for the circular trajectory.

5. Draw an arc clockwise from the old x and v to a new set of values on the phase plane. (Figure 36.1) How long an arc? Try about 15 degrees and calculate x_{center} for the new x and v at the end of the arc. If x_{center} has changed a lot, so that the next arc would make a sharp corner with the current arc, you should erase and start again with a shorter arc. On the other hand, if one makes the arcs *too* small, then "eyeballing" errors—the graphical equivalent of round-off errors—will limit the accuracy.

6. With a protractor, keep track of the length of each arc: from undamped linear systems, we know that the elapsed time for each arc θ is $\Delta t = \theta/\omega_{ref}$. Add that to the previous time and label the new state with its value of t. [Because we are keeping track of elapsed time, we can generalize the phase-plane-delta method to non-homogeneous equations by letting F be a function of t as well: $F(x, v, t)$.]

7. Use this final state x and v to start over with Step 4. With enough repetitions of this process, a trajectory can be plotted.

Proof: We have developed the method conceptually, without much rigor. Clearly it works for nearly linear problems. Should we worry about the effect of $\dot{x}_{center}$ and $\ddot{x}_{center}$? From Equation 36.4 we know that the slope of the trajectory is exactly

$$S_{traj} = \frac{-F(x,v)}{(v/\omega_{ref})\,\omega_{ref}^2}$$

The slope of the radial line from the abscissa at x_{center} to the state values of x and v is

$$S_{radius} = \frac{(v/\omega_{ref})}{x - x_{center}} = \frac{(v/\omega_{ref})}{x - \left(\frac{\omega_{ref}^2 x - F(x,v)}{\omega_{ref}^2}\right)} = \frac{(v/\omega_{ref})\,\omega_{ref}^2}{F(x,v)} = \frac{-1}{S_{traj}}$$

Even for large non-linearities, the initial arc of the phase-plane-delta method gives exactly the correct direction to the trajectory; the only disadvantage of using it for strong non-linearities is that the arcs must be kept short. Furthermore, we know that the speed of the state point along the trajectory V_{traj} has the component v in the x-direction, and the component $v \cdot S_{traj}$ in the y-direction, so that

$$\begin{aligned} V_{traj} &= \sqrt{v^2 + (vS_{traj})^2} = \sqrt{v^2 + v^2\left(\frac{-F(x,v)}{(v/\omega_{ref})\,\omega_{ref}^2}\right)^2} \\ &= \sqrt{v^2 + \left(\frac{-F(x,v)}{\omega_{ref}}\right)^2} \end{aligned}$$

Compare this with the phase-plane-delta method's

$$
\begin{aligned}
V_{traj} &= \omega_{ref} \cdot L_{radius} = \omega_{ref}\sqrt{(v/\omega_{ref})^2 + (x - x_{center})^2} \\
&= \omega_{ref}\sqrt{(v/\omega_{ref})^2 + \left(x - \frac{\omega_{ref}^2 x - F(x,v)}{\omega_{ref}^2}\right)^2} \\
&= \sqrt{v^2 + \left(\frac{F(x,v)}{\omega_{ref}}\right)^2}
\end{aligned}
$$

Evidently the phase-plane-delta method also gives exactly the correct initial rate of progress along the trajectory.

Note: Many historical sources will define the instant center $\delta_{phaseplane} \equiv -x_{center}$ and plot it along the negative x-axis. Some authors will also interchange the coordinates so that the motion is counter-clockwise.

Example: *Friction damping* often has a static-friction break-away value, and a nearly constant dynamic-friction value. This is experienced by drivers: at the moment of coming to a stop, the brakes lock up and the nose of the car dips, unless the driver is skillful. The dynamic value may not be exactly constant, but may slightly decline with increasing speed; at very high speed, automobile brakes are slightly less effective than at moderate speeds. For analysis, friction damping is often simplified to *Coulomb damping*

$$
F_{\text{damper}} = c\frac{\dot{x}}{\left|\dot{x}\right|} = c \cdot \operatorname{sgn}\left(\dot{x}\right)
$$

where c has the dimensions of Newton or lbf, and the function "signum" has a value of unity with the sign of the argument. Inserting a Coulomb damper into a simple system, the governing equation is

$$
\begin{aligned}
\ddot{x} + \frac{cg_c}{m} + \frac{kg_c}{m}x &= 0 \text{ for } \dot{x} > 0 \\
\ddot{x} - \frac{cg_c}{m} + \frac{kg_c}{m}x &= 0 \text{ for } \dot{x} < 0
\end{aligned}
$$

Comparing this with Equation 36.2, we see that

$$
F(x,v) = \frac{cg_c}{m}\operatorname{sgn}(v) + \frac{kg_c}{m}x
$$

Choosing $\omega_{ref}^2 = kg_c/m$ and applying Equation 36.6 we get

$$
f(x,v) = \frac{cg_c}{m}\operatorname{sgn}(v)
$$

Therefore, Equation 36.7 yields

$$\begin{aligned} x_{center} &= \frac{-c}{k}\mathrm{sgn}\,(v) \\ &= \frac{-c}{k} \text{ for } v > 0 \\ &= \frac{+c}{k} \text{ for } v < 0 \end{aligned}$$

Our circular trajectories will have a fixed center on the negative x-axis whenever the velocity is positive (the upper half-plane) and a fixed center on the positive x-axis whenever the velocity is negative (the lower half-plane). Starting at some initial displacement x_o, we draw a lower half-circle with its center at c/k, ending up at $-(x_o - 2c/k)$. We continue with an upper half-circle with its center at $-c/k$, ending up at $(x_o - 4c/k)$, and so on. We observe that

- As the trajectory spirals inward, the amplitude decreases by a constant amount $4c/k$ for each revolution, for a straight-line decay envelope—not the exponential decay we would have had with linear damping.
- The time elapsed for each 180-degree arc is π/ω_{ref}. Because a revolution takes a total of 360 degrees for this problem, the period is $2\pi/\omega_{ref}$ and the frequency is ω_{ref}, unaffected by damping—this is an unusual result because most non-linear or damped trajectories are composed of arcs that sum up to slightly more or less than 360 degrees. Recall that linear damping, in Chapter 9, slowed down the oscillations by a factor of $\sqrt{1-\zeta^2}$.
- This system is "piecewise linear" and the projection of the trajectory on the x-axis is pieced together from cosine functions.
- When the trajectory terminates on the x-axis between $-c/k$ and $+c/k$, it cannot proceed clockwise any further, and gets stuck in this "deadband."

Friction dampers were popular in early race cars, because they are easily adjustable. In modern practice, the straight-line decay is deemed insufficient for damping out large oscillations, and the deadband passes along irritating forces from small perturbations that stay within the lock-up region. Manufacturers of hydraulic shock absorbers go to some lengths to minimize the amount of static friction or "sticktion" in their products.

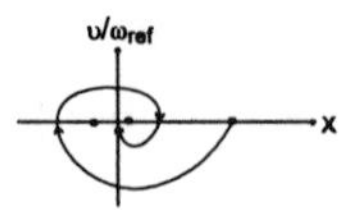

Exercise 36.1 *Repeat the process with a model which includes static friction. Show that the basic decay envelope is unaffected by sticktion, but that the deadband is widened.*

Example: In machinery we sometimes encounter masses which are attached to springs with a certain amount of "play." If the slack region amounts to 2ϵ, the spring force can be described by the function

$$\begin{aligned} F_{\text{spring}} &= k(x-\epsilon) \qquad \text{for } x > \epsilon \\ &= 0 \qquad \text{for } -\epsilon < x < \epsilon \\ &= k(x+\epsilon) \qquad \text{for } x < -\epsilon \end{aligned}$$

and the governing equation is

$$\begin{aligned} \ddot{x} + \frac{kg_c}{m}(x-\epsilon) &= 0 \text{ for } x > \epsilon \\ \ddot{x} + \frac{kg_c}{m}(x-x) &= 0 \text{ for } -\epsilon < x < \epsilon \\ \ddot{x} + \frac{kg_c}{m}(x+\epsilon) &= 0 \text{ for } x < -\epsilon \end{aligned}$$

Choosing $\omega_{ref}^2 = kg_c/m$ we get

$$\begin{aligned} x_{center} &= \epsilon \text{ for } x > \epsilon \\ &= x \text{ for } -\epsilon < x < \epsilon \\ &= -\epsilon \text{ for } x < -\epsilon \end{aligned}$$

Our circular trajectories will have a fixed center at on the positive x-axis when we are to the far right of the phase plane, and a fixed center on the negative x-axis when we are to the far left; this is another piecewise linear system. In the central region, the center of the trajectory moves with x, so we have to draw tiny arcs and keep adjusting the center of our compass. We discover that it is a good idea to choose x_{center} not on the basis of x at the beginning of the arc, but on the average value of x during each little arc. If we had not figured it out before, we will soon notice that $\dot{x}$=constant in that region. Starting at some large positive initial displacement x_o, we draw a quarter-circle with its center at ϵ, continue on a horizontal line for $-\epsilon < x < \epsilon$,.and then start a half-circle around $x_{center} = -\epsilon$. We observe that

- The trajectory returns to the original state (or within the expected graphical error of it), as we should expect from an undamped system. To investigate the entire phase plane, we have to start different trajectories with a range of initial displacements.

- It takes more than 360 degrees of arcs to return to the positive x-axis. The two outer half-circles contribute the first 360 degrees, for an elapsed time of $2\pi/\omega_{ref}$. Adding up the little arcs in the inner region adds an additional time $4\epsilon/v_{\max} = 4\epsilon/x_{\max}\omega_{ref}$. The total period is for $x_{\max} > \epsilon$

is

$$\begin{aligned} \tau &= \frac{2\pi}{\omega_{ref}} + \frac{4\epsilon}{x_{\max}\omega_{ref}} = \left(2\pi + \frac{4\epsilon}{x_{\max}}\right)\sqrt{\frac{m}{kg_c}} \\ f &= \frac{\sqrt{kg_c/m}}{2\pi + 4\epsilon/x_{\max}} \end{aligned}$$

so that ω approaches ω_{ref} for $x_{\max} \gg \epsilon$, but tends toward zero if $x_{\max}$ is reduced towards ϵ.

- This system is "piecewise linear" and the projection of the trajectory on the x-axis is pieced together from sine functions and straight lines.
- If the initial conditions are $-\epsilon < x_o < \epsilon$ and $\dot{x}_o = 0$, there is no energy in the system and it stays stationary.

This example illustrates that non-linear restoring forces cause the natural frequency to be a function of maximum amplitude.

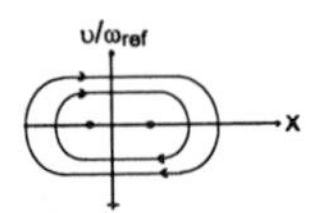

Exercise 36.2 *Investigate the natural frequency of a mass which is held against a support spring by gravity, for amplitudes which include the mass bouncing clear of the spring for part of the cycle.*

Example: Some *autonomous systems* have been modeled by means of a damping term which can be positive or negative, depending on the amplitude

$$\frac{m}{g_c}\ddot{y} + c\left(y^2 - a^2\right)\dot{y} + ky = 0$$

where c has the dimensions N-s/m^3 or lbf-sec/ft^3. As always, the system can be investigated more easily if the number of parameters is reduced by dividing by the first coefficient

$$\ddot{y} + \frac{cg_c}{m}\left(y^2 - a^2\right)\dot{y} + \frac{kg_c}{m}y = 0$$

rescaling time to dimensionless $t^* \triangleq \sqrt{kg_c/m} \cdot t$ and renormalizing to get rid of the last coefficient

$$y'' + \frac{c}{\sqrt{km/g_c}}\left(y^2 - a^2\right)y' + y = 0$$

and rescaling the displacement to $x \triangleq y/a$ to obtain

$$\begin{aligned} x'' + \left(\frac{ca^2}{\sqrt{km/g_c}}\right)\left(x^2 - 1\right)x' + x &= 0 \\ x'' + \mu\left(x^2 - 1\right)x' + x &= 0 \qquad (36.9) \end{aligned}$$

This is the *van der Pol* equation, and contains only the single dimensionless parameter $\mu \triangleq ca^2/\sqrt{km/g_c}$. The damping coefficient is positive when $|x| > 1$ so that we expect the trajectories to spiral inward when we are far out on the right or left of the phase plane. We have negative damping when we are near the vertical axis, so we expect the trajectories to spiral outward in this central region. Letting $\omega_{ref}^2 = 1$ in the t^* time-scale (or kg_c/m in real time t), we apply the phase-plane-delta method

$$\begin{aligned} F(x,v) &= \mu\left(x^2-1\right)v + x \\ f(x,v) &= \mu\left(x^2-1\right)v \\ x_{center} &= \mu\left(1-x^2\right)v \end{aligned}$$

To investigate this equation in a hurry, we start by trying a large value of the parameter, $\mu = 1$, so that the inward- or outward-spiraling will proceed rapidly. We find that

- If we start with a small initial value $x_o > 1$, the trajectory will rapidly spiral away from the center towards a oddly-shaped limit cycle crossing the x-axis near $x = 2$.
- If we start with a large initial value $x_o > 1$, the trajectory will rapidly spiral inward towards the same limit cycle crossing the x-axis near $x = 2$.
- If we increase the value of the system parameter μ, the shape of the limit cycle gets sharper corners; the positive and negative "damping" forces outpace the forces spring and inertia forces and we approach a relaxation cycle.
- If we decrease the value of μ, the limit cycle becomes more nearly circular, and it takes a long time to approach it unless we know to start with $x_o \approx 2$. When μ is very small, approximate algebraic methods (which we will introduce in future chapters) might be more efficient than graphical ones.

Some mathematicians like to think of the origin in a linear problem as an "attractor" keeping the state-variable vector (x, v) within rein (and, in the case of damping, pulling the state variable towards itself); and the limit cycle of an autonomous oscillator as a "strange attractor" for the state of the system.

Exercise 36.3 *Investigate a mechanical system with the governing equation*

$$\frac{m}{g_c}\ddot{y} + c\left(\dot{y}^2 - V^2\right)\dot{y} + ky = 0$$

by discovering the normalizations which will reduce it to the form

$$x'' + \mu\left(\left(x'\right)^2 - 1\right)x' + x = 0$$

and sketching-out the response when the system parameter $\mu = 1$.

Example: As mentioned above, we can investigate equations with excitation by letting F be a function $F(x, v, t)$. For example, the simple pendulum with a horizontally oscillating pivot $x_{\text{pivot}} = X_o \sin(\omega_{ex} t)$ is governed by

$$\begin{aligned}\frac{mL^2}{g_c}\ddot{\theta} + \frac{mgL}{g_c}\sin\theta &= \frac{mL\cos(\theta)}{g_c}\omega_{ex}^2 X_o \sin(\omega_{ex} t) \\ \ddot{\theta} + \left(\frac{g}{L}\right)\sin\theta &= \left(\frac{g}{L}\right)\left(\frac{\omega_{ex}^2 X_o}{g}\right)\cos(\theta)\sin(\omega_{ex} t)\end{aligned}$$

and can be rescaled to $t^* \triangleq \sqrt{g/L}\cdot t$

$$\begin{aligned}\theta'' + \sin\theta &= \left(\frac{\omega_{ex}^2 X_o}{g}\right)\cos(\theta)\sin\left(\frac{\omega_{ex}}{\sqrt{g/L}}t^*\right) \\ \theta'' + \sin\theta &= \xi\cos(\theta)\sin(\rho t^*) \qquad (36.10)\end{aligned}$$

where $\xi \triangleq \omega_{ex}^2 X_o/g$ is a dimensionless measure of the pivot-acceleration input, and $\rho \triangleq \omega_{ex}/\sqrt{g/L}$ is a dimensionless measure of the excitation frequency. Choosing $\omega_{ref} \triangleq \sqrt{g/L}$, the center of our trajectory arcs is

$$x_{center} = -(\sin\theta - \theta - \xi\cos(\theta)\sin(\rho t^*))$$

We can investigate the problem in several stages:

1. The homogeneous-equation solutions ($\xi = 0$) can be surveyed by plotting trajectories from various initial conditions in the phase plane. We discover that there are attractors, around which the trajectories can circle (or oval), located on the θ-axis at $\theta = 0, \pm 2\pi, \pm 4\pi$, etc.; they are separated by "saddle points" at $\theta = \pm\pi, \pm 3\pi$, etc., from which trajectories can proceed either in the "south-west" or "north-east" direction. Infinitesimal differences in θ_o can set off the system around a different attractor: this property of some non-linear systems is called "sensitivity to initial conditions." If we also permit initial velocities, we find that there are solutions in which the system keeps spinning in the one direction or the other: to the right in the upper part of the phase-plane, and vice versa.

2. We can contemplate the effect of adding damping: the trajectories in the upper half-plane will tend to slope downward, and vice versa; sooner or later every trajectory will be captured by an attractor, and spiral towards it. Small differences in initial conditions far away in the upper or lower phase plane, may cause the system to fork one way or the other, and end up with a different attractor. Our ability to predict the final state of a system requires more and more precision if the initial velocity is high; ultimately, we cannot do it. The principle of sensitivity to initial conditions applies here, as it also applies to roulette wheels, and to predicting the weather more than a few days ahead of time.

3. Instead of damping, we can add the excitation in our governing equation. Every time we draw a phase-plane-delta arc, we measure it and calculate the passage of time, so that we can compute the new x_{center} from the new displacement, velocity, and time. If ξ is very small and $\rho \neq 1$, we may have a near-linear response near $\theta = 0$. For larger ξ, the trajectories become more erratic, and may occasionally tip over to the domain of another attractor, at $\theta = \pm 2\pi$. As we plot trajectories, we find that this popping back-and-forth between attractors is "chaotic" and not regular in any statistical sense. For engineers, who make their living by predicting either deterministic or else at least statistical behavior of physical systems, this is disconcerting.

Graphical representations are easy to interpret, and the phase-plane-delta can be applied to a very wide range of problems. However, this last example illustrates how tedious it might become for problems with excitation. At some point, numerical simulation becomes easier.

Problem 36.4 *Investigate the decay behavior of a system with cubic damping*

$$\frac{m}{g_c} \ddot{x} + c \dot{x}^3 + kx = 0$$

Hint: start by writing a dimensionless form of the equation, expressing the dependent parameter

$$\pi_1 \triangleq \frac{x}{x_o}$$

in terms of the independent parameter

$$\pi_2 \triangleq x_o^2 \frac{c \cdot k^{1/2}}{m^{3/2}}$$

Pick a value for π_2 and plot the solution in the phase plane.

Problem 36.5 *Investigate the natural frequency of a system with a softening spring*

$$\frac{m}{g_c} \ddot{x} + k\left(x - \varepsilon x^3\right) = 0$$

Find the independent parameter(s), the dependence of frequency on the amplitude, and the maximum amplitude of stable oscillations.

Problem 36.6 *Investigate the decay of a system in which the spring has a hysteresis loop*

$$\frac{m}{g_c} \ddot{x} + k\left(x + \varepsilon \cdot sgn\left(\dot{x}\right)\right) = 0$$

Sketch the force-displacement curve of the spring, develop the dimensionless parameters, and find the decay envelope.

Problem 36.7 *Investigate the motion of a rotational mass-spring system which is driven by an escapement. There is a moderate amount of linear damping. The escapement adds a constant amount of energy whenever the system passes from negative displacement to positive displacement* θ.

Chapter 37

ANALYTICAL SOLUTION and elliptic integrals

A few non-linear vibrations have well-known solutions in the form of standard functions. Among them is the pendulum and the system with a cubic non-linearity.

37.1 Integration in the Phase Plane

The free vibration of undamped linear system is represented by

$$\frac{m}{g_c}\ddot{x} + f(x) = 0$$

As in the previous chapter, we eliminate time t from the equation by substituting $v \cdot dv/dx$ for $\ddot{x}$; we can then separate variables v and x:

$$\frac{m}{g_c} v dv = -f(x)\, dx$$

For Initial Conditions of $v = v_{(0)}$ and $x = x_{(0)}$, we can integrate both sides of the equation:

$$\frac{m}{g_c}\int_{v_{(0)}}^{v} v dv = -\int_{x_{(0)}}^{x} f(x)dx$$

resulting in

$$\frac{1}{2}\frac{m}{g_c}\left(v^2 - v_{(0)}^2\right) = -\int_{x_{(0)}}^{x} f(x)dx$$

This is a conservation-of-energy statement which can be solved for the state-variable path v as a function of x.

Next, we can reintroduce time by substituting dx/dt for v, and solve for t as a function of x.

37.2 Linear Example

First, we will try out direct integration on a simple problem to which we already know the answer: an undamped linear system represented by

$$\ddot{x} + \omega_n^2 x = 0$$

As in the previous chapter, we eliminate time t from the equation by substituting $v \cdot dv/dx$ for $\ddot{x}$; we can then separate variables v and x:

$$vdv = -\omega_n^2 x dx$$

For Initial Conditions of $v = 0$ and $x = x_{(0)}$, we can integrate both sides of the equation:

$$\int_0^v vdv = -\omega_n^2 \int_{x(0)}^x xdx$$

resulting in

$$\frac{1}{2}v^2 = \omega_n^2 \frac{1}{2}\left(x_{(0)}^2 - x^2\right)$$

Taking the square root,

$$v = \mp\omega_n \sqrt{x_{(0)}^2 - x^2}$$

We will choose the *negative* square root because we expect a negative velocity, decreasing x from the initial displacement $x_{(0)}$. This relationship defines the trajectory in the phase plane, an ellipse with half-axes of $x_{(0)}$ and $x_{(0)} \cdot \omega_{ref}$.

We now reintroduce time by substituting dx/dt for v; we can then separate the variables x and t and integrate both sides:

$$\int_0^t dt = t = \frac{-1}{\omega_n} \int_{x(0)}^x \frac{dx}{\sqrt{x_{(0)}^2 - x^2}}$$

where x within the integral is the dummy variable.

Looking in a table of integrals, we find:

$$\int \frac{d\xi}{\sqrt{a^2 - \xi^2}} = \arcsin\frac{\xi}{a}$$

Therefore,

$$t = \frac{-1}{\omega_n}\left(\arcsin\frac{x}{x_{(0)}} - \arcsin\frac{x_{(0)}}{x_{(0)}}\right)$$
$$\frac{x}{x_{(0)}} = \sin\left(\mp\omega_n t + \frac{\pi}{2}\right) = \cos\left(\omega_n t\right)$$

If we want to know the period of the oscillation, we take four times the time to get from the abscissa to the negative ordinate:

$$t_n = \frac{-4}{\omega_n}\int_{x(0)}^0 \frac{dx}{\sqrt{x_{(0)}^2 - x^2}} = \frac{-4}{\omega_n}(\arcsin 0 - \arcsin 1) = \frac{2\pi}{\omega_n}$$

37.3 Non-Linear Springs

A symmetrically non-linear spring can be modeled by $F_{spring} = k(x + \varepsilon x^3)$; positive ε gives us a hardening spring, and negative ε a softening spring. The undamped vibrational system is represented by

$$\ddot{x} + \omega_{ref}^2(x + \varepsilon x^3) = 0$$

where $\omega_{ref}^2 \equiv kg_c/m$ and ε has the dimensions of L^{-2}. As in the previous chapter, we eliminate time t from the equation by substituting $v \cdot dv/dx$ for $\ddot{x}$; we can then separate variables v and x:

$$vdv = -\omega_{ref}^2(x + \varepsilon x^3)dx$$

For Initial Conditions of $v = 0$ and $x = x_{(0)}$, we can integrate both sides of the equation:

$$\int_0^v vdv = -\omega_{ref}^2 \int_{x_{(0)}}^x (x + \varepsilon x^3)dx$$

resulting in

$$\frac{1}{2}v^2 = \omega_{ref}^2 \left(\frac{1}{2}\left(x_{(0)}^2 - x^2\right) + \frac{1}{4}\varepsilon\left(x_{(0)}^4 - x^4\right) \right)$$

Factoring out $x_{(0)}^2 - x^2$ and taking the square root,

$$v = \mp\omega_{ref}\sqrt{\left(x_{(0)}^2 - x^2\right)}\sqrt{1 + \frac{\varepsilon}{2}\left(x_{(0)}^2 + x^2\right)}$$

We will choose the *negative* square root because we expect x to decrease from the initial displacement $x_{(0)}$. This relationship defines the trajectory in the phase plane for the first fourth of the oscillatory cycle; the other quadrants will be symmetrical. (If ε were zero, it would define an ellipse with half-axes of $x_{(0)}$ and $x_{(0)} \cdot \omega_{ref}$.)

We now reintroduce time by substituting dx/dt for v; we can then separate the variables x and t and integrate both sides:

$$\int_0^t dt = t = \frac{-1}{\omega_{ref}} \int_{x_{(0)}}^x \frac{dx}{\sqrt{\left(x_{(0)}^2 - x^2\right)}\sqrt{1 + \frac{\varepsilon}{2}\left(x_{(0)}^2 + x^2\right)}}$$

where x within the integral is the dummy variable.

We recognize that this resembles an Elliptic Integral of the First Kind:

$$F = \int_0^x \frac{d\xi}{\sqrt{1-\xi^2}\sqrt{1-\lambda^2\xi^2}} = \int_0^\varphi \frac{d\psi}{\sqrt{1-\lambda^2\sin^2\psi}}$$

We can put our integral into the first of these forms by writing $\xi \equiv$

We can put our integral into the second form of the Elliptic Integral of the First Kind by substituting $x_{(0)} \cos\psi$ for the dummy variable x, and $x_{(0)} \cos\varphi$ for the final value of x in the upper limit of the integration. (This expression resembles the solution for the linear problem $x = x_{(0)} \cos \omega_n t$.)

$$\begin{aligned} t &= \frac{-1}{\omega_{ref}} \int_{x_{(0)}\cos\psi = x_{(0)}}^{x_{(0)}\cos\psi = x} \frac{d\left(x_{(0)}\cos\psi\right)}{\sqrt{\left(x_{(0)}^2 - \left(x_{(0)}\cos\psi\right)^2\right)}\sqrt{1 + \frac{\varepsilon}{2}\left(x_{(0)}^2 + \left(x_{(0)}\cos\psi\right)^2\right)}} \\ &= \frac{-1}{\omega_{ref}} \int_{\psi=\arccos 1}^{\psi=\arccos\left(x/x_{(0)}\right)} \frac{-x_{(0)}\sin\psi d\psi}{x_{(0)}\sqrt{1-\cos^2\psi}\sqrt{1+\frac{\varepsilon}{2}x_{(0)}^2\left(1+\cos^2\psi\right)}} \\ &= \frac{1}{\omega_{ref}} \int_{\psi=0}^{\psi=\varphi} \frac{d\psi}{\sqrt{1+\frac{\varepsilon}{2}x_{(0)}^2 + \frac{\varepsilon}{2}x_{(0)}^2\cos^2\psi}} \\ &= \frac{1}{\omega_{ref}} \int_0^{\varphi} \frac{d\psi}{\sqrt{\left(1+\frac{\varepsilon}{2}x_{(0)}^2\right)\left(1+\frac{\frac{\varepsilon}{2}x_{(0)}^2}{1+\frac{\varepsilon}{2}x_{(0)}^2}\cos^2\psi\right)}} \end{aligned}$$

where $\varphi = \arccos(x/x_{(0)})$. Substituting $\lambda^2 \equiv \varepsilon x_{(0)}^2/(2 + \varepsilon x_{(0)}^2)$

$$t = \frac{1}{\omega_{ref}\sqrt{\left(1+\frac{\varepsilon}{2}x_{(0)}^2\right)}} \int_0^{\varphi} \frac{d\psi}{\sqrt{1+\lambda^2\sin^2\psi}}$$

which is the standard form for the incomplete Elliptic Integral if ε is a negative number—in other words, for a softening spring. The values are listed as the elliptic cosine $cn(\varphi, \lambda))$ in tables of Jacobian elliptic functions. The period follows from the complete Elliptic Integral

$$t_n = \frac{4}{\omega_{ref}\sqrt{\left(1+\varepsilon x_{(0)}^2\right)}} \int_0^{\pi/2} \frac{d\psi}{\sqrt{1-\lambda^2\sin^2\psi}}$$

37.4 Pendulum

A simple pendulum can be modeled by

$$\ddot{\theta} + \omega_{ref}^2 \sin x = \theta$$

where $\omega_{ref}^2 \equiv g/L$. In our usual linearization, we truncate to only the first term of the expansion $\sin\theta = \theta - \theta^3/6 + \theta^5/120 - \ldots$; from the previous section, we could also keep the second term, making $\varepsilon = -1/6$; now we will try to get an exact solution. As in the previous chapter, we eliminate time t from the equation by substituting $v \cdot dv/dx$ for $\ddot{x}$; we can then separate variables v and x:

$$vdv = -\omega_{ref}^2 \sin(x)dx$$

For Initial Conditions of $v = 0$ and $x = x_{(0)}$, we can integrate both sides of the equation:

$$\int_0^v vdv = -\omega_{ref}^2 \int_{x_{(0)}}^x \sin(x)dx$$

resulting in

$$v^2 = 2\omega_{ref}^2 \left(\cos x - \cos x_{(0)}\right)$$

Taking the square root,

$$v = \mp\omega_{ref}\sqrt{2\left(\cos x - \cos x_{(0)}\right)}$$

We will choose the *negative* square root because we expect x to decrease from the initial displacement $x_{(0)}$. This relationship defines the trajectory in the phase plane for the first fourth of the oscillatory cycle; the other quadrants will be symmetrical. (If ε were zero, it would define an ellipse with half-axes of $x_{(0)}$ and $x_{(0)} \cdot \omega_{ref}$.)

We now reintroduce time by substituting dx/dt for v; we can then separate the variables x and t and integrate both sides:

$$\int_0^t dt = t = \frac{-1}{\omega_{ref}} \int_{x_{(0)}}^x \frac{dx}{\sqrt{2\left(\cos x - \cos x_{(0)}\right)}}$$

where x within the integral is the dummy variable.

Exercise 37.1 *Using elliptical functions in a mathematics reference, plot the frequency of a pendulum as a function of maximum amplitude, relative to the small-amplitude value.*

Chapter 38

PSEUDO-LINEARIZATION and equivalent damping

If the non-linearities and the damping are both small, we can look for the sinusoidal solution which most nearly fits the governing equation throughout one cycle.

38.1 Krylov-Bogoliubov's First Approximation

A very general normalized equation for a non-linear system is

$$\ddot{x} + F(x, v) = 0 \tag{38.1}$$

where $F(x, v)$ may encompass spring-like dependence on displacement x, damping-like dependence on v, or a combination of these. We can make Equation 38.1 *look* like a linear equation by arbitrarily *choosing* a nominal reference frequency ω_{ref}, and rewriting it as

$$\ddot{x} + \omega_{ref}^2 x + \mu f(x, v) = 0 \tag{38.2}$$

where $\mu f(x, v) \equiv F(x, v) - \omega_{ref}^2 x$, a small number if the non-linearities and the damping are both small and if the reference frequency ω_{ref} is well chosen. (You may let $\mu = 1$; we have introduced it in order to be able to look at the dependence of the solution on the magnitude of this term.) If $\mu f(x, v)$ is indeed small relative to the other terms (or at least nearly constant), it may be treated as a slight modification of a linear equation. The solution of the linear equation would be

$$\begin{aligned} x &= A\cos\left(\omega_{ref}t+\theta\right) \\ v &= \dot{x} = -\omega_{ref}A\sin\left(\omega_{ref}t+\theta\right) \end{aligned}$$

We will assume that the solution to the non-linear equation has this same form, except that A and θ are not quite constant, but are slowly varying functions of time t:

$$\begin{aligned} x &= A_{(t)}\cos\left(\omega_{ref}t+\theta_{(t)}\right) \\ v &= -\omega_{ref}A_{(t)}\sin\left(\omega_{ref}t+\theta_{(t)}\right) \end{aligned} \tag{38.3}$$

Whatever the solution to the non-linear equation is, we can force this solution to fit it by adjusting the amplitude $A_{(t)}$ or the phase angle $\theta_{(t)}$. As a matter of fact, we have too many means of adjustment, because either one of these functions of time can do the job (except at the maxima of x, where only $A_{(t)}$ will work, and at the zero-values of x, where only $\theta_{(t)}$ will work). Therefore we can place another constraint on the magnitudes of $A_{(t)}$ and $\theta_{(t)}$: we note that the derivative of $x = A_{(t)}\cos\left(\omega_{ref}t+\theta_{(t)}\right)$ is:

$$\dot{x} = \frac{dA}{dt}\cos\left(\omega_{ref}t+\theta_{(t)}\right) - \left(\omega_{ref}+\frac{d\theta}{dt}\right)A_{(t)}\sin\left(\omega_{ref}t+\theta\right)$$

We would like this to be equal to our expression $v = -\omega_{ref}A_{(t)}\sin\left(\omega_{ref}t+\theta_{(t)}\right)$; this is only possible if

$$\frac{dA}{dt}\cos\left(\omega_{ref}t+\theta\right) - \left(\omega_{ref}+\frac{d\theta}{dt}\right)A\sin\left(\omega_{ref}t+\theta\right) = -\omega_{ref}A\sin\left(\omega_{ref}t+\theta\right)$$

which reduces to

$$\frac{dA}{dt}\cos\left(\omega_{ref}t+\theta_{(t)}\right) - \left(\frac{d\theta}{dt}\right)A_{(t)}\sin\left(\omega_{ref}t+\theta_{(t)}\right) = 0 \tag{38.4}$$

We would like our postulated solution Equation 38.3 to satisfy the governing Equation 38.2 and this constraint Equation 38.4. For the former, we need

$$\ddot{x} = \dot{v} = -\omega_{ref}\frac{dA}{dt}\sin\left(\omega_{ref}t+\theta_{(t)}\right) - \left(\omega_{ref}+\frac{d\theta}{dt}\right)\omega_{ref}A_{(t)}\cos\left(\omega_{ref}t+\theta_{(t)}\right)$$

Writing Φ for $\omega_{ref}t+\theta_{(t)}$, we now have our solution terms

$$\begin{aligned} x &= A_{(t)}\cos\Phi \\ v &= -\omega_{ref}A_{(t)}\sin\Phi \\ \ddot{x} &= -\omega_{ref}\frac{dA}{dt}\sin\Phi - \left(\omega_{ref}+\frac{d\theta}{dt}\right)\omega_{ref}A_{(t)}\cos\Phi \\ \mu f(x,v) &= \mu f\left(A_{(t)}\cos\Phi, -\omega_{ref}A_{(t)}\sin\Phi\right) \end{aligned}$$

Substituting into Equation 38.2 we get

$$-\omega_{ref}\tfrac{dA}{dt}\sin\Psi - \left(\omega_{ref}+\tfrac{d\theta}{dt}\right)\omega_{ref}A_{(t)}\cos\Psi + \omega_{ref}^2 A_{(t)}\cos\Psi$$
$$+\mu f\left(A_{(t)}\cos\Psi, -\omega_{ref}A_{(t)}\sin\Psi\right) = 0$$

which simplifies to

$$-\omega_{ref}\frac{dA}{dt}\sin\Phi - \left(\frac{d\theta}{dt}\right)\omega_{ref}A_{(t)}\cos\Phi = -\mu f\left(A_{(t)}\cos\Phi, -\omega_{ref}A_{(t)}\sin\Phi\right)$$

Together with Equation 38.4, we now have two equations for the derivatives of A and θ:

$$\left(-\omega_{ref}\sin\Phi\right)\tfrac{dA}{dt} + \left(-\omega_{ref}A_{(t)}\cos\Phi\right)\tfrac{d\theta}{dt} = -\mu f\left(A_{(t)}\cos\Phi, -\omega_{ref}A_{(t)}\sin\Phi\right)$$
$$\left(\cos\Phi\right)\tfrac{dA}{dt} + \left(-A_{(t)}\sin\Phi\right)\tfrac{d\theta}{dt} = 0$$

which we can separate into

$$\frac{dA}{dt} = \frac{\mu}{\omega_{ref}}\cdot f\left(A_{(t)}\cos\Phi, -\omega_{ref}A_{(t)}\sin\Phi\right)\cdot\sin\Phi$$
$$\frac{d\theta}{dt} = \frac{\mu}{\omega_{ref}A_{(t)}}\cdot f\left(A_{(t)}\cos\Phi, -\omega_{ref}A_{(t)}\sin\Phi\right)\cdot\cos\Phi$$

If we satisfied these equations, we would have an exact solution; but we are going to solve them only approximately by assuming that the derivatives of A and θ are steady throughout one cycle of Ψ going from zero to 2π. The optimal results for steady values come from

$$\frac{dA}{dt} \cong \frac{\mu}{\omega_{ref}}\cdot\frac{1}{2\pi}\int_0^{2\pi} f\left(A\cos\Phi, -\omega_{ref}A\sin\Phi\right)\cdot\sin\Phi d\Phi$$
$$\frac{d\theta}{dt} \cong \frac{\mu}{\omega_{ref}A}\cdot\frac{1}{2\pi}\int_0^{2\pi} f\left(A\cos\Phi, -\omega_{ref}A\sin\Phi\right)\cdot\cos\Phi d\Phi$$

Because A and θ are assumed to vary slowly, we approximate A (and θ) by a constant value on the right side of each equation. Although it is paradoxical to assume values for the derivatives of A and θ on the left side of the equation, and an unchanging value on the right side, we can justify it by showing that the integrals are weak functions of the steady drift in the values of A and θ.

Solving these integrals, we obtain the rate of decay of the oscillation dA/dt, as well as the frequency of oscillations $(\omega_{ref} + d\theta/dt)$.

38.2 Non-Linear Springs

A non-linear spring might have an element law approximated by $F_{spring} = kx \pm x^3$, where a plus-sign would represent a hardening spring, and a minus-sign, a softening spring. In terms of Equation 38.2, this leads to $\mu f(x,v) = \pm\mu x^3$, where $\mu = \epsilon g_c/m$. Inserting this into our integrals with $x = A\cos\Phi$,

$$\begin{aligned}\frac{dA}{dt} &\cong \frac{\mu}{\omega_{ref}} \cdot \frac{1}{2\pi}\int_0^{2\pi} \left((\pm A\cos\Phi)^3\right) \cdot \sin\Phi d\Phi \\ \frac{d\theta}{dt} &\cong \frac{\mu}{\omega_{ref}A} \cdot \frac{1}{2\pi}\int_0^{2\pi} \left((\pm A\cos\Phi)^3\right) \cdot \cos\Phi d\Phi\end{aligned}$$

which leads to $dA/dt = 0$;

$$\begin{aligned}\frac{d\theta}{dt} &\cong \frac{\pm\mu A^2}{\omega_{ref}} \cdot \frac{1}{2\pi} \cong \frac{\mu A^2}{\omega_{ref}} \cdot \frac{1}{2\pi}\int_0^{2\pi} \cos^4\Phi d\Phi \\ &\cong \frac{\pm\mu A^2}{\omega_{ref}} \cdot \frac{1}{2\pi}\left[\frac{3\Phi}{8} + \frac{\sin 2\Phi}{4} + \frac{\sin 4\Phi}{32}\right]_0^{2\pi} \\ &\cong \frac{\pm\mu A^2}{\omega_{ref}} \cdot \frac{3}{8}\end{aligned}$$

so that the actual frequency of free oscillations is $\omega_{ref} + \frac{3}{8}\frac{\mu A^2}{\omega_{ref}}$ for the hardening spring, and $\omega_{ref} - \frac{3}{8}\frac{\mu A^2}{\omega_{ref}}$ for the softening spring.

More generally, a non-linear spring might be represented by $F_{spring} = kx + \epsilon\,|x|^n\,sgn(x)$. In terms of Equation 38.2, this leads to $\mu f(x,v) = \mu\,|x|^n \text{sgn}(x)$, where $\mu = \epsilon g_c/m$. For hardening springs, ϵ and μ are positive, for softening springs, negative. Inserting this into our integrals with $x = A\cos\Phi$,

$$\begin{aligned}\frac{dA}{dt} &\cong \frac{\mu}{\omega_{ref}} \cdot \frac{1}{2\pi}\int_0^{2\pi} (|A\cos\Phi|^n\,\text{sgn}(A\cos\Phi)) \cdot \sin\Phi d\Phi \\ \frac{d\theta}{dt} &\cong \frac{\mu}{\omega_{ref}A} \cdot \frac{1}{2\pi}\int_0^{2\pi} (|A\cos\Phi|^n\,\text{sgn}(A\cos\Phi)) \cdot \cos\Phi d\Phi \\ &\cong \frac{\mu}{\omega_{ref}A} \cdot |A|^n \cdot \frac{2}{\pi}\int_0^{\pi/2} \cos^{n+1}\Phi d\Phi\end{aligned}$$

which leads to $dA/dt = 0$ and

$$\begin{aligned}\text{for } n &= 0,\ \frac{d\theta}{dt} \cong 0 \\ \text{for } n &= 1,\ \frac{d\theta}{dt} \cong \frac{\mu}{2\omega_{ref}} \\ \text{for } n &= 2,\ \frac{d\theta}{dt} \cong \frac{4\mu A}{3\pi\omega_{ref}} \\ \text{for } n &= 3,\ \frac{d\theta}{dt} \cong \frac{3\mu A^2}{8\omega_{ref}}\end{aligned}$$

Remembering that μ and ϵ can be positive or negative, we see that the frequency of free vibration of the linear system $n = 1$ is approximately

$$\omega = \left(\omega_{ref} \pm \frac{\mu}{2\omega_{ref}}\right) = \sqrt{\frac{kg_c}{m}} \pm \frac{\epsilon}{2\sqrt{km/g_c}}$$

which is the first two terms of the exact solution

$$\omega = \sqrt{\frac{(k \pm \epsilon)\, g_c}{m}} = \sqrt{\frac{kg_c}{m}\left(1 \pm \frac{\epsilon}{k}\right)} = \sqrt{\frac{kg_c}{m}}\left(1 \pm \frac{1}{2}\frac{\epsilon}{k} + \frac{1}{6}\left(\frac{\epsilon}{k}\right)^2 \pm \ldots\right)$$

confirming the first-order-approximation nature of the Krylov-Bogoliubov method.

The general result we have obtained is that the frequency of oscillation of hardening springs increases with amplitude; for softening springs it decreases with amplitude—and at some amplitude it may even go to zero, indicating that an unstable point on the phase plane is reached.

The method we have used can be adapted to any spring characteristic that we can express in a polynomial or in some other function, but works best if that characteristic consists of a small modification of a linear element law.

38.3 Non-Linear Damping

A non-linear damper might have an element law approximated by $F_{damper} = \epsilon v^3$. In terms of Equation 38.2, this leads to $\mu f(x, v) = \mu v^3$, where $\mu = \epsilon g_c/m$. Inserting this into our integrals with $v = -\omega_{ref} A \sin\Phi$,

$$\begin{aligned}
\frac{dA}{dt} &\cong \frac{\mu}{\omega_{ref}} \cdot \frac{1}{2\pi}\int_0^{2\pi} \left((-\omega_{ref} A \sin\Phi)^3\right) \cdot \sin\Phi d\Phi \\
\frac{d\theta}{dt} &\cong \frac{\mu}{\omega_{ref} A} \cdot \frac{1}{2\pi}\int_0^{2\pi} \left((-\omega_{ref} A \sin\Phi)^3\right) \cdot \cos\Phi d\Phi
\end{aligned}$$

which leads to

$$\begin{aligned}
\frac{dA}{dt} &\cong -\mu\omega_{ref}^2 A^3 \cdot \frac{1}{2\pi}\int_0^{2\pi} \sin^4\Phi d\Phi \\
&\cong -\mu\omega_{ref}^2 A^3 \cdot \frac{1}{2\pi}\left[\frac{3\Phi}{8} - \frac{\sin 2\Phi}{4} + \frac{\sin 4\Phi}{32}\right]_0^{2\pi} \\
&\cong -\mu\omega_{ref}^2 A^3 \cdot \frac{3}{8}
\end{aligned}$$

and $d\theta/dt \cong 0$; this first-order approach does not capture the slight influence of damping on the period of free oscillation.

More generally, a non-linear damper might be represented by

$$F_{damper} = \epsilon \left|v\right|^n sgn(v)$$

In terms of Equation 38.2, this leads to

$$\mu f(x,v) = \epsilon |v|^n \, sgn(v)$$

where $\mu = \epsilon g_c/m$. Inserting this into our integrals

$$\begin{aligned}
\frac{dA}{dt} &\cong \frac{\mu}{\omega_{ref}} \cdot \frac{1}{2\pi}\int_0^{2\pi} \left(|-\omega_{ref} A \sin\Phi|^n \operatorname{sgn}(-\omega_{ref} A \sin\Phi)\right) \cdot \sin\Phi d\Phi \\
&\cong -\mu\omega_{ref}^{n-1} \cdot |A|^n \cdot \frac{1}{\pi}\int_0^{\pi} \sin^{n+1}\Phi d\Phi \\
\frac{d\theta}{dt} &\cong \frac{\mu}{\omega_{ref} A} \cdot \frac{1}{2\pi}\int_0^{2\pi} \left(|-\omega_{ref} A \sin\Phi|^n \operatorname{sgn}(-\omega_{ref} A \sin\Phi)\right) \cdot \cos\Phi d\Phi
\end{aligned}$$

which yields $d\theta/dt = 0$, so that free-oscillation $\omega \cong \omega_{ref}$ and

$$\begin{aligned}
\text{for Coulomb damping} &: \quad n = 0, \ \frac{dA}{dt} \cong \frac{-2\mu}{\pi\omega} \\
\text{for linear damping} &: \quad n = 1, \ \frac{dA}{dt} \cong \frac{-1}{2}\mu A \\
\text{for quadratic damping} &: \quad n = 2, \ \frac{dA}{dt} \cong \frac{-4}{3\pi}\mu A^2 \omega \\
\text{for cubic damping} &: \quad n = 3, \ \frac{dA}{dt} \cong \frac{-3}{8}\mu A^3 \omega^2
\end{aligned}$$

The Coulomb damping result is exactly the result we obtained before. We can compare the second of these with the exact solution for linear damping, where

$$\omega = \sqrt{1 - (\mu/2\omega_{ref})^2}\,\omega_{ref} \cong \omega_{ref}$$

and

$$dA/dt = Ae^{-\mu t/2} = A\left(1 - (\mu/2)\,t + \left(\mu^2/8\right)t^2 - \cdots\right) \cong A - (\mu A/2)\,t$$

Exercise 38.1 *Express these relations in terms of the logarithmic decrement*

$$\delta = \ln \frac{A - \frac{\pi}{\omega} \cdot \frac{dA}{dt}}{A + \frac{\pi}{\omega} \cdot \frac{dA}{dt}}$$

Exercise 38.2 *What is the value of dA/dt for $F_{damper} = \alpha v + \beta v^2 sgn(v)$?*

38.4 Self-Sustaining Oscillations

If the damping is positive for large amplitudes of x or v, but negative for small amplitudes, it is possible to get a limit cycle rather than decay to a point. For example, we might have a problem described by the van der Pol equation

$$\frac{m}{g_c}\ddot{x} + c\left(x^2 - a^2\right)\dot{x} + kx = 0$$

which we can reduce to the dimensionless form

$$x'' + \mu \left(x^2 - 1\right) x' + x = 0$$

We now can solve all equations which can be reduced to this form once-and-for-all. Applying the Krylov-Bogoliubov First Approximation, we let $\mu = \varepsilon$ and

$$\begin{aligned} f(x,v) &= \left(x^2-1\right)v = \left(A^2\cos^2\Phi - 1\right)\left(-\omega_{ref}A\right)\sin\Phi \\ &= -\omega_{ref}A^3\sin\Phi\cos^2\Phi + \omega_{ref}A\sin\Phi \end{aligned}$$

Inserting this into the Krylov-Bogoliubov Equations,

$$\begin{aligned} \left(\frac{dA}{dt}\right)_{avg} &\cong \frac{\mu}{\omega_{ref}} \cdot \frac{1}{2\pi}\int_0^{2\pi} \left(-\omega_{ref}A^3\sin^2\Phi\cos^2\Phi + \omega_{ref}A\sin^2\Phi\right) d\Phi \\ &\cong \frac{\mu}{\omega_{ref}} \cdot \frac{1}{2\pi}\left(-\omega_{ref}A^3\frac{2\pi}{8} + \omega_{ref}A\frac{2\pi}{2}\right) \cong \mu\left(\frac{-A^3}{8} + \frac{A}{2}\right) \\ \left(\frac{d\theta}{dt}\right)_{avg} &\cong \frac{\mu}{\omega_{ref}A} \cdot \frac{1}{2\pi}\int_0^{2\pi} \left(-\omega_{ref}A^3\sin\Phi\cos^3\Phi + \omega_{ref}A\sin\Phi\cos\Phi\right) d\Phi \\ &\cong 0 \end{aligned}$$

We see that the vibrational frequency $\omega_{ref} + d\theta/dt$ is unchanged (in the first approximation), and the $dA/dt = 0$ when $A = 0$ (an unstable point) or when $A^2 = 4$. Converting back to the original coordinates, the limit cycle has an amplitude of $y_{\max} = a \cdot x_{\max} = 2a$, and the angular frequency of the self-sustaining oscillations is $\omega = \sqrt{kg_c/m}$. All this is a first approximation, applicable to small ε; to handle larger non-linearities, we would want look at second- and third-order effects.

38.5 Forced KBM

Krylov-Bogoliubov can be used for steady-state periodic excitation if we include the function of time, the so-called KBM method. In practice, this means restricting ourselves to response at the excitation frequency and solving for the equilibrium amplitude

$$\frac{dA}{dt} = 0 \cong \frac{\mu}{\omega_{ex}} \cdot \frac{1}{2\pi}\int_0^{2\pi} f\left(A\cos\Phi, -\omega_{ex}A\sin\Phi, t\right) \cdot \sin\Phi d\Phi$$

where the dummy variable Φ corresponds to $\omega_{ex}t+\theta$. The problem that this creates is that the time function $t = \left(\Phi - \theta\right)/\omega_{ex}$, where θ is unknown. Therefore we have to solve this equation simultaneously with one for $d\theta/dt = 0$. Trying this on for the linear-damping problem

$$\ddot{x} + \left(\left(2\zeta\omega_n\right)\dot{x} + \left(\omega_n^2\right)x - \left(\frac{F_o}{k}\right)\left(\omega_n^2\right)\sin(\omega_{ex}t)\right) = 0$$

we arrange it as

$$\ddot{x} + \omega_{ex}^2 x + \mu f(x, v, t) = 0$$

where

$$\mu f(x, v, t) = (2\zeta\omega_n) v + \left(\omega_n^2 - \omega_{ex}^2\right) x - \left(\frac{F_o}{k}\right) \left(\omega_n^2\right) \sin(\omega_{ex} t)$$

Therefore, the steady-state solution requires

$$\begin{aligned}
\frac{dA}{dt} &= 0 \cong \frac{1}{\omega_{ex}} \cdot \frac{1}{2\pi} \int_0^{2\pi} \begin{pmatrix} (2\zeta\omega_n)(-\omega_{ref} A \sin\Phi) \\ + \left(\omega_n^2 - \omega_{ex}^2\right)(A\cos\Phi) \\ - \left(\frac{F_o}{k}\right)\left(\omega_n^2\right)\sin(\Phi - \theta) \end{pmatrix} \cdot \sin\Phi d\Phi \\
&\cong \frac{1}{\omega_{ex}} \cdot \frac{1}{2\pi} \int_0^{2\pi} \begin{pmatrix} (2\zeta\omega_n)(-\omega_{ex} A \sin\Phi) \\ + \left(\omega_n^2 - \omega_{ex}^2\right)(A\cos\Phi) \\ - \left(\frac{F_o}{k}\right)\left(\omega_n^2\right)(\sin\Phi\cos\theta - \cos\Phi\sin\theta) \end{pmatrix} \cdot \sin\Phi d\Phi \\
&\cong \frac{1}{\omega_{ex}} \cdot \frac{1}{2\pi} \left(- (2\zeta\omega_n)(\omega_{ex} A) - \left(\frac{F_o}{k}\right)\left(\omega_n^2\right)(\cos\theta) \right) \cdot \int_0^{2\pi} \sin^2\Phi d\Phi
\end{aligned}$$

and

$$\begin{aligned}
\frac{d\theta}{dt} &= 0 \cong \frac{1}{\omega_{ex} A} \frac{1}{2\pi} \int_0^{2\pi} \begin{pmatrix} (2\zeta\omega_n)(-\omega_{ex} A \sin\Phi) \\ + \left(\omega_n^2 - \omega_{ex}^2\right)(A\cos\Phi) \\ - \left(\frac{F_o}{k}\right)\left(\omega_n^2\right)(\sin\Phi\cos\theta - \sin\Phi\cos\theta) \end{pmatrix} \cdot \cos\Phi d\Phi \\
&\cong \frac{1}{\omega_{ex} A} \cdot \frac{1}{2\pi} \left(\left(\omega_n^2 - \omega_{ex}^2\right) A + \left(\frac{F_o}{k}\right)\left(\omega_n^2\right)\sin\theta \right) \cdot \int_0^{2\pi} \cos^2\Phi d\Phi
\end{aligned}$$

so that

$$A = -\left(\frac{F_o}{k}\right)\left(\frac{\omega_n}{2\zeta\omega_{ex}}\right)\cos\theta = -\left(\frac{F_o}{k}\right)\left(\frac{\omega_n^2}{\omega_n^2 - \omega_{ex}^2}\right)\sin\theta$$

and

$$\cot\theta = \frac{1 - \frac{\omega_{ex}^2}{\omega_n^2}}{2\zeta\frac{\omega_{ex}}{\omega_n}}$$

Problem 38.3 *Investigate the equation*

$$x'' + \mu\left(\left(x'\right)^2 - 1\right)x' + x = 0$$

for small values of μ. Find the limit-cycle amplitude by finding the condition where dA/dt averages zero.

Chapter 39

SERIES EXPANSIONS and subharmonics

39.1 Perturbation Method

The classical method for solving a non-linear equation is to express the solution as a linear solution with corrective terms. Consider the pendulum problem

$$\ddot{\theta} + \frac{g}{l} \sin\theta = 0$$

Remembering the Maclaurin series expansion of $\sin\theta = \theta - \theta^3/6 + \theta^5/120 - \ldots$, we can write

$$\ddot{\theta} + \omega_o^2\theta + \mu \cdot F(\theta) = 0$$

where the non-linearity is

$$F(\theta) = \sin\theta - 1 = -\theta^3/6 + \theta^5/120 - \ldots$$

and μ is the perturbation parameter which gives us the opportunity of introducing the non-linearity in small doses. Let the solution be expressed in the form

$$\theta = \theta_o(t) + \mu \cdot \theta_1(t) + \mu^2 \cdot \theta_2(t) + \ldots$$

We start with an initial-displacement solution without the non-linearity

$$\theta_o = A_o \cos\omega t$$

where ω is itself a function of the amplitude and the non-linearity

$$\omega^2 = \omega_o^2 + \mu b_1(A) + \mu^2 b_2(A) + \ldots$$

If we use these to substitute in the governing equation,

$$\begin{gathered}\left(\ddot{\theta}_o(t) + \mu\cdot \ddot{\theta}_1(t) + \mu^2\cdot \ddot{\theta}_2(t) +\right) \\ + \left(\omega^2 - \mu b_1(A) - \mu^2 b_2(A) + \ldots\right)\left(\theta_o(t) + \mu\cdot\theta_1(t) + \mu^2\cdot\theta_2(t)+\right) \\ +\mu(\sin\theta - 1) = 0\end{gathered}$$

which we can sort out by the order of μ

$$\begin{aligned}\left(\ddot{\theta}_o(t)\right) + \left(\omega^2\right)\left(\theta_o(t)\right) &= 0 \\ \mu\left(\left(\ddot{\theta}_1(t)\right) + \left(\omega^2\cdot\theta_1(t) - b_1(A)\cdot\theta_o(t)\right) + (\sin\theta - 1)\right) &= 0 \\ \mu^2\left(\left(\ddot{\theta}_2(t)\right) + \left(\omega^2\cdot\theta_2(t) - b_1(A)\cdot\theta_1(t) - b_2(A)\cdot\theta_o(t)\right)\right) &= 0\end{aligned}$$

and solve the first equation, etc.

As we add more terms, the algebra becomes overwhelming. However, we can investigate some basic problems of harmonics using only a few terms.

39.2 Duffing's Equation

When we have non-linear response, we get harmonic distortions of the input in the output. If we solve for the solution of

$$\dot{x} + ax + bx^3 = f\cdot\cos\omega_{ex}t$$

we know from our perturbation solutions that harmonic terms of the form $\cos 3\omega_{ex}t$ will appear. By reciprocity, can we also have subharmonics?

39.3 Subharmonics

To examine the possibility of subharmonics, let us examine the postulated solution

$$x = A_{sub}\cos\frac{\omega_{ex}t}{3} + A_1\cos\omega_{ex}t$$

$$x = -A_{sub}\frac{\omega_{ex}^2}{9}\cos\frac{\omega_{ex}t}{3} - A_1\omega_{ex}^2\cos\omega_{ex}t$$

$$\begin{aligned}x^3 &= \begin{pmatrix} A_{sub}^3\cos^3\frac{\omega_{ex}t}{3} \\ +3A_{sub}^2A_1\cos^2\frac{\omega_{ex}t}{3}\cos\omega_{ex}t \\ +3A_{sub}A_1^2\cos\frac{\omega_{ex}t}{3}\cos^2\omega_{ex}t \\ +A_1^3\cos^3\omega_{ex}t\end{pmatrix} \\ &= \begin{pmatrix}\frac{3}{4}\left(A_{sub}^2 + A_{sub}A_1 + 2A_1^2\right)A_{sub}\cos\frac{\omega_{ex}t}{3} \\ +\frac{1}{4}\left(A_{sub}^3 + 6A_{sub}^2A_1 + 3A_1^3\right)\cos\omega_{ex}t\end{pmatrix}\end{aligned}$$

Substituting this solution and its second derivative into the governing equation, we get

$$\left(\begin{array}{c} -A_{sub}\frac{\omega_{ex}^2}{9}\cos\frac{\omega_{ex}t}{3} + aA_{sub}\cos\frac{\omega_{ex}t}{3} + b\ldots \\ -A_1\omega_{ex}^2\cos\omega_{ex}t + aA_1\cos\omega_{ex}t + b\ldots \end{array} \right) = \left(\begin{array}{c} 0+ \\ f\cdot\cos\omega_{ex}t \end{array} \right)$$

so that we have two algebraic equations to satisfy:

$$\begin{aligned} \left(\left(a - \frac{1}{9}\omega_{ex}^2\right) + \frac{3}{4}b\left(A_{sub}^2 + A_{sub}A_1 + 2A_1^2\right)\right)A_{sub} &= 0 \\ \left(\left(a - \omega_{ex}^2\right)A_1 + \frac{1}{4}b\left(A_{sub}^3 + 6A_{sub}^2A_1 + 3A_1^3\right)\right) &= f \end{aligned}$$

If $A_{sub} = 0$, the first equation is satisfied and the second requires $\left(a - \omega_{ex}^2\right)A_1 + \frac{3}{4}bA_1^3 = f$. If $A_{sub} \neq 0$, we eliminate ω_{ex}between the two equations and get

$$A_{sub}^3 - A_1\left(21A_{sub}^2 + 27A_{sub}A_1 + 51A_1^2 + 32\frac{a}{b}\right) = \frac{f}{b}$$

which we solve for A$_{sub}$

$$A_{sub} = \frac{-A_1}{2} \pm \sqrt{\frac{4}{27b}\left(\omega_{ex}^2 - 9a\right) - \frac{7}{4}A_1^2}$$

where A_{sub} must be a real number, so the conditions of its existence must include

$$\begin{aligned} \frac{4}{27b}\left(\omega_{ex}^2 - 9a\right) &\geq \frac{7}{4}A_1^2 \\ \omega_{ex}^2 &\geq 9\left(a + \frac{21}{16}A_1^2b\right) \end{aligned}$$

The equality is where the subharmonic just can exist; at that point, then,

$$\omega_{ex}^2 \approx 9a$$

and

$$A_{sub} = \frac{-A_1}{2} \pm 0$$

which we can introduce into our polynomial to find A_1 as a function of f:

$$\begin{aligned} \frac{343}{32}bA_1^3 + 8aA_1 + f &= 0 \\ A_1 &\approx f/8a \end{aligned}$$

Thus we find that a subharmonic oscillation is plausible when the subharmonic coincides with "resonance." What we don't know is whether this solution is stable.

39.4 Stability

Let us recast the subharmonic solution by writing

$$\ddot{x} + ax + bx^3 = f \cdot \cos 3\omega_{resp} t$$

which we know to have a subharmonic solution near $\omega_{resp} = \sqrt{a}$. Let us disturb x by adding or subtracting a small increment ε, so that

$$\ddot{x} + \ddot{\varepsilon} + a(x+\varepsilon) + b(x+\varepsilon)^3 = f \cdot \cos 3\omega_{resp} t$$

which can be expanded and the original equation subtracted, leaving

$$\ddot{\varepsilon} + a\varepsilon + b\left(3x^2\varepsilon + 3x\varepsilon^2 + \varepsilon^3\right) = 0$$

Leaving off higher powers of ε and inserting the response we have postulated for x, we get a $\cos 2\omega_{resp} t$ term because of the square, and the equation takes the form

$$\ddot{\varepsilon} + \left(a + 2\left(1 - \frac{a}{\omega_{resp}^2}\right)\cos 2\omega_{resp} t\right)\varepsilon = 0$$

This has the form of the Mathieu equation. If this ε grows and diverges, the solution of which this is a disturbance will be unstable. But if the Mathieu equation for the disturbance is stable, the subharmonic solution is stable. For the values obtained here, it turns out to be stable.

Exercise 39.1 *Find the magnitude of the harmonic distortion term* $\cos 3\omega_{ex} t$ *in the response of Duffing's equation*

$$\dot{x} + ax + bx^3 = f \cdot \cos \omega_{ex} t$$

Chapter 40

NUMERICAL SIMULATION and chaos

We have looked at graphical methods (Chapter 36), analytical solutions (Chapter 37), approximate methods (Chapter 38), and series expansions (Chapter 38). When all else fails, we carry out numerical simulations.

40.1 Euler's Method

In Section 16.3.1, we represented functions as lists of numbers representing the amplitude at discrete time intervals, and developed a central-difference method which had potential stability problems. For non-linear equations, it is safer and easier to express the governing equations

$$\ddot{x} + F(x, v, t) = 0 \tag{40.1}$$

in terms of the state variables x and v, replacing $\ddot{x}$ with $\dot{v}$

$$\frac{dx}{dt} \triangleq v \tag{40.2}$$

$$\frac{dx}{dt} = -F(x, v, t) \tag{40.3}$$

If we number the input and output data points with subscripts, designated as t_n, x_n, and v_n, we can substitute the discrete expressions for x and v near the point-in-time n

$$\begin{aligned} \frac{x_{n+1} - x_n}{\Delta t} &\cong v_{n+\frac{1}{2}} \\ \frac{v_{n+1} - v_n}{\Delta t} &\cong -F(x_{n+\frac{1}{2}}, v_{n+\frac{1}{2}}, t_{n+\frac{1}{2}}) \end{aligned}$$

Because some of the points are not centered on even values of n, we require Δt to be small and fudge out our expressions, replacing the in-between values with nearby values

$$\begin{aligned} \frac{x_{n+1} - x_n}{\Delta t} &\cong v_n \\ \frac{v_{n+1} - v_n}{\Delta t} &\cong -F(x_n, v_n, t_n) \end{aligned}$$

so that we can solve them for new $n+1$ values in terms of old n-values

$$x_{n+1} \cong x_n + v_n \cdot \Delta t \tag{40.4}$$

$$v_{n+1} \cong v_n - F(x_n, v_n, t_n) \cdot \Delta t \tag{40.5}$$

We can easily write this Euler algorithm in terms of a recursive programming language like *FORTRAN* or *BASIC*

1. Enter Initial Condition values of the state variables x_{old} and v_{old}

2. Start the procedure

    - Compute $F_{old} = F(x_{old}, v_{old}, t_{old})$ and replace any previous value.
    - Obtain $x_{new} = x_{old} + v_{old} \cdot \Delta t$ and $v_{new} = v_{old} + F_{old} \cdot \Delta t$.
    - Print-out x_{new} and v_{new} and display them graphically on a phase plane.
    - Replace $x_{old} = x_{new}$ and $v_{old} = v_{new}$.
    - Return to start of procedure.

3. Stop the procedure after a sufficient number of Δt increments.

Because we fudged by using "old" values to predict behavior throughout the time-step, we need to use very small steps for accuracy. Alternatively, after we have tested the basic procedure, we can replace the simple Euler algorithm with more complex Runge-Kutta algorithms, which are arranged to have very small residual errors.

Problem 40.1 *Investigate the decay behavior of a system with cubic damping*

$$\frac{m}{g_c} \ddot{x} + c \dot{x}^3 + kx = 0$$

Hint: start by writing a dimensionless form of the equation in terms of $y \triangleq x/x_o$

$$\ddot{y} + \left(\frac{c \cdot k^{1/2}}{m^{3/2}}\right)\dot{y}^{3} + y = 0$$

Problem 40.2 *Investigate the natural frequency of a system with a softening spring*

$$\frac{m}{g_c}\ddot{x} + k\left(x - \varepsilon x^3\right) = 0$$

Find the independent parameter(s), the dependence of frequency on the amplitude, and the maximum amplitude of stable oscillations.

Problem 40.3 *Investigate the decay of a system in which the spring has a hysteresis loop*

$$\frac{m}{g_c}\ddot{x} + k\left(x + \varepsilon \cdot sgn\left(\dot{x}\right)\right) = 0$$

Sketch the force-displacement curve of the spring, develop the dimensionless parameters, and find the decay envelope.

Problem 40.4 *Investigate the motion of a rotational mass-spring system which is driven by an escapement. There is a moderate amount of linear damping. The escapement adds a constant amount of energy whenever the system passes from negative displacement to positive displacement* θ*.*

Problem 40.5 *Investigate the effect of torque excitation on a pendulum, normalized to*

$$\theta + 0.1\,\theta + \sin\theta = a \sin\left(\frac{2\pi t}{\tau_{ex}}\right)$$

making Δt *an even fraction of the excitation period* τ_{ex}*. Can you find values of excitation strength a and period* τ_{ex} *which will sustain (1) steady oscillation, (2) chaotic switching to other "attractors," and (3) steady rotation in one direction or another.*

Chapter 41

VIBRATION CONTROL active and semi-active

In high-technology vehicles, from automobiles to spacecraft, we now see application of control systems to overcome vibration. The general concept includes

- measuring "output" vibrations by means of an accelerometer, straingage, or other sensor;
- generating a control input, either
  - an electromechanical or hydraulic force excitation, or else
  - a parametric variation, e.g., of an elastic system component; and
- a control system which "feeds back" information in a "closed loop" from the output to the input.

The control strategy is to minimize the vibrational response resulting from uncontrollable inputs, by adding control inputs which cancel as much of the response as possible. For linear systems, the theory is well developed and explained in textbooks and references.[1] In practice, evaluating the vibrational response is not always clear-cut.

41.1 Suspension Performance Criteria

Judging the performance of an automotive suspension involves several different considerations.

[1]e.g., *The CRC Handbook of Mechanical Engineering,* Frank Kreith, Editor, CRC Press LLC, Boca Raton, Florida, 1998, ISBN 0-8493-9418-X.

41.1.1 Isolation

The basic job of an automotive suspension is to minimize the transmission of road irregularities and wheel motion y to the seats. If x is the vertical displacement of a car, the passengers' perception of *ride quality* is linked to the rate-of-change of acceleration d^3x/dt^3, called "jerk," transmitted to the seats. Steady accelerations are not easily distinguished from gravity, unless they are large compared to gravity.

If the only criterion for optimizing the springs and dampers were minimizing jerk in the presence of small road irregularities, then linear theory would apply, and we would use Equation 12.6 to predict periodic-excitation response

$$\frac{\dddot{x}_{\max}}{\dddot{y}_{\max}} = \frac{\sqrt{1+\left(2\zeta\frac{\omega_{ex}}{\omega_n}\right)^2}}{\sqrt{\left(1-\frac{\omega_{ex}^2}{\omega_n^2}\right)^2+\left(2\zeta\frac{\omega_{ex}}{\omega_n}\right)^2}} \tag{41.1}$$

and the Laplace-transform approach of Chapter 17 for transient response.

41.1.2 Amplitude Limits

We have to anticipate that the linear range occasionally might be exceeded; in a car suspension, this is called "bottoming out." Because wheel travel is limited by interference with bodywork, etc., the unsprung axle or wishbones will encounter helper springs or rubber bump stops. This will make the system non-linear.

In a well-designed suspension, hitting the bump stops would occur rarely, but would cause memorable jolts. Rather than incorporating these singular spikes into the average value of jerk, we may count these events, and make reducing the number of bottoming-out jolts a second criterion of suspension performance.

If we have a statistical description of the input from the road (Chapter 18), we could use our linear-spring model to predict the output, to whether (and how much) it exceeds the limits—but we would have difficulty telling how *often* that happens.

41.1.3 Traction

When we expand our car model to include the mass of the wheel and the elasticity of the tire, we end up with at least a two-mass system (Chapter 20). In addition to our ride quality criterion based on the jerk of the sprung mass $\dddot{x}$, and the displacement limit between sprung and unsprung masses, we are also interested in the deflection of the tire and the force on the road. If they become too small, traction will be lost. In older, live-axle cars, this showed up dramatically on rough roads in the form of sideways hop and chatter in high-speed turns. Therefore, uniformity of tire deflection is a third criterion.

41.1.4 Power

As we noted in Chapter 25, even on smooth roads, damping in the tire is kept low to reduce rolling resistance. On rough roads, shock absorbers between the sprung and unsprung masses also absorb and dissipate energy, which must be supplied by propulsive power. For economic reasons, we would like to keep this power absorption and dissipation moderate.

41.1.5 Weight and Cost

To a large extent, performance criteria trade off against each other: softening the spring to reduce the natural frequency reduces the transmissibility of small road disturbances, but increases the number of times the suspension bottoms out. On the other hand, increasing wheel travel permits all of the performance measures to be improved simultaneously. Historically, passenger cars with one or two inches longer suspension travel have done well at high speeds on dirt roads, and excelled in Saharan rallies. In most cars, wheel travel is limited by practical and economic considerations.

41.2 Passive System

Optimizing even a passive system is a complex task when multiple criteria are involved. The classical approach is to write a single criterion which is a weighted sum of the other criteria. However, if even one of the criteria is non-linear, the results depend not only on the system, but also on the amplitude of the input disturbance.

The *input disturbances* are not easily characterized. On an interstate highway, they may be very small, coming from tar-strips between concrete slabs. In a city, they may include potholes and gutters. On the cobblestone roads of a picturesque European city, they will be high in frequency. On Australian dirt roads pounded into regular waves by chattering truck axles, they will contain critical frequencies. On older roads built on landfill, they may consist of large, slow undulations. In short, the mix of frequencies and amplitudes makes it difficult to find an all-around optimum, and most vehicles are designed for only a part of the range of possible road surfaces.

41.3 Adaptive System

As we suggested in Chapter 12, it is possible to build systems which adapt to varying road conditions. Some ride-height-adjusting systems change the spring constant, but most simple systems change the damping coefficient, for example by opening or closing an orifice in a hydraulic shock absorber. If there are inputs which cause large resonant response, then a high damping factor is clearly desirable; when road inputs are predominantly at high frequencies, then minimal damping will give the best results.

When the choice of damping is up to the driver, then his or her sensibilities will lead to his or her decisions. Can we invent a system to make the choice automatically? For example, if the relative displacement between sprung and unsprung masses is sensed, repeated large amplitudes might command an increase in the damping; a long interval without large amplitudes might permit a gradual reduction in damping. The system will not do as well as the driver, who not only senses the vibrational response, but also has the ability to look ahead at the road and anticipate future input.

The characteristic feature of an adaptive system is that the time constant for adjustment is long, longer than a cycle of oscillation.

41.4 Semi-Active Control

The next step forward is to make adjustments to the damper quickly, within a cycle of oscillation. We can generate forces as a controlled input, whenever there is a relative velocity between the sprung and unsprung masses, but in only one direction, the direction which retards the velocity. This restriction limits the application of classical control theory, and pushes us towards using numerical simulation (Chapter 40). It is difficult to state generically how much of an improvement over an adaptive system a semi-active control can offer, or how much it falls short of fully active control: only by experiments and simulations of particular control schemes can those questions be answered.

Example: Recall the quarter-car model of Chapters 20 and 25, where x_1 and x_2 are the displacements of the unsprung and sprung masses, respectively; we can write the equations

$$\frac{m_1}{g_c}\ddot{x}_1 + c\,\dot{x}_1 - c\,\dot{x}_2 + k_{tire}x_1 + k_{spr}x_1 - k_{spr}x_2 = \mathfrak{f}(t) - \frac{m_1 g}{g_c} \quad (41.2)$$

$$\frac{m_2}{g_c}\ddot{x}_2 - c\,\dot{x}_1 + c\,\dot{x}_2 - k_{spr}x_1 + k_{spr}x_2 = \frac{-m_1 g}{g_c} \quad (41.3)$$

which we can express in Hamilton's canonical form as

$$\dot{x}_1 = v_1 \quad (41.4)$$

$$\dot{x}_2 = v_2 \quad (41.5)$$

$$\dot{v}_1 = \frac{g_c}{m_1}(-cv_1 + cv_2 - k_{tire}x_1 - k_{spr}x_1 + k_{spr}x_2 + \mathfrak{f}(t)) - g \quad (41.6)$$

$$\dot{v}_2 = \frac{g_c}{m_2}(cv_1 - cv_2 + k_{spr}x_1 - k_{spr}x_2) - g \quad (41.7)$$

and convert to Euler algorithms in the manner of Equations 40.4 and 40.5.

This basic formulation now can be expanded to include non-linear effects and control:

- For complex road conditions, we insert reference road profiles $\mathfrak{f}(t)$, or even actual measured road data.
- For travel limitations, we make k_{spr} be a function of the amplitude; perhaps it jumps to a higher rubber-bump-stop value when $(x_1 - x_2)$ exceeds a certain value.
- For tire characteristics, we make k_{tire} a function of x_1, or at least replace $k_{tire}x_1$ with zero whenever the tire leaves the ground.
- For non-linear damping, we make c a function of the velocity $(v_1 - v_2)$, making sure that it has the correct sign.
- For semi-active control, we make c a function of whatever control quantities we measure, making sure that it has the correct sign.

The output will give us information on our performance measures: $\dddot{x}_2$ for ride quality, $(x_1 - x_2)$ for wheel-travel limits; x_1 for ground contact and traction, etc.

Adjusting the damper, perhaps by means of a solenoid valve, requires information signals, but not substantial power input. This is the motivation for using semi-active control. However, indirectly there is still absorption and dissipation of energy in the damper, stolen from propulsive power.

41.5 Active Control

If we are willing to supply electromechanical or hydraulic power to actuators between the sprung and unsprung masses, we can operate full-fledged control systems to smooth the ride, limit wheel excursions, compensate for tilt and sway, etc. How well we can meet our multiple criteria in the presence of random disturbances depends on how much information we can provide to the control system:

- accelerometers on the sprung mass
- accelerometers on the unsprung masses
- displacement sensors between sprung and unsprung mass
- distance sensors under the car, pointed at the road, for ride height.

The last item offers the opportunity for open-loop control.

Once we have decided on a control strategy, we can simulate it by adding the actuator force $\mathfrak{F}$ to our numerical algorithms, Equations 41.6 and 41.7

$$\dot{v}_1 = \frac{g_c}{m_1}\left(-cv_1 + cv_2 - k_{tire}x_1 - k_{spr}x_1 + k_{spr}x_2 + \mathfrak{f}(t) + \mathfrak{F}\right) - g \quad (41.8)$$

$$\dot{v}_2 = \frac{g_c}{m_2}\left(cv_1 - cv_2 + k_{spr}x_1 - k_{spr}x_2 - \mathfrak{F}\right) - g \quad (41.9)$$

Exercise 41.1 *We have used a quarter-car model for our example. How many degrees of freedom are involved in a whole car?*

Problem 41.2 *Program the quarter-car model of the example, and modify it to incorporate cubic damping. Test the performance for several amplitudes of excitation at lower-than-resonant, resonant, and higher-than-resonant frequencies, and for a ramp input.*

Problem 41.3 *Program the quarter-car model of the example, and modify it to incorporate a simple semi-active control algorithm of your own invention. Test the performance for several periodic inputs, and for at least two ramp inputs of different severity.*

Chapter 42

FLOW-INDUCED VIBRATIONS and flow instabilities

A large number of vibration problems arise from flow through or around structures. In many cases they may be due to pulsations in the flow due to a pump or due to the blade-passage frequency of an upstream blower or fan. In other cases they arise from the non-linear nature of the governing equations of fluid flows, which are difficult to solve, but can be shown to have properties leading to paradoxical behavior.[1] There are several different physical mechanisms involved, each of which has different characteristics.

42.1 Turbulent Excitation

The most-studied source of turbulence is the instability in **boundary layers**, the slowed-down regions at surfaces bounding a flow. When a boundary layer has developed beyond a certain point, individual eddies begin at the wall and propel themselves outward, initially gaining energy from the shearing flow. As they move further away from the wall, they begin to dissipate, breaking up into smaller and smaller eddies. As a result, velocity measurements show a wide spectrum of random fluctuations in a layer of the fluid near the wall. In a pipe, the boundary layers ultimately merge, and form a solid core of turbulence within the pipe. Theory and experiment show that the phenomenon depends on a dimensionless similitude parameter called the Reynolds number which is

[1] Garrett Birkhoff, *Hydrodynamics, A Study in Logic, Fact and Similitude,* Princeton University Press, 1950; reprinted 1955 by Dover; reprinted 1978 by Greenwood Publishing Group, ISBN 0-31320118-8.

defined as

$$\text{Re} \triangleq \frac{v_{ref} L_{ref}}{(\mu/\rho)} \tag{42.1}$$

where v_{ref} is a reference velocity, for example, the velocity well away from the wall, or a velocity computed from the shear and other parameters at the wall; L_{ref} is a reference dimension of the flow geometry, for example, the length over which the boundary layer has developed, or some thickness measure of the boundary layer; μ is the viscosity of the fluid; and p is the density of the fluid. Because the choice of reference quantities is somewhat arbitrary, the values of Reynolds numbers are meaningful only for comparing similar situations.

Exercise 42.1 *What must the dimensions of viscosity μ be?*

The turbulence causes pressure fluctuations at the wall. If the wall is flexible—a structural panel or a tensioned web—it will cause a broad-band random excitation. Each characteristic mode of the wall will receive irregular excitation as described in Chapter 18. In most engineering situations, the response to parallel flow will be moderate, but will steadily increase as the Reynolds number increases.

Turbulence is also created in **shear layers** away from a wall, where fluid streams mix. Regular "flow structures" can be observed to form, and break down into smaller eddies which impinge on structures downstream. However, the most severe turbulence arises in the wake of **screens** placed across the flow, or in the wakes of other unstreamlined bodies.

42.2 Vortex Shedding

The simplest bluff body is a circular **cylinder**, such as a string in an Aeolian harp. When a wind blows across the string, it will experience an alternating lift force at a particular excitation frequency $f_{v.s.}$. The dimensionless frequency/velocity parameter is called the Strouhal number

$$S \triangleq \frac{f_{v.s.} D}{v_\infty} \tag{42.2}$$

where D is the diameter of the cylindrical string and v_∞ is the velocity of the approaching flow well away from the string. If the string is tuned so that one of its frequencies is in resonance with the excitation, $f_n = f_{v.s.}$, it will vibrate and transmit sound by way of the supporting soundboard. For a wide range of Reynolds numbers,

$$10^2 > \frac{v_\infty D}{(\mu/\rho)} > 10^5$$

the excitation frequency is proportional to the wind velocity, so that the Strouhal number is a constant

$$S \approx \frac{1}{5}$$

Complementary to the alternating lift, the wake undulates; the cylinder sheds a "street" of alternating vortices. Theodor von Kårmån showed that a **vortex street** forms a stable configuration if it has a width of approximately D and repeats every $5D$.

Vortex shedding applies to a wide range of structures in cross-flow, from the cables the Navy uses to tow paravanes (kite-like underwater probes "flown" beside a ship) to smokestacks in the wind. A smokestack has the natural frequency of a vertical cantilever. If the transverse excitation from the wind reaches the natural frequency of that cantilever, the smokestack may go into resonance, and ultimately tumble down. Sometimes a dimensionless velocity is defined in terms of the structural frequency

$$v^* \triangleq \frac{v_\infty}{f_n D} \tag{42.3}$$

and the critical wind velocity is

$$v^*_{v.s.} \approx 5$$

Exercise 42.2 *What is the excitation frequency on a four-foot-wide smokestack in a* 50 *m.p.h. wind if* $S \approx 0.2$*? What is the diameter Reynolds number at that velocity? Is the chosen value of the Strouhal number appropriate, or do we need to look up references for better information? Useful information:* (μ/ρ) *for atmospheric air is approximately* 0.17×10^{-3} *ft²/sec or* 0.016×10^{-3} *m²/s.*

Exercise 42.3 *What is the first critical velocity for water flowing across a "thermowell," a hollow cantilever containing a temperature probe, with a diameter of one centimeter and a natural frequency of* 30 *Hertz, if* $v^*_{v.s.} \approx 5$*? Is the chosen value of the critical velocity appropriate, or do we need to look up references for better information? Useful information:* (μ/ρ) *for cool water is approximately* 0.011×10^{-3} *ft²/sec or* 1.0×10^{-6} *m²/s.*

How large the resonant amplitude will turn out to be depends on the mass behind the excitation forces on the one hand, and on the damping of the structure on the other. Both are traditionally combined into the mass-damping parameter

$$\frac{m}{\rho D^2} \times 2\pi\zeta \tag{42.4}$$

where m is the mass-per-unit-length of the cylinder which resists vibration, and ρD^2 is proportional to the mass (also per-unit-length of tube) of the fluid which is active in generating vibrational forces. ζ is the damping ratio which dissipates vibration; because it is experimentally determined from the logarithmic decrement, it is customary to write $2\pi\zeta$. To have destructive vibrations, it is necessary to meet two conditions:

1. that the excitation resonates with the structure

$$\begin{aligned} f_n &= f_{v.s.} \\ v^* &= v^*_{v.s.} \end{aligned}$$

2. and that the response amplitude is damaging.

The excitation forces are not entirely independent of structural motion:

- the excitation may synchronize and "lock-in" with the structural vibration if $f_{v.s.} \approx f_n$; but
- the excitation forces fall off with increasing response, and amplitudes exceeding one or two diameter are generally not observed.

Non-circular cylinders and other bluff bodies also experience vortex shedding. In general, it is necessary to search the literature for similar geometrical situations for experiments or field experience with comparable similitude numbers.

Surprisingly, closely packed cylindrical arrays, such as **tube rows** in duct-mounted heat exchangers, and **tube bundles** in shell-and-tube heat exchangers, experience similar vibration excitation. Although we cannot visualize vortex streets, we use the term vortex shedding because of these resemblances:

- For a wide range of Reynolds numbers, the excitation occurs at fixed Strouhal numbers, which are different from 0.2, but can be looked up in tables, plots, or maps[2] as a function of tube packing (pitch-to-diameter ratio) for different geometrical arrangements (in-line or staggered).
- The excitation exists even in the absence of tube response; it may manifest itself as an acoustic resonance in the cavity of a gas heat-exchanger.
- The response amplitude depends on the mass-damping parameter; problem amplitudes are observed mostly in heat exchanger having liquid on the shell side.
- Lock-in can be observed near resonance.

Measured tube amplitudes usually show a gradually increasing floor of random oscillations due to incoming turbulence. At a critical flow, the amplitude will jump to a much larger value. In liquid flows, that amplitude is likely to be

[2] Robert D. Blevins, *Flow-Induced Vibration,* Second Edition, Van Nostrand Reinhold, New York, 1990, ISBN 0-442-20651-8.

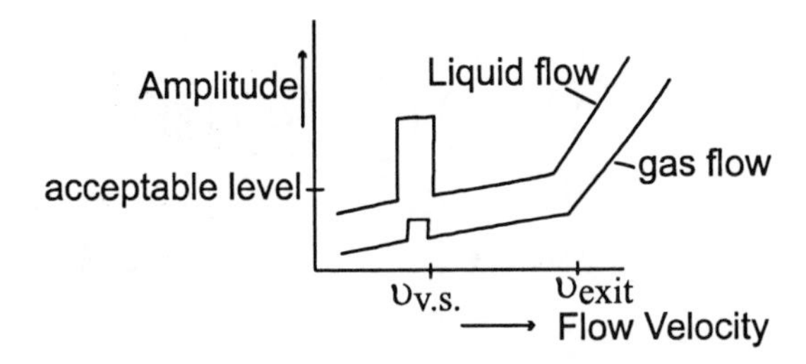

Figure 42.1: Amplitude versus Flow

damaging, either because tubes bang together in mid-span, or because tubes wear at the points where they pass through baffle and support plates, or even because fatigue stresses appear at the tubesheets. In gas flows, the amplitude may be tolerable, and the vibration will abate if the flow is increased past the critical velocity. At some point, a critical velocity for an entirely different type of flow-induced vibration may be reached, and the amplitudes begin to rise steeply, this time without limit. (Figure 42.1)

42.3 Flutter

The classic examples of flutter involve two degrees of freedom. In a **wing**, the two motions are twisting and up-and-down flexing. In practice, the two motions are coupled, because neither the center of gravity of the wing, nor the center of flexure of the spar, are in the same place as the center of aerodynamic lift (which is about one-fourth of the chord length behind the leading edge of the wing). If the two motions are about 90 degrees out-of-phase, energy is transferred from the flow to the oscillation: when the wing is twisted to a greater angle of attack and greater lift, it is in the midst of upward motion, and vice versa. The oscillation becomes self-sustaining at some critical speed, and with a further increase of speed lead to failure. The onset of flutter can be often delayed to a higher critical velocity by moving the center of mass forward.

Connors' **fluid-elastic whirling** in tube rows[3] resembles wing flutter because

- there are two ninety-degree-out-of-phase motions: the odd-numbered tubes moving up- or down-stream, and the even-numbered tubes moving laterally; (Figure 42.2)
- the changing geometry creates lift- and drag-forces which extract energy from the flow to maintain the vibration; and

[3]Peter M. Moretti, "Caught in a Cross Flow – The Paradox of Flow-Induced Vibrations," *Mechanical Engineering,* Vol. 108, No. 12 (Dec. 1986), pp. 56–61, American Society of Mechanical Engineers (ASME); see also correction in Feb. 1987 issue, p. 2.

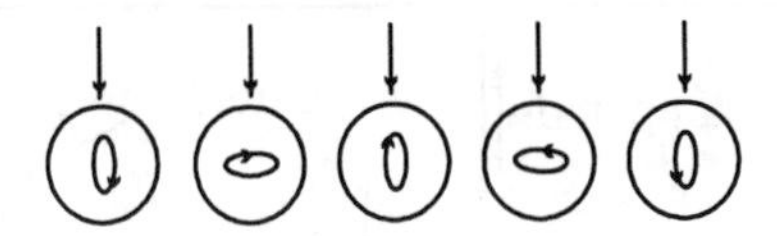

Figure 42.2: Fluid-Elastic Whirling

- there is a critical velocity beyond which the oscillation increases without limit, until damage occurs.

The critical velocity in tube rows and arrays is a function of the mass and damping parameters; it typically takes the form

$$\frac{v_{crit}}{f_n D} = \beta\sqrt{\frac{m}{\rho D^2} \times 2\pi\zeta} \tag{42.5}$$

The dimensionless constant β depends on the tube arrangement and pitch-to-diameter ratio, typically ranging from about 3.0 upward; values for regular arrangements such as 90° square in-line, and 30°, 45°, or 60° staggered, have been published in the literature (see Chapter 43).

Weaver's **fluid-elastic instability** in tube bundles[4] can occur with only one tube moving, and therefore is a likely candidate for the initial onset of fluid-elastic vibration. As a one-degree-of-freedom system, it is thought to derive its instability from a time-delay in the flow. The most likely source of the time delay is in the boundary-layer's development and separation on the tube.[5]

42.4 Jet Switching

The flow behind a closely-spaced tube row is not symmetrical; the jets emerging between the tubes veer to one side or the other, and coalesce into larger jets.[6] The deflected jets are linked to reaction lift- and drag-forces on the tubes. Roberts' **jet-switching** mechanism results from motions of the tubes which, once initiated, change the pairing of the jets to generate oscillation-sustaining forces. The critical velocity depends on the combined mass-damping parameter, roughly following the same form as Equation 42.5. Flow-visualization shows that the jet deflections can exist simultaneously with vortex shedding; and that

[4]Peter M. Moretti, "Flow-Induced Vibrations in Arrays of Cylinders," *Annual Review of Fluid Mechanics*, Vol. 25 (1993), pp. 99–114, ISBN 0-8243-0725-9.

[5]Jia-Qi Cai, *A Theoretical Study of Time Delays on a Single Cylinder and a Cylinder in an Array Oscillating in Uniform Flows*, Ph.D. Dissertation, Oklahoma State University, Stillwater, July 1995.

[6]P.M. Moretti & M. Cheng, "Instability of Flow Through Tube Rows," *Trans. ASME, J. Fluids Engineering*, Vol. 109, No. 2 (June 1987), pp. 197–198.

similar switchable asymmetries exist in deep tube bundles.[7] Since the critical-velocity equation has a similar form for jet-switching as for fluid-elastic theories, it is difficult to distinguish between mechanisms. For regular tube arrangements, it is sufficient to plot experimental data in the functional form

$$\frac{v_{crit}}{f_n D} = \mathfrak{F}\left(\frac{m}{\rho D^2}, 2\pi\zeta\right) \qquad (42.6)$$

The idea of jet-switching relates to a number of other multi-stable flow situations, from the instability in two-dimensional diffusers—the flow prefers to follow on or the other of the walls if they are angled steeply—to Coanda jets which attach themselves to the surface from which they emerge, but can break free under certain circumstances. One startling example is the rough shaking experienced by an airplane when the wing goes into (intermittent) stall. A related phenomenon is the **galloping** of power lines when they are coated by winter ice, presenting an asymmetrical profile to a crosswind.

42.5 Closure

Flow-induced vibrations have been surveyed because they are important, because they show how practical problems cut across specialties like structures and fluid dynamics, and because they demonstrate the importance of searching the literature to find experimental results before embarking on design and prediction.

The use of dimensionless equations also makes it clear how to fix "trouble jobs" in the field. Examining the equations for vortex shedding and for fluid-elastic phenomena in heat exchangers shows that both respond to the same *general* strategies:

- lowering the velocity, by limiting operating parameters or by changing the flow arrangement;
- raising the natural frequency, by introducing more support plates; and/or
- increasing the damping, by adding clamping elements etc.

However, if one understands the physical mechanism, it is sometimes possible to apply *specific* measures, such as strakes applied to smokestacks to modify vortex shedding.

Problem 42.4 *In the library and on the Internet, search for information on the failure of the Tacoma Narrows Bridge in 1940. Summarize the debate on the nature of the failure: was it vortex-shedding or an aerodynamic instability?*

[7] M. Cheng and P.M. Moretti, "Flow Instabilities in Tube Bundles," ASME Pressure Vessel and Piping Publication PVP Vol.154 *Flow-Induced Vibration,* Honolulu, July 23–27, 1989, pp. 11–15.

What might have been a fix for the problem if it had been understood before the failure?

Chapter 43

LITERATURE SEARCHES

A design, development, or research engineer often must search for information on existing know-how to solve a problem efficiently. The most current information is generally found not in textbooks, but in technical journals. The search for this information involves several techniques: initial search, systematic search, backward search, and forward search.

43.1 Initial Search

Before you can do a successful systematic search, you have to survey the terrain and figure out a search strategy. This is the hardest and most discouraging part of the search—because progress is sporadic—but will pay off sooner or later. It involves prowling around in libraries or on the Internet:

- Ask colleagues whether they know of recent literature on your subject, in the hope of finding a book or a paper which helps you identify the **technical community** or **conferences** dealing with your problem.
- Search library catalogs, on-site or via the Internet,[1] and book sellers[2] for books from which you can learn the technical background, current **terminology,** and **subject headings** used in describing your problem.
- Skim through the subject area in index journals and bibliographies. The *Engineering Index* in bound volumes goes back to 1884, and for recent years is available electronically and on CD-ROM as the *Compendex* database. *Science Citations Index—Subject Index* goes back to 1955. You may stumble across an important paper, but the main objective at this stage is to identify **key terms** used in the titles, and/or recurring **authors** in your field, and/or core **journals** publishing on your subject.

[1] *e.g.*, `http://www.library.okstate.edu/liblink.htm`
[2] *e.g.*, `http://www.crcpress.com/`

- Look for bibliographies within textbooks, articles in the *Annual Review of Fluid Mechanics,* etc., feature review articles in *Applied Mechanics Reviews,* and other abstracts journals; if you are lucky and find your subject, you will have a whole list of references.

- Skim through the tables of contents of recent journals in the appropriate general area; this can be done most easily through *Current Contents.* Appropriate journals might include *Trans. ASME Journals, AIAA Journal, J. Sound & Vibration,* etc.

By the end of the initial search, you might have found some interesting papers; you will certainly have become familiar with words and names that might make good search terms.

43.2 Systematic Search

The techniques used in the initial search required you to skim through a lot of material, because the classification of that material into subject headings is rather general. The next step is to focus on specific terms, keywords, authors, and/or journals.

Since the early 1970s it has been possible to carry out computer searches of literature databases (such as *Compendex*[3] in engineering) much more precisely—not only by author or tabulated keyword, but also by any word used in the title or abstract. This capability was first brought to libraries over telephone modems by providers like *Dialog*[4] (and to individuals by *Knowledge Index*). More recently, many libraries acquired databases on CD-ROM for convenient on-site searching. *Compendex* is now also available on the Internet for subscribers to *Engineering Information Village.*[5] In addition, you should search for dissertations.

Finding the right search terms is a trial-and-error process. The database search engine maintains a thesaurus listing all the available search terms. You will save a lot of wasted effort by using the command for looking up search words so that you can examine the ones next to them in the thesaurus, to make sure you don't miss slight variants of the words. At this point, you want to be sure you have broadened the search sufficiently to include almost all pertinent papers.

You should also find out whether you can use asterisks or question marks as "wild-card" letters which allow alternative endings to search words, so you don't miss a paper just because it uses the plural of a search word.

If you use only a single search term, you will often get far too many "hits" for you to look at. If that happens, you should refine the search by using the code for

[3] `http://ei.telebase.com/m/cpx.htm`

[4] `http://www.dialog.com/`

[5] `http://www.ei.org/`

the Boolean "and" command. For example, if you are interested in "flow-induced vibrations in heat-exchanger tube bundles," you might search "vibration and tube" so that only papers using *both* of these terms in the front matter will be listed in the output; obviously you should try several other combinations and examine which one gives the best yield of truly pertinent papers. On the other hand, if you include all possible terms "flow and vibration and heat-exchanger and tube and bundle," you will probably get hardly any hits at all, and miss many papers that don't happen to use all of these words. A good search phrase yields most of the pertinent papers, without too much chaff. There will always be some chaff: a search phrase that yields papers on active-suspension systems for automobiles and railcars, may also get you a paper on the settling-out of solutions and suspensions of chemicals in railway tank-cars.

Your Initial Search in Section 43.1 should have alerted you to the fact that different research groups use different language for describing the same problem. One man's "tube bundle" is another man's "cylindrical array;" you can include both by using the code for the Boolean operator "or" combined with parenthesis symbols. For example, an improved search phrase might read "(vibrations or whirling) and (tubes or arrays)."

When you have developed a suitable search combination, you will obtain a manageable set of pertinent papers. Reading the titles will narrow it down some; downloading and reading selected abstracts will narrow it down further.

The systematic search process will give you a solid body of references; but it will not include classic papers because few electronic databases go back to before 1970; it will also not include the latest papers that have not been typed into the database yet.

43.3 Obtaining Papers

If you are at a university library, you can usually find the journals in the stacks, or request the papers through inter-library loan. If you use the Internet, database information providers can refer you to Document Delivery services which will mail, fax, or download papers to you for a fee.

43.4 Backward Expansion

When you have obtained one or more appropriate papers, you can examine their reference lists or bibliographies for promising titles, in order to search backwards for older papers. This is your opportunity to find the key papers in the field, classic papers that made a big impact on the technical community and laid the foundation for current research. Key papers may also be found in the reference or bibliography sections of specialized textbooks and monographs.

43.5 Forward Expansion

When you have identified key papers from the Systematic Search and the Backward Expansion, you can search forward to the best and most recent papers:

- Since the 1950s, reverse lists of references have been compiled and listed in the *Science Citations Index—Citations Index,*[6] which is available at libraries and through services like *Dialog.* Look for papers which cite your key papers among their references. Since new papers are listed here only by author and journal, you need to go on and look up the titles in the *Science Citations Index—Author Index.* If the titles sound promising, you can then look up or order these papers.

- You should also check the most recent papers by those authors which have shown up favorably in your search activities, by looking up their names in the *Science Citations Index—Author Index.*

Only when you have gone through all of the search steps can you be reasonably certain that you have located the most important papers.

43.6 Documentation

Throughout your searches you must keep up with the important job of documenting the results, so that you won't have to return and repeat the process as questions arise! (Every professor is familiar with the plaint of a student holding a copy of a paper or chapter that he can't cite because he failed to record the source.) You should have in hand:

- a list of all related papers you found;

- short summaries of the more pertinent papers, noting whether they cover theory, experiment, and/or computer simulation; and

- a summary of the state-of-the-art, with conclusions, graphs, and formulas.

This is much easier if you have kept 3×5 index cards of *all* references *throughout* the process, writing down author, year, title, journal, issue, and page numbers in standard style on the front of each card, and a note about the publication's content on the back.

Many professionals enter all their search results into a computerized database to make sure that they can properly cite any references they have used. In addition to commercial programs for this purpose, there is the application BIBTEX which is used by the LATEX systems mentioned in Chapter 1.

[6] `http://www.isinet.com/`

43.7 Staying Current

Once a basic search has been completed, keeping up with new developments is relatively easy; you can:

- rerun your search phrase and look for new "hits;"
- check the latest *Science Citation Index—Citations Index* for new papers citing your sources;
- check the latest *Science Citation Index—Author Index* for new papers from active authors; and
- check core journals for new papers in *Current Contents*.[7]

This is an ongoing process for as long as you work in an area.

Exercise 43.1 *How many publications by the author of this text can you find, from 1965 to the present? Does it make any difference whether you use the first name or only initials? How could you weed out papers by a namesake with the same last name and first initial? If you have more than one search resource at your disposal, which one gives you the best return?*

Problem 43.2 *Find up-to-date papers on methods for computing the power spectrum of stationary random vibrations from short data records. Partial answer: Look for phrases like discrete Fourier transform (DFT), Cooley-Tukey fast Fourier transform (FFT), auto-regressive (AR) model, maximum entropy spectral analysis, Burg algorithm, Marple least-square method, smoothness priors long AR model, and digital filtering.*

Problem 43.3 *Find a few classic papers on the "wavelet" approach for obtaining trends in non-stationary random vibrations. Hint: Some of them antedate electronic databases, so you may have to look at the lists of references in current papers on wavelets.*

Problem 43.4 *Do a literature survey on waving motions of flags. Hint: You may have to look for alternate wording, e.g., "instability of an elastic strip hanging in an airstream." Partial answer: Some classic papers are international: Fairthorne, ARC Report No. 1345, 1930; Thoma, (German) ZAMM, 1930; Sparenberg, Proc. Nederland Acad. Sci., 1962; Uno, J. Textile Mach. Soc. Japan, 1973; Watanabe, Proc. First. Internat. Conf. Web Handling, 1991. The latest papers are likely to cite these classic papers.*

---

[7] `http://www.isinet.com/products/products.html`

Index